About Island Press

Island Press is the only nonprofit organization in the United States whose principal purpose is the publication of books on environmental issues and natural resource management. We provide solutions-oriented information to professionals, public officials, business and community leaders, and concerned citizens who are shaping responses to environmental problems.

In 1994, Island Press celebrated its tenth anniversary as the leading provider of timely and practical books that take a multidisciplinary approach to critical environmental concerns. Our growing list of titles reflects our commitment to bringing the best of an expanding body of literature to the environmental community throughout North America and the world.

Support for Island Press is provided by Apple Computer, Inc., The Bullitt Foundation, The Geraldine R. Dodge Foundation, The Energy Foundation, The Ford Foundation, The W. Alton Jones Foundation, The Lyndhurst Foundation, The John D. and Catherine T. MacArthur Foundation, The Andrew W. Mellon Foundation, The Joyce Mertz-Gilmore Foundation, The National Fish and Wildlife Foundation, The Pew Charitable Trusts, The Pew Global Stewardship Initiative, The Rockefeller Philanthropic Collaborative, Inc., and individual donors.

About the Center for Plant Conservation

The Center for Plant Conservation (CPC) was founded in 1984 as a national, nonprofit organization to facilitate and coordinate off-site plant conservation within the U.S. botanic gardens and arboreta as a complement to on-site conservation. Its goal is to prevent the extinction of the rarest plant taxa native to the United States. A principal objective is to ensure that these imperiled plants are securely maintained as living collections within what is now a network of twenty-five botanic gardens and arboreta nationwide. Called Participating Institutions, today they conserve many of the rarest plants of their regions. In 1995, the network housed 480 plants—called the National Collection of Endangered Plants—in protective cultivation or in seed storage. About half of these plants have been sponsored by concerned individuals and organizations, thereby ensuring their long-term protection within the CPC garden network. Associated programs in research, education, and display are developed by the Participating Institutions and the Center's national office, which is located at the Missouri Botanical Garden in St. Louis. The national office is also responsible for computerized record keeping of the National Collection, for coordinating the development of national off-site plant conservation guidelines and policies, and for developing a national support group, The Friends of CPC.

Restoring Diversity

We dedicate this book to all those
whose life work is to protect and restore diversity.

Restoring Diversity

Strategies for Reintroduction of Endangered Plants

Edited by Donald A. Falk
Constance I. Millar
Margaret Olwell

Foreword by Reed F. Noss

Center for Plant Conservation
Missouri Botanical Garden

ISLAND PRESS
Washington, D.C. • Covelo, California

Grateful acknowledgment is expressed by the authors for permission to use the following material:

Excerpt from THE FAMILY REUNION copyright © 1939 by T. S. Eliot and renewed 1967 by Esme Valerie Eliot, reprinted by permission of Harcourt Brace & Company.

Figures CS5-1 and CS5-2 appeared previously in "Recovery Planning and Reintroduction of the Federally Threatened Pitcher's Thistle (*Cirsium pitcheri*) in Illinois," Natural Areas Journal 13:164–76; and in *Recovery of Endangered Species*, eds. M. Bowles, and C. Whelan, Cambridge University Press, 1994. Reprinted with the permission of Cambridge University Press.

No copyright claim is made in "The Regulatory and Policy Context," "FOCUS: Reintroducing Endangered Hawaiian Plants," "FOCUS: *Pinus Torreyana* at the Torrey Pine State Reserve, California," "Rare Plant Mitigation: A Policy Perspective," "*Pediocactus Knowltonii*," and "Texas Snowballs (*Styrax Texana*) Reintroduction," works produced by employees of the U. S. government.

Library of Congress Cataloging in Publication Data

Restoring diversity : strategies for reintroduction of endangered plants /
    editors, Donald A. Falk, Constance I. Millar, Margaret Olwell.
        p.   cm.
    Includes bibliographical references (p.) and index.
    ISBN 1-55963-296-8 (cloth). — ISBN 1-55963-297-6 (paper).
    1. Plant reintroduction.   2. Endangered plants.   3. Restoration ecology.
I. Falk, Donald A.   II. Millar, Constance I.   III. Olwell, Margaret.
    QK86.4.R47   1996   95–18936
    581.5′29—dc20   CIP

Contents

Acknowledgments

A multi-year, multi-faceted project such as this requires the efforts of many talented, dedicated people, only a few of whom are publicly identified with the outcome. Here we would like to acknowledge the efforts of the people and institutions who made this work possible.

The catalytic funding for this project was provided by The Joyce Foundation of Chicago, to whom we express our deepest thanks. Additional assistance was provided by the American Association of Botanic Gardens and Arboreta, ARCO Foundation, Bureau of Land Management, Illinois Department of Conservation, International Union for the Conservation of Nature (Reintroduction Specialists Group), Missouri Department of Conservation, Natural Areas Association, Society for Ecological Restoration, U.S. Fish and Wildlife Service, and the U.S.D.A. Forest Service. We are extremely appreciative of the support provided by these institutions and hope that their ongoing contributions to the conservation of biological diversity will be well recognized.

We are deeply grateful to the Center's host institution, the Missouri Botanical Garden, and in particular its director, Peter Raven, for consistent and enthusiastic support of CPC generally and of this project in particular. Without his support, this project would have never come to fruition.

Special thanks are due to the staff, board, and Science Advisory Council of the Center for Plant Conservation, who organized and implemented the program. Staff members Marie Bruegmann and Gregory Wieland played the key roles in organizing the conference, with assistance from staff members Jeanne Cablish, Sheila Kilgore, Michael O'Neal, Grace Padberg, interim director Mick Richardson, and Science Advisory Council Chair Barbara Schaal (Washington University). We also thank CPC Executive Director Brien Meilleur for providing valuable assistance during the preparation and editing of this book.

Organization of the original study project and symposium, and initial work on the book, took place while two of us (Falk and Olwell) were on staff at the Center for Plant Conservation. Since major funding was acquired while we were still at the Center, and CPC provided much logistical support for all aspects of symposium and book development, we wish to acknowledge this relationship prominently. We thank the CPC, as well as Society for Ecological Restoration, and U.S.D.A. Forest Service, for granting us the time to complete the present work.

The twenty-five Participating Institutions of the Center for Plant Conservation constitute some of the best expertise in the biology of rare plant species to be found anywhere. Their staffs, who manage the National Collection of

Endangered Plants, have an intimate understanding of the biology of rare species. Together, these institutions constitute an important resource for conservation and research; they also provide much of the inspiration for this book.

The Rare Plant Reintroduction Project was guided by an outstanding steering committee: Ken Berg (Bureau of Land Management), John Fay (U.S. Fish and Wildlife Service), Donna House (The Nature Conservancy), Lloyd Loope (National Park Service), Charles McDonald (U.S. Fish and Wildlife Service), Linda McMahan (Berry Botanic Garden), Steve Packard (The Illinois Nature Conservancy), Bruce Pavlik (Mills College), Robert Unnasch (The Nature Conservancy), and Peter White (University of North Carolina, North Carolina Botanical Garden). Their ideas shaped this project from its inception and helped us to focus on the most important issues.

Many others contributed materially to the fulfillment of this project and we would like to offer them thanks: William Curtis (*St. Louis Post-Dispatch* and *BioScience*), Emilia (Parra) Falk (U.S.D.A. Forest Service), William Jordan (University of Wisconsin Arboretum), Daniel Janzen (University of Pennsylvania), Carol Lippincott (University of Florida, formerly Fairchild Tropical Garden), Thomas Lovejoy (Smithsonian Institution and National Biological Survey), Michael Maunder (Royal Botanic Gardens U.K.), Gary Paul Nabhan (Arizona-Sonora Desert Museum Native Seeds/Search), Mark Schwartz (Illinois Natural History Survey), John Schwegman (Illinois Department of Conservation), William Stevens (*New York Times*), Christopher Topik (U.S.D.A. Forest Service), Susan Wallace (VanBloem Gardens), and Donald Waller (University of Wisconsin).

From the moment we engaged them in this project, the staff of Island Press has supported our efforts. Special thanks to Joe Ingram, Chuck Savitt, Barbara Youngblood, and especially to our editor, Barbara Dean, who contributed materially to shaping the book in its early stages. Island Press continues to be a vital and creative force in conservation, and we are proud to have developed this work with their support.

The contributors to *Restoring Diversity* deserve special acknowledgment, for it is upon their insight, scholarship, and dedication that the quality of the book ultimately rests. We appreciate the time and care that have gone into the preparation of the individual chapters and hope that the book will help to bring recognition to their outstanding contributions to the science and art of restoration.

Coediting a book is a lot like a marriage. The good ones are based on patience, communication, give-and-take, shared priorities, and a sense of humor (not necessarily in that order). To our readers, we aver that this is a stronger, better book for the three minds that shaped it. To each other we say thank you for being good collaborators in this enterprise. We can't imagine more convivial partners.

D.A.F.
C.I.M.
M.O.

Foreword

In these times of crumbling natural ecosystems, of rabid political resistance to conservation, it may seem utopian or even downright stupid to dream of restoring ecological diversity at any meaningful scale. It is hard enough to hold on to the few scraps of natural area that remain. Wild land now buried under asphalt and concrete seems impossible to revive. Species that have gone extinct cannot be brought back, at least not with current technologies. Few rare species show signs of recovering, even with the expenditure of millions of conservation dollars in their behalf. How, then, can massive ecological restoration and species reintroduction be technically or politically feasible?

Perhaps restoring diversity is most of all about restoring hope. Some vestiges of hope among us serve as seeds for renewed optimism. Despite all the bad news and failed programs for endangered species and ecosystems, conservation biologists and restoration ecologists persist in their missions. They stubbornly cling to the belief that much of the damage our species has wrought can be repaired and will be repaired. And there are a few encouraging signs of recovery. Some regions of the United States, largely deforested during the nineteenth and early twentieth centuries, are regaining their tree cover, although it may be centuries before the richness of the original old growth is seen again. Some species, such as the American alligator, have miraculously bounced back from the brink of extinction. Others have reversed long-term declines. What can we learn from these positive examples? Did we do something right, or were the species more adaptable than we gave them credit for? How can we enhance prospects of recovery for a much larger array of endangered plants and animals?

In *Restoring Diversity*, reintroduction of endangered plants is the theme around which a variety of recovery efforts and restoration projects is described. The editors point to the challenge of linking restoration ecology to the main body of ecological theory. They state that ecological restoration and species reintroductions must be an integral part of conservation strategy. I agree, and I share much of their optimism (at least on my good days I do). We have no shortage of potential case studies. Opportunities for restoring biodiversity are everywhere, if only because damage is everywhere. Even in our increasingly crowded world, we have a moral duty to make room for other species.

The reader will note that virtually all the case studies described in this book are small restoration projects for individual rare species. That has been

the state of the art. Will lessons from small-scale experiments transfer readily to regional experiments? Can we help biomes and continents heal as we restore populations of rare plants site by site? The editors rightly note the importance of context: "[T]here can be no successful reintroduction of any organism, or restoration of any community, unless these individual efforts are part of healthy, intact, functioning, and diverse large-scale ecosystems." If restorationists and conservation biologists design their reintroduction projects with this broad context in mind, perhaps the ultimate result of their efforts will be more than the sum of its parts. Peter White, in his chapter, describes how the spatial context of reintroduction sites will affect their ability to recover. Similarly, Richard Primack explains why restorationists must consider spatial and temporal scales far beyond their individual sites or the lifespans of the plants they work with. I emphatically underscore these points. Please, restorationists, don't get lost in the minutiae of your sites and your study organisms. These details are important, but so is the bigger picture, the broader mission. Ask how your project can contribute to the ultimate goals of reassembling tattered floras and faunas and allowing the evolutionary process to continue flowering.

That biologists and others are trying their damndest to restore populations and communities is a sign that we have not lost all faith in the future of life. We doggedly persist, planting seeds here, removing drainage tiles there, knocking down dams, yanking out exotics, closing roads, and watching Nature come back. Let us do more of these things, and let us carry on even when all reasonable grounds for optimism have vanished. No matter how futile our efforts often seem, restoring diversity is the right thing to do.

Reed F. Noss
Oregon State University

Introduction

One of the penalties of an ecological conscience is that one lives alone in a world of wounds.

Aldo Leopold, *Round River*

The Context for Reintroduction

The world of conservation is concerned largely with healing the wounds to which Leopold referred many years ago. Most conservationists would undoubtedly prefer to spend their time enjoying and studying the endless fascination of undamaged natural systems. But the daily practice of conservation takes place in a larger context of loss, damage, fragmentation, and all the other vestiges of late-twentieth-century civilization's impact on the planet.

Fields such as conservation biology and restoration ecology serve primarily to help us understand the consequences of this impact and to describe the undisturbed systems that remain as reference points for the healing process. The first line of defense, as always, remains the protection of these benchmark ecosystems: their composition, structure, and dynamic processes over space and time. However, to do the work of conservation in a wounded world, conservation practitioners, advocates, and regulators are increasingly confronted with the need to intervene in the functioning of ecosystems. The extent of this intervention as practiced today—as a component, in the current lexicon, of ecosystem management—would have been almost unthinkable little more than a generation ago, and it certainly would not have fallen under the banner of conservation. Land managers seed and weed; they start and suppress fires; they use herbicides, pesticides, lawn-mowers, chain saws, dredges, drip torches, electric fences, dynamite, bulldozers, helicopters, computers, and satellite Global Positioning units. These are the modern conservationist's tools of the trade.

The need for such intervention leads to wonderfully oxymoronic concepts characteristic of our age, such as the "managed natural area." For better or worse, even some of our larger parks and protected areas are becoming managed systems. Perhaps in only a few, increasingly extraordinary places, such as the great boreal forests of North America and Eurasia, can we still protect systems at the largest scale without active management. However, it may be more difficult for isolated fragments of landscape to behave in a "natural" manner and to replicate large-scale ecological processes that include the effects of millennia of human habitation.

The primary reason for this unprecedented intervention is, of course, the continuing and accelerating loss and degradation of global biological diversity (World Watch Institute 1994; Raven, Berg, and Johnson 1993; Wilson 1992; Woodwell 1993). In many cases, the choice is not between the survival or degradation of a pristine landscape but the more painful choice between two less desirable alternatives: accepting the loss of species and habitat or living in a managed world. One may debate whether landscape intervention and management are a tragic fall from the presumed innocence of nature (McKibben, 1989) or whether the very notion of unpeopled wilderness is a romantic European concept (Anderson 1993). In either case, intervention appears more and more necessary and inevitable. And as we debate, the losses continue.

One of the emerging tools of biological management is the reintroduction of organisms that have become locally extirpated from their natural habitat. In the United States, reintroduction is used increasingly by federal, state, and private conservation agencies and even plays a role in the implementation of the Endangered Species Act. For instance, nearly one-fourth of all U.S. plants listed under the act include reintroduction in their recovery prescription. This includes thirty-five species for which reintroduction is already in process or planned (Falk and Olwell 1992). Well-known reintroduction programs of such animals as the California condor, the black-footed ferret, and the red wolf have raised public awareness and acceptance of such measures; they have also raised new questions about the costs, biological effectiveness, and ethics of such projects.

Not all reintroduction work takes place in a conservation context, however. The relocation of populations of sensitive species is often proposed by developers, including public agencies, who find these organisms in the way of the construction or resource-extraction projects they wish to undertake. In California alone, there are now hundreds of pending applications by private developers to relocate populations of rare species to more convenient locations, under the banner of compensatory mitigation. Particularly plagued by this trend, wetlands, have been including coastal tide marshes and estuaries, freshwater bogs, swamps, cienegas, vernal pools, and many other sensitive

and important habitats. This application of restoration techniques is not hypothetical; it is being proposed and used with rare species every day.

Thus, whether we are talking about reintroduction in a conservation context—in which all parties accept the overall objective to enhance biodiversity—or in a compensatory mitigation context—in which this consensus may be absent—it seems fair to say that the age of biological management is upon us, whether we are prepared for it or not.

Unfortunately, by and large we are not. Despite hundreds of ongoing projects, the biological understanding for relocating or reintroducing species, populations, and communities is poorly developed. Links between restoration and the main body of ecological theory have barely been forged. For example, population-level introductions can be viewed as empirical tests of models of dispersal, invasion, and competition—factors that may influence greatly how reintroductions work ecologically. Likewise, every introduction of organisms into the environment is an opportunity to observe the founder effect empirically, especially with regard to changes in gene frequencies over time (Lewin 1989). Similar links need to be made with models of reproductive ecology, ecotypic variation, population biology, metapopulation dynamics, food webs and energetics, patch dynamics, successional pathways, habitat fragmentation, landscape heterogeneity, and many other concepts (Pickett and Parker 1994; Soulé 1987). In fact, we argue that restoration will be fully able to contribute to the conservation of biological diversity only as it becomes understood in the light of ecology. In the meantime, every restoration project that proceeds without gathering such baseline data is a lost opportunity for learning.

At the policy and strategic levels, endangered species reintroductions are similarly proceeding without adequate guidance. Few national guidelines exist for the reintroduction of endangered plants into the landscape, despite the substantial amount of current activity. Essential strategic considerations, such as the relationship between conserving existing habitats and establishing new ones, are consequently not reflected in many projects. The confluence of these two streams—the ecological practice of reintroduction and its strategic policy context—describes the conceptual framework for this book.

In thinking about reintroduction and restoration over the past several years, we have found that the issues separate into several concentric layers. At the most immediate and pragmatic level, the practice of reintroduction confronts certain *technical questions*, concerned largely with the feasibility of reintroducing and establishing living organisms. How can we successfully establish species A on site X? Should planting be done by broadcasting seed or by planting rooted seedlings? Which nutrients need to be provided during the

establishment phase? How can we stabilize the substrate of a highly disturbed site? Much of the restoration literature over the past decade has focused on such matters, and justifiably so.

These technical issues grade imperceptibly into a set of *ecological questions* about reintroduction and restoration. Which species are appropriate for a given location? How many individuals need to be introduced to form the basis for a sustainable population? For each species, what level of genetic variation should be incorporated, and how often should its status be assessed? How should outplanted material be distributed spatially in relation to existing populations and communities? How often, and how soon after outplanting, should the site be returned to a natural disturbance regime? Should pollinators, dispersal agents, and mycorrhizal symbionts be actively reintroduced, or can we rely on natural recolonization? Perhaps most important, how do we define success in ecological terms?

Unavoidably to the modern conservationist, these ecological questions are in turn embedded in a context of *strategic* and *political questions*, that can strongly influence the eventual outcome. Who owns the land? How do they intend to manage it? What is their relationship to (and role in) the proposed introduction project? Is the reintroduction effort entirely within a conservation context (such as a protected natural area), or does it occur within some other setting (such as compensatory mitigation for a population to be destroyed elsewhere)? How will ongoing management and monitoring be funded, and who will guarantee quality? What is the long-term security of the reintroduction site as protected habitat? Without a clear strategy in this realm, many reintroductions risk being ecologically ephemeral at best.

We believe that experience with these questions demonstrates convincingly the importance of being attentive to context. Ecologically, the central and recurrent issue of *Restoring Diversity* is the integration of site and population work into the larger wholes of landscapes and ecosystems. While we focus on endangered species because of their urgency, our context is always community and ecosystem integrity and function. We hope to draw attention to the larger biological context in which individual reintroductions take place. To be more precise, we state that no organism can be successfully reintroduced, no community can be successfully restored, unless these individual efforts are part of healthy, intact, functioning, and diverse large-scale ecosystems. This compels attention to community and ecosystem processes as well as landscape-scale spatial patterns of heterogeneity and change. We persist doggedly with that perspective, while exploring the techniques and ecology pertinent to individual sites and populations.

The same contextual thinking guides our approach to the strategic and political surroundings of reintroduction. Much of this work—including salvage of individuals and populations about to be destroyed and creation of new

communities—takes place outside the usual conservation frame, in a context of substantial new harm to the environment. The circumstances—mining, highway construction, commercial development—may vary, but the impacts are familiar: drained wetlands, asphalt prairies, more species gone forever. The relationship of restoration to these losses is ironic, even paradoxical, because the formula for destroying one site all too often includes a prescription for creating a "new" and supposedly comparable habitat or population elsewhere. The questions thus arise: On what basis may mitigation proposals be evaluated? Do the new sites meet the standards of diversity and stability that we would expect for other reintroductions? Although compensatory mitigation may resemble conservation-oriented reintroduction at the site level, the perspective changes when we look at such projects in a larger context. Strategically, mitigation represents the most controversial application of the techniques of restoration.

The CPC Study Project

These and other questions underlay the decision by the Center for Plant Conservation (CPC) to organize a study project, conference, and publication on the biology and strategy of rare plant reintroduction. As the only national organization devoted exclusively to conserving endangered native plants, CPC was in a logical position to take the lead in exploring these issues. Moreover, because of the nature of its central program—which includes the National Collection of Endangered Plants, maintained by a cooperative network of botanic gardens, arboreta, seed banks, and other facilities—CPC needed to clarify its own policies on several important questions, including its response to requests for plant material to be used in reintroduction projects.

Following a pathway used previously to explore questions of genetic diversity in rare plants (Falk and Holsinger 1991), CPC set out in 1990 to discover and organize the best thinking on the subject. With the support of its host institution, the Missouri Botanical Garden, CPC formed a steering committee, secured funding, and began a series of workshops on various aspects of rare plant reintroduction and ecological restoration. These efforts culminated in April 1993 with a national conference held in St. Louis, the first and most comprehensive meeting to date on the subject. Invited participants included academic biologists, agency staff, and activists concerned with the anthropogenic endangerment of plant diversity and the possible role that reintroduction might play in conservation. The organization of the meetings reflected a desire to approach the subject analytically, with sessions on biology, planning, and political context. In addition, a special session of case studies provided an opportunity to examine some of the larger issues in an applied

setting. Each invited presenter was assigned a specific topic. The strong attendance, excellent presentations, and coverage of the meeting in the scientific and popular press (Allen 1994; Roberts 1993; Stevens 1993) encouraged us to produce a book that explored these important issues further.

Organization of the Contents

Restoring Diversity reflects the approach to the underlying issues we outlined earlier, with sections on ecology, the planning context, critical policy issues surrounding compensatory mitigation, and illustrative case studies. The book is organized into five parts. We begin with the overall policy and environmental context, examining the setting in which reintroduction may be called for as a conservation tool. Next, we explore a series of biological issues that influence whether reintroductions are likely to be successful and, indeed, how success itself may be defined. The third part is devoted entirely to the use of reintroduction in the context of compensatory mitigation, which brings together matters of policy, biology, and uncertainty in the most intense situations. Each part concludes with a "focus" example, illustrating how some of the considerations discussed play out in real-world applications. In Part IV contributors describe a series of case studies that illustrate the extent to which every reintroduction project is unique. Finally, we have included model guidelines for rare plant reintroductions (Part V), which incorporate and combine the ideas contained in the book's chapters with our own experience. The guidelines build on a great deal of earlier, thoughtful policy work by a number of individuals and agencies, which we have tried to synthesize in a form useful to practitioners and regulators alike in evaluating and applying the tools of reintroduction.

Above all, we have sought to touch upon the many issues that influence feasibility and significance of returning endangered plant species to the landscape. We devoted considerable time to the sequencing of topics and the interweaving of the chapters, seeking coherence and tight organization in the final work. To achieve this objective in the published work, the papers presented at the conference were supplemented with several additional manuscripts solicited and written specifically for this project. We trust that the final result holds together conceptually and thematically for the reader.

The Terminology of Restoration

Lastly, a word about words. Because our field is new and evolving rapidly, the lexicon of conservation is, understandably, a semantic jungle. Soon after they

are introduced into general circulation, phrases such as *sustainable develop-ment* and *ecosystem management* often acquire so many simultaneous mean-ings that they cease to stand for anything at all.

The present topic is no exception. For instance, much heat (but little light) has been generated by debates over the distinction between *introduction* and *reintroduction*. The ends of this continuum are probably clear: the most con-servative action, such as augmenting a recently damaged population with propagules drawn from the same site, can comfortably be called reintroduc-tion, while the release of a member of a phylogenetic lineage never known from the continent might illustrate an extreme case of a new introduction. Generally, *introduction* involves putting something new into an ecosystem, while *reintroduction* involves replacing something that is very recently lost or that remains present. Unfortunately, this seemingly simple distinction turns out to be considerably less useful and more difficult to apply in practice. As with many continuua, most cases fall somewhere in the middle; black and white, one might say, are just extreme shades of gray. Few people actually ad-vocate introducing Siberian tigers to the Andes or endemic Hawaiian silver-swords to Borneo; such actions are clearly beyond the pale from the perspec-tive of this book. Much more problematic (and typical) are cases in which a species has not been recorded in a particular watershed for several decades, or where the degree of genetic differentiation among populations is unclear.

We believe that a "terminological" approach to defining biological man-agement is doomed to failure unless the terms are all made explicit with re-spect to *scale* in three dimensions: biological hierarchy, space and time. In particular, we find that the boundary between introduction and reintroduc-tion is impossible to determine unless scaling parameters have been estab-lished. Conversely, once scale is made explicit, the distinction frequently be-comes obvious.

As a thought exercise, consider the outplanting of fifty cuttings of a riparian endemic species X, known to have active metapopulation dynamics and rapid turnover rates, taken from source site S and proposed for outplanting at re-ceptor site R in an adjacent drainage in the same watershed. How is one to as-certain if this is an introduction or a reintroduction—or, indeed, if there is any real difference? The usual criteria for making this distinction—such as *historical range* or *local genotype*—actually beg the question they purport to answer. The tools of scale can help to clarify the issue by providing a consis-tent conceptual framework for analysis. For example, an examination of *spa-tial scale* might reveal that species X occurs in the bioregion, watershed, and drainage, but is not previously documented from the particular ten-meter re-ceptor site R. Thus, from a spatial perspective, this action is a reintroduction at the spatial scales of bioregion, watershed, and drainage, but possibly an in-troduction of a new element at the scale of a particular patch of habitat. On

the *time scale*, investigating the period of occurrence at each spatial scale, the species might be documented in the bioregion during the past ten thousand years, and in the watershed and drainage consistently for the past thousand years; in the receptor site R drainage, the average persistence of a particular patch might be ten years. The outplanting might thus be a temporal reintroduction at time scales of a decade or longer but a temporal introduction at finer (shorter) scales of temporal resolution. Finally, on the *biological scale*, electrophoretic analysis might reveal measurable genetic differentiation between all known sites of species X. Thus we might say that the proposed action is a reintroduction at the species, genus, family, and higher taxonomic levels, but an introduction at the level of ecotype or population.

The reader will undoubtedly have observed in the preceding example that the probability of an action becoming a new introduction instead of a reintroduction increases at finer levels of resolution on each scale. In other words, there is a relationship with what landscape ecologists call grain. This useful characteristic enables the restorationist to discern that the distinction between introduction and reintroduction is dependent upon the grain of resolution in the evaluative criteria in each scale. Hence, any project can be defined either way; coarse-grain criteria will, other things being equal, define more actions as reintroductions, while fine-grain criteria will tend to mark more projects as introductions. While this outcome may be disconcerting to some, it is the clearest way to work the problem without its being obscured by terminology and a priori categories.

Several other terms occur inevitably in writing about (re)introduction and are used throughout this book. *Enhancement* (also called augmentation, reinforcement, and restocking) refers to the addition of individuals to an existing population, with the aim of increasing population size or diversity and thereby improving viability. This is most conservatively done using genetic material drawn from the receptor site itself. *Translocation* describes the act of moving plants from an *in situ* location to any other site. A special category of translocation is *rescue* (or salvage), which implies that individuals have been saved from destruction by being moved elsewhere (although "elsewhere" is not necessarily protected habitat and is often *ex situ*). Conversely, *outplanting* (release) refers to movement of plants from an *ex situ* location to an *in situ* location, including restoration sites. Finally, we employ the term *compensatory mitigation* to describe situations in which an existing population or community are destroyed in exchange for creating or establishing a replacement.

The Society for Ecological Restoration defines *restoration* as the process of repairing damage to the diversity and dynamics of indiginous ecosystems. *Revegetation* and *reclamation* are related activities, but there's an important distinction: these are generally used to describe actions taken on severely degraded sites (such as mines, landfills, tailing piles, or construction zones)

where the immediate objective is to prevent further physical deterioration by establishing any vegetation at all, regardless of provenance or long-term ecological function. Such measures may serve as a prelude, or first step, to ecological restoration. (For a useful discussion of these and other terms, see Botanic Gardens Conservation International 1995.)

There are some terms we would like to excise along the way: *failure* is one of them. As ecologist Don Waller of the University of Wisconsin noted in his conference presentation, there are no true failures in ecological research, only unexpected outcomes. Pavlik (Chapter Six) similarly observes that "our current inability to construct a robust definition of success is due largely to our past unwillingness to document failure." In any field as new as restoration ecology, there should be no expectation that every outplanting trial, germination test, or successional management protocol will produce exactly the results we anticipate. For example, recognition of the importance of mycorrhizae to plant metabolism and growth was largely the result of many "failed" growth experiments, both in greenhouses and in the field. The "failure" of transplanted figs to reproduce alerted biologists to the role the fig wasp played in the trees' reproduction. Where rare species are concerned, the knowledge base is far too small to allow for anything other than an attitude of utter humility. The only thing we can count on is that the outcomes of these actions can tell us a great deal if we listen attentively.

In these and other cases, the only real failure would be rigidity in our thinking. Ecological restoration is characterized to a marvelous degree by uncertainty, risk, and unpredictability—all reminders of the merits of what Wes Jackson of the Land Institute calls an "ignorance-based worldview." We implore the would-be restorationist to design experiments as thoughtfully as possible and then to watch the results carefully with an eye for learning what the plants are saying.

REFERENCES

Allen, W. H. 1993. "Reintroduction of Endangered Plants." *BioScience* 44 (2): 65–68.

Anderson, K. 1993. *Before Wilderness: Environmental Management by Native Californians.* Menlo Park, CA: Ballena Press.

Botanic Gardens Conservation International. 1995. *A Handbook for Botanic Gardens on the Reintroduction of Plants to the Wild.* Joint publication with the IUCN Species Survival Commission, Reintroduction Specialist Group. BGCI, Richmond, Surrey, U.K.

Falk, D. A., and K. E. Holsinger. 1991. *Genetics and Conservation of Rare Plants.* New York: Oxford University Press.

Falk, D. A., and M. Olwell. 1992. "Scientific and Policy Considerations in Restoration and Reintroduction of Endangered Species." *Rhodora* 94 (879): 287–315.

Lewin, R. 1989. "How to Get Plants into the Conservationist's Ark." *Science* 244: 32–33.

McKibben, W. 1989. *The End of Nature*. New York: Doubleday.

Pickett, S. T. A., and V. T. Parker. 1994. "Avoiding the Old Pitfalls: Opportunities in a New Discipline." *Restoration Ecology* 2 (2): 75–79.

Raven, P. H., L. R. Berg, and G. B. Johnson. 1993. *Environment*. Ft. Worth, Tex.: Harcourt Brace Jovanovich.

Roberts, L. 1993. "Wetlands Trading Is a Loser's Game, Say Ecologists." *Science* 260: 1890–1892.

Soulé, M. E., ed. 1987. *Viable Populations for Conservation*. Cambridge, U.K.: Cambridge University Press.

Stevens, W. K. 1993. "Botanists Contrive Comebacks for Threatened Plants." *New York Times*, May 11, p. C1.

Wilson, E. O. 1992. *The Diversity of Life*. Cambridge, Mass.: Harvard University Press.

Woodwell, G. M., ed. 1993. *The Earth in Transition: Patterns and Processes of Biotic Impoverishment*. New York: Cambridge University Press.

Worldwatch Institute. 1994. *A Worldwatch Institute Report on Progress Toward a Sustainable Society*. New York: Norton.

Restoring Diversity

The Environmental and Policy Context for Reintroduction

As we noted in the Introduction, context is everything. In *Restoring Biodiversity*, this suggests the importance of attentiveness to a wide range of considerations that surround the specific act of reintroducing rare plants to the landscape. In other words, the ultimate determinant of success in ecological restoration may be the success with which individual acts at the site or population level are integrated into the larger trends and dynamics of landscapes and ecosystems. This does not imply, of course, that one can simply take local, immediate conditions for granted. These too are part of the context of reintroduction and include the technical feasibility of getting plants and communities established, the availability of source material with which to work, and the prospects for cooperation from local officials and the community. Rather, we suggest that the restorationist must focus simultaneously on the big picture and the small, on the grand design and the fine details, so as to replicate the way in which ecosystems work and improve our chances of success.

The first two chapters in this section are designed to illustrate some of the larger contextual considerations that influence the reintroduction of endangered plant species. As an opening problem statement, and to illuminate the overall need for population management and reintroduction, Morse (Chapter One) provides an overview of the endangerment of plant diversity in North America. As he observes, rare species constitute 15 to 22 percent of the entire continental vascular flora (a percentage that reveals, to our discredit, that we have done no better in protecting our natural diversity than have other countries with far fewer resources). Despite the heroic efforts of conservation groups and public agencies and important national legislation such as the Endangered Species Act (ESA), every year sees the disappearance of significant populations, communities, and habitats. Such losses do not affect rare plants exclusively, of course, but these organisms are often sensitive indicators of ecological deterioration and environmental quality. In addition to more obvious events, such as species-level extinction, many rare species are suffering from the invisible depletion of their numbers and erosion of the genetic diversity that enables them to adapt to environmental change. Thus, Morse suggests, the ongoing and increasing endangerment of hundreds, perhaps thousands, of native plant species creates the conditions in which more direct intervention—including reintroduction—may be required.

The following chapter by Kutner and Morse (Chapter Two) explores in more detail one of the most significant emerging contextual forces, global climate change. Drawing on an analysis conducted by The Nature Conservancy, they describe the potential impact of climate change on species that are presently rare and endangered. These impacts include latitudinal and elevational shifts in vegetation zones, reductions in suitable protected habitat, and daunting obstacles to population-level migration. Climate change is also

predicted to have significant effects on ecological processes, such as the timing and severity of storms, fires, and flood events. Species occupying specialized habitat—such as wetlands, topographic anomalies, or zones with unique edaphic conditions—may face a particularly difficult transition: they may be unable to persist in their current habitat if climatic conditions change significantly, but also unable to colonize other patches of suitable habitat across both natural and anthropogenic dispersal barriers. Moreover, the rate of vegetation change, one of the least-discussed aspects of climate change impacts, may exceed the ability of species to "migrate" across the landscape. The net effect of these forces will likely be felt most sharply by rare species. For many, introducing new populations may be a vital part of strategies for their conservation.

The biological context for individual reintroductions includes their connection into patterns and processes at the landscape and bioregional scale. While reintroduction projects are often physically carried out at small physical, biological, and temporal scales, their ultimate success—and thus their significance to biodiversity conservation efforts—rests on the integration of individual projects into heterogenous patterns of distribution at larger scales and over longer periods of time. In Chapter Three, White explores the challenges of connecting reintroduction projects at larger scales, such as linking reintroduction to landscape-scale processes and patterns of diversity in natural systems. The spatial context of individual reintroduction sites, for example, may influence their subsequent biological behavior, including dispersal, colonization, community interactions, exposure to disturbance events, and other dynamics influenced by habitat patch size and isolation. Compounding this complexity is the multi-layered biological hierarchy at work in any system, with individual plants, populations, metapopulations, and communities each interacting across the landscape in different ways. Moreover, as he notes, models of landscape-scale equilibrium are evolving rapidly, as recognition grows that different parameters or community components may achieve equilibrium at different spatial and temporal scales. In examples drawn from the floristically rich southern Appalachians, White demonstrates convincingly that reintroductions must be designed and managed at the landscape and regional scale if they are to achieve lasting significance.

One of the basic contextual dimensions for conserving biological diversity is the legislative arena, addressed by McDonald's description in Chapter Four of the regulatory and policy setting for introduction and reintroduction of endangered species. In addition to its better-known provisions for listing and protecting existing populations of endangered and threatened species, the Endangered Species Act (ESA) allows for reintroduction as a component in species recovery strategies. In fact, recovery plans for nearly one-quarter of all

federally listed plant species include reintroduction as a proposed action, including many for which it is a criterion for down-listing or de-listing. In this context, the protection status of new populations (often replacing others that have been inadvertently destroyed) is a pivotal issue; as McDonald notes, the ESA permits reintroduction of endangered species as experimental populations into unoccupied historical habitat. Reintroduced populations that are vital to the immediate survival of the species are designated as essential and receive full protection of the act, while other experimental populations are designated nonessential. In evaluating the strategic context for reintroduction as a long-term conservation tool, these distinctions may become enormously consequential.

Our FOCUS example illustrating how these contextual pieces fit together is the threatened flora of Hawaii, discussed by Mehrhoff (Chapter Five). With a native flora of nearly one thousand flowering plant species, 89 percent of which are endemic to the island chain and 32 percent of which are endangered or extinct, Hawaii is arguably "the endangerment capital of the United States." The presence of nearly as many naturalized exotic species as there are natives makes conservation even more difficult, as does the persistence of forty-two native plant species with no more than ten individuals existing in the wild. Hawaii brings together the imperatives for conservation in the most pressing manner imaginable, creating an intense biological and strategic context for restoration. Although transplant experiments have been attempted with several dozen endangered native species, the long-term success of these efforts is still highly uncertain. The challenges faced in Hawaii, although difficult in the extreme, mirror and perhaps presage those that restorationists will find elsewhere.

CHAPTER 1

Plant Rarity and Endangerment in North America

Larry E. Morse

———————•◦•———————

Actions focused on the conservation of particular plant species have become a substantial portion of all conservation activities in North America. Yet plant diversity continues to be degraded or lost, even in protected areas. While populations of rare plant species are maintained and managed in a variety of ways, there is an increasing need for and interest in the use of species reintroduction and population restoration as part of broadly integrated conservation strategies (Falk 1987).

In this chapter, I consider threats to rare and common plant species, review current conservation strategies, address the need for reintroduction and restoration, and offer some cautions. My discussion focuses on vascular plant species of North America north of Mexico; the even more intense plant conservation challenges in Hawaii are reviewed by Mehrhoff (Chapter Five).

I do not particularly consider here nonvascular plants. While some are protected by the same strategies used for vascular plants, completely new conservation strategies may be required for other kinds of organisms traditionally treated as plants, particularly algae and fungi.

Rarity and Endangerment in the North American Flora

Rarity and *endangerment* are not synonymous. When Columbus landed in the Americas five centuries ago, presumably there were many rare species among the native plants of the Western Hemisphere. Some of these may have become more common in the following centuries, while others have become more rare (Millar and Libby 1991). *Rarity* is an expression of the pattern of distribution and abundance of a species at a specified time. *Endangerment* (also known as threat) refers to factors (generally anthropogenic) that may make a species more susceptible to decline or extinction.

In general, a rare species is more readily threatened than a common one. However, even very common species can be jeopardized by such threats as exotic pests or diseases (Campbell and Schlarbaum 1994); the decline of the American chestnut (*Castanea dentata*) in the United States several decades ago is now just one case among dozens. Commercial exploitation can also reduce a common species' abundance substantially, as has happened to various cacti (Cactaceae) and the American ginseng (*Panax quinquefolius*).

Concepts of Rarity

One approach to categorizing the several kinds of rarity is that developed by Deborah Rabinowitz. As presented by Kruckeberg and Rabinowitz (1985), rarity can be expressed by the interaction of three factors: geographic range size, habitat specificity, and local population size. The rarest species have all three factors: small geographic ranges, narrow habitat specificity, and small local populations, as is the case for Harper's beauty (*Harperocallis flava*), known only from a few small populations in intermittent wetlands in a tiny area of the Florida panhandle (Clewell 1985). Fiedler and Ahouse (1992) expand upon the work of Rabinowitz, addressing hierarchies of causes of rarity.

Rarity criteria can be applied to a species throughout its range, or in a specified geographical area. Virtually all widespread species are rare somewhere, usually at the peripheries of their geographical or ecological ranges. Thus it is important to distinguish between plants that are rare globally (everywhere they occur) and plants that are locally rare in a specified area, even though they are common elsewhere. On occasion, a globally rare species can be so abundant in a small area that it would not ordinarily be considered rare at that location. For example, the globally rare seaside alder (*Alnus maritima*) is locally abundant along the Nanticoke River in Delaware and eastern Maryland (Stibolt 1981).

Numbers and Kinds of Rare Species

Due to differences in criteria used, there are currently several lists of rare or endangered vascular plant taxa in the United States. About seven thousand species are represented on one or more of these lists, out of the total of about seventeen thousand native U.S. species (Falk 1991; Kartesz 1992; Morse, Kartesz, and Kutner 1995). Also, about 90 mainland and 110 Hawaiian vascular plant species are extinct or possibly so (Russell and Morse 1992). Since these various numbers are based on different criteria for rarity, threat, or vulnerability, they serve here only to emphasize the magnitude of the plant conservation challenge in North America. Yet, many species are on most or all of these lists, showing there is concern for their continued existence regardless of the particular criteria used.

The precarious state of the native Hawaiian flora is discussed by Mehrhoff (Chapter Five). Of Hawaii's 980 native species of flowering plants, about 50 are known from 10 or fewer individuals and an additional 160 species from fewer than 1,000 individuals.

The U.S. Fish and Wildlife Service (1993, 1994) has formally listed 470 U.S. vascular plant taxa, as well as one lichen species and is considering about 2,000 additional plant taxa for listing, including a few dozen lichens and bryophytes. The Center for Plant Conservation has identified about 780 U.S. species as most likely to be vulnerable to extinction within a decade and tracks approximately 4,100 taxa of high conservation priority (Falk 1991). The Nature Conservancy, working with the various State Natural Heritage Programs, has identified about 4,900 native U.S. species to date as globally rare. Most of the 1,150 flowering plants and pteridophytes native to Hawaii are included in this total; even relatively common Hawaiian endemics are nevertheless globally rare.

The Conservancy and Heritage data show that rarity varies not only with geography but also with taxonomy. States with large numbers of globally rare vascular plant species include Arizona, California, Florida, Nevada, New Mexico, Texas, and Utah. Several plant families have large numbers of rare species in the U.S. mainland: the Asteraceae or Compositae (about 380), Fabaceae or Leguminosae (210), Brassicaceae or Cruciferae (160), Scrophulariaceae (160), and the Poaceae or Gramineae (100). On the other hand, the following families with more than 100 native mainland U.S. species have relatively few (10 or fewer) globally rare species: Adiantaceae, Amaranthaceae, Iridaceae, Juncaceae, Salicaceae, Verbenaceae, and Violaceae.

North America's rare plants fall into many categories, as shown by the following examples.

Presumably Ancient Lineages

It is difficult to know with certainty that a species itself has descended essentially unchanged since ancient times. Two kinds of indirect evidence are commonly used to show that a species may represent an ancient lineage. In some cases, a distinctive genus is known from pre-Pleistocene fossils. For example, the giant sequoia (*Sequoiadendron giganteum*) belongs to a genus known as fossils as old as the Cretaceous; the modern species is extant in fewer than a hundred groves in a 250-mile-long portion of California's Sierra Nevada (Hartesvelt et al. 1975). While strongly protected and carefully managed at most of these sites, the species is nevertheless vulnerable to climate change, exotic insects or diseases, and comparable kinds of broad threats.

In other cases, phylogenetic peculiarity, combined with a disjunct or relict distribution pattern, suggests a long history. Corkwood (*Leitneria floridana*),

for example, is a taxonomically peculiar species that is the only known member of its order (Cronquist 1981); it occurs in a few dozen scattered sites in the Gulf Coastal Plain, where it has been severely impacted by wetlands drainage for agricultural development.

RECENTLY EVOLVED SPECIES

While difficult to document, some very rare species are believed to have evolved quite recently (in geologic terms) and may be capable of adapting and expanding in future centuries or millennia if they can evade extinction. The clearest examples are allopolyploids, such as the well-studied Malheur wire-lettuce (*Stephanomeria malheurensis*) of Oregon (Gottlieb 1979; Guerrant, Chapter Eight), the Vermont maidenhair fern (*Adiantum viridimontanum*) of Vermont (Paris 1991), and Murray's birch (*Betula murrayana*) of Michigan (Barnes and Dancik 1985), a controversial fertile octoploid of complex origin. Ranker and Arft (1994) discuss the conservation significance of fertile hybrid–derived polyploid species. Chromosomal rearrangements offer another mechanism for rapid isolation and speciation; for example, as in the *Clarkia unguiculata* species complex of California (Holsinger 1985).

CLONAL RELICTS

A small number of plants are presently known only as individual, long-persisting, self-incompatible clones, spreading vegetatively but forming no viable seed. Among the best-known instances of this phenomenon are the box-huckleberry (*Gaylussacia brachycera*) and the Canby's mountain-lover (*Pachistima canbyi*) of the southern Appalachians (Coville 1919; Braun 1961).

SUBSTRATE-RESTRICTED RARITIES

Hundreds of rare plant species are restricted to regionally unusual geologic substrates, such as sandstone, limestone, granite, gypsum, or serpentine. Of these, the numerous endemics of the western serpentine outcrops are best studied, with recent reviews by Kruckeberg (1984) and Brooks (1987). Yet other rare species are confined to habitats related to unusual landforms, such as the shores of the Great Lakes, where such plants as the dwarf lake iris (*Iris lacustris*) are found (Voss 1972). Such species are unlikely to survive and compete in other kinds of habitats and therefore must be protected within their present landscapes.

MICROCLIMATE-RESTRICTED RARITIES

In many parts of the continent, locally endemic or regionally rare plants are confined to areas microclimatically warmer, cooler, wetter, or drier than the general climate of the surrounding landscape. For example, the Kate's Moun-

tain clover (*Trifolium virginicum*) and several other plants occur almost exclusively on the hot, south-facing slopes of the mid-Appalachian shale barrens (Keener 1983). Similarly, Oconee bells (*Shortia galacifolia*) is known only from steep-walled valleys of the Blue Ridge escarpment in the Carolinas and Georgia (Hardin 1977), and the Florida yew (*Taxus floridana*) and the Florida torreya (*Torreya taxifolia*) only from cool seepage areas in steep ravines near the Apalachicola River in the Florida panhandle and adjacent Georgia (Platt and Schwartz 1990). In addition to such microclimatically restricted, globally rare plants, numerous locally rare disjuncts or range-edge occurrences are also in microclimatically unusual settings, such as hemlock (*Tsuga canadensis*) and associates in cool, moist gorges in southern Ohio (Braun 1961).

Declines in More Common Species

For more common species, concerns are often raised about possible reductions in geographic distribution, habitat diversity, population size, or genetic variation. Geographical considerations focus mainly on the loss of disjunct or peripheral populations, especially when the affected occurrences are the last ones in a particular state or province. The widespread interest in this consideration is shown by the numerous efforts to protect plants that are rare in a particular U.S. state but common elsewhere. About forty-five hundred of the species that are monitored by the Heritage Programs as rare in one or more states in their range are nevertheless common elsewhere. About fifteen hundred species are apparently extirpated from at least one state but extant elsewhere, including about two hundred globally rare species.

Geographical disjunction is often coupled with unusual habitats. For example, the Hudson River (New York) populations of the heart-leaved plantain (*Plantago cordata*) are the only remaining places where this plant occurs along a freshwater-intertidal estuarine shore, rather than along a small stream on limestone or dolomite, as is more characteristic of inland populations of this species (Tessene 1969).

In contrast to the many instances of protection for geographically disjunct plant populations, there seem to be few cases in which habitat peculiarity within a species' main range is cited as a primary justification for conservation of a particular plant population. For example, the subalpine populations of the Aleutian maidenhair fern (*Adiantum aleuticum*) are considered significant for conservation in Idaho (Moseley and Groves 1990), although the species is common at lower elevations in the state. In Santa Cruz County, California, conservation attention is being given to the sandhills populations of ponderosa pine (*Pinus ponderosa*) due to the unusual substrate (Marangio and Morgan 1987). California's McNab cypress (*Cupressus macnabiana*) generally occurs on serpentine; the species' few non-serpentine populations

are considered significant for conservation (C. Millar, personal communication, August 1994).

Noticeable reduction in abundance of relatively widespread species has sometimes led to specific conservation measures to maintain exemplary stands as protected areas, such as the national parks established in the western United States for the coast redwood (*Sequoia sempervirens*), the saguaro (*Carnegia gigantea*), and the organpipe (*Stenocereus thurberi*). However, for most widespread species, disjunct stands (rather than range-center occurrences) are most likely to receive individual protection and management.

The conservation of within-species genetic variation in common (or rare) species currently occurs in two quite different ways. Formally recognized subspecies and varieties can be considered for conservation in the same ways as full species (although generally at a lower priority). Nontaxonomic genetic variation is more difficult to address, but can be approached through the use of molecular data or other genetic information to identify unusual populations or individuals (see Millar and Libby 1991; Holsinger and Gottlieb 1991). The impacts of habitat fragmentation, invasive exotics, and other landscape-level phenomena on the genetic structure of plant populations are still poorly known.

Habitat Loss and Degradation

The changes to the American landscape in the past five centuries need not be reviewed in detail here; we cannot turn back the clock and preserve large undisturbed tracts of the presettlement landscape. Instead, in most instances, we must do our best to conserve biological diversity with the landscapes, habitats, and plant populations we now have, optimizing the effectiveness of scarce conservation resources under continuing development pressures (Falk 1991).

Habitat loss and habitat degradation (including inappropriate land-management practices) are usually cited as major causes of plant species decline, extirpation, or extinction in the United States. Effects of habitat fragmentation on species biology and ecosystem integrity are also being increasingly studied (see Harris 1984; Angermeier and Karr 1994; Noss and Cooperrider 1994).

Losses of aquatic and wetland habitats have been particularly well documented. Dahl, Johnson, and Frayer (1991) report that 2.6 million acres of wetlands in the conterminous United States were lost from the mid-1970s to the mid-1980s, with 54 percent of this loss due to agricultural conversion. Of the estimated 221 million acres of wetlands present in colonial times, over half have been lost. Even greater percentages of loss are reported by Noss and Cooperrider (1994) for eastern forests (96 percent logged by 1920), tallgrass prairies east of the Mississippi (over 98 percent lost) and for riparian forests

and various other habitat types long subject to agricultural development. In California alone, ten million acres of natural habitat had been converted to agricultural or urban uses before 1950 and an additional five million acres converted between 1950 and 1980 (Jensen, Torn, and Harte 1993).

In recent years, the impacts of invasive exotic (nonindigenous) species on natural habitats have received considerable attention. Many hundreds of plant species desirable in agricultural, sylvicultural, or horticultural contexts, as well as non-native weeds, have become thoroughly naturalized in the United States, and thousands of additional species are now locally established outside cultivation and may pose future risks. Merely designating a presently natural area as a preserve will not keep human visitors, birds, floods, and the wind from bringing in other species in the future. McClintock (1987), McKnight (1991), and the Office of Technology Assessment (U.S. Congress, 1993) provide current reviews.

Current Plant Conservation Strategies

Reintroductions and population restorations are increasingly important tools in the broad range of plant conservation strategies currently employed in the United States. Plant conservation occurs on two major tracks. First is *habitat conservation* itself—the protection of exemplary occurrences of forests as forests, prairies as prairies, and swamps as swamps. Most, if not all, of the common species of plants in the United States occur in at least a few places where they receive protection at this habitat level. Occasional monitoring of health and stability of these widespread species is appropriate to give early warning of possible threats from pests, diseases, economic exploitation, exotic competitors, or other factors.

Restoration involving dominant and characteristic species is increasingly significant in expanding and buffering natural remnants of broadly depleted habitats such as prairies, eastern old-growth forests, oak savannas, and coastal dune vegetation. More difficult is the establishment of new examples of such ecosystems in areas where they are long gone. While safe guesses can usually be made concerning the common plants to use, rarer plants can only be added to such assemblages by inference or speculation unless historical records exist (see White, Chapter Three).

The second and more challenging kind of plant conservation is *species-by-species protection* of the rarer kinds of plants, requiring site-specific knowledge of individual occurrences of these less common species. Since rare plants often grow together in peculiar habitats, a number of rare species can often be protected as ensembles at well-chosen sites, such as the serpentine outcrop at Ring Mountain, California, where the Tiburon mariposa (*Calochortus*

tiburonensis) and several other locally endemic species occur together (Kruckeberg 1984; Strahan and Wolley 1987).

There are other elements of North American plant conservation programs:

INFORMATION MANAGEMENT
Native plant conservation at the species or the vegetation-community level requires systems of considerable, carefully organized information for use in environmental review and conservation planning (Morse and Henifin 1981; Jenkins 1985). For example, each of the fifty U.S. states now has a Natural Heritage Inventory Program, where comprehensive data are assembled on the taxonomic classification, status, occurrences, and conservation needs of the rare plants (and animals and natural communities) of the state. Five Canadian provinces also have similar programs. Initially, existing information from scientific literature, museum collections, and local experts is summarized, mapped, and computer-indexed. Precise locations, population status, land ownership, and management needs of the occurrences of the higher-priority species are then determined through fieldwork by staff, contractors, and volunteers. Priorities are revised as new sites for a species are found or as historically known sites are lost.

Since each Nature Heritage Inventory Program uses the same information-management methodology, data can be assembled and summarized to provide continental and range-wide perspectives on the local and global significance of the various plant species at any given site. Developed initially to provide guidance for conservation planning, the Heritage Programs are now also heavily involved in environmental review and in providing information to managers of lands owned by local, state, tribal, and national governments. Further background on the Heritage Programs is provided by Jenkins (1985), Morse (1987), Master (1991), Stolzenberg (1992) and Stein et al. (1995).

LANDOWNER CONTACT, AWARENESS, AND INTEREST
The cooperation of landowners is essential to long-term success in plant conservation. Far more sites are lost to ignorance than to deliberate vandalism. For privately owned lands, numerous formal and informal protection tools are available, including registration, easements, and management agreements. Acquisition by a conservation organization or land trust can provide long-term protection to important tracts of land. Designations, dedications, and management agreements are similarly useful for significant areas on government lands. Reviews of these land-protection methods are provided by Hoose (1981) and the National Research Council (1993).

PRESERVE DESIGN
Over its history, The Nature Conservancy, a private nonprofit conservation organization, has protected over eight million acres of land significant to

species or habitat conservation, including about fourteen thousand individual tracts of land. Thousands of other places have been protected by other organizations, and by government agencies (local, state, and federal) and tribal authorities.

Appropriate design is essential to the long-term success of such preserves, large or small. While a tiny, arbitrarily bounded preserve may be better than none at all for a while, a well-designed preserve needs to take into account such factors as the ecological needs of the priority species being protected, the nature of present and anticipated adjacent land use, and the possible need for a landscape unit large enough to maintain natural processes (such as flash floods or fire) upon which a species or its habitat may depend.

Preserve designs often include *primary areas*, containing the plant(s) and other features being protected, with suitable habitat for expansion; and *secondary areas*, which provide surrounding buffer land against extrinsic threats. At a landscape scale, primary (core) and secondary (buffer) areas have been identified in planning and managing various biosphere reserves under the auspices of the United Nations (Franklin 1977; von Droste zu Hulshoff and Gregg 1985). Landscape factors have also been considered in designing hundreds of small and large Nature Conservancy preserves. More complicated designs can include a number of primary preserves within a local setting (such as a watershed or mountain ridge) in which inappropriate land uses are discouraged, a planning strategy being used by The Nature Conservancy for larger or more complex preserves (Sawhill 1991). Jensen (1987), Jenkins (1989), Jensen, Torn, and Harte (1993), and Noss and Cooperrider (1994), among others, discuss the range of considerations involved in long-term conservation planning at the landscape level.

LEGAL PROTECTION OF SPECIES
The extent to which the legal system supports plant species conservation varies greatly from nation to nation and, to some degree, from state to state. In the United States, several federal laws have encouraged conservation of rare plants and animals and their habitats. Best known of these are the Endangered Species Act of 1973, which provides strong protection for a few hundred species; the National Environmental Policy Act (NEPA), which requires environmental review of many development and land-management proposals; and the various federal programs for wetlands protection.

PUBLIC LAND MANAGEMENT
Policies of the major federal land-management agencies provide further conservation measures for many hundreds of rare plant species in the United States (see McDonald, Chapter Four). About 29 percent of the U.S. land area is managed by federal agencies. Of these 671 million acres of federal lands, 270 million acres are managed by the Bureau of Land Management, 191

million acres by the U.S. Forest Service, 90 million acres by the U.S. Fish and Wildlife Service, 80 million acres by the National Park Service, 25 million acres by the Department of Defense, and 15 million acres by other federal agencies (National Fish and Wildlife Foundation 1992).

In 1993, the Heritage Programs and The Nature Conservancy examined land-ownership data for the 344 species (and subspecies) of plants federally listed at that time (National Heritage Data Center Network 1993). About 50 percent of these plants have at least one current occurrence known from federal lands, and about 12 percent have all or nearly all of their occurrences on federal lands. More than sixty listed plants are recorded from U.S. Forest Service lands alone, with slightly lower counts from Department of Defense and Bureau of Land Management lands.

Federal agencies are increasingly addressing the needs of rare plant species in developing their land-management plans. Under its "sensitive species" concept, the U.S. Forest Service takes into management consideration over a thousand rare plant species and attempts to sustain them without need for formal federal listing (Topik 1994). Many units of the National Park Service, U.S. Fish and Wildlife Service, and Bureau of Land Management also routinely include the needs of rare plant species in their land-management planning. In the past several years, the Department of Defense has also taken a strong interest in environmentally sensitive management of U.S. military lands, which contain a surprisingly large number of occurrences of rare plants. In May 1994, several U.S. federal agencies formed the Federal Native Plant Conservation Committee to promote better coordination among federal plant conservation programs (Canfield 1994).

Ex situ Samples

Long-term, off-site maintenance of living genetic samples (*ex situ* conservation) is an increasingly important strategy that complements the protection and management of existing natural occurrences of rare plants. *Ex situ* conservation provides important security against loss or depletion of the remaining natural stands of a rare species and also serves as a source for plant material for scientific studies, horticultural experimentation, and public education, as well as for introductions and reintroductions. Options for germplasm maintenance include cultivation of a population sample (not merely a few specimen plants) in a botanical garden or arboretum; maintenance of a seed or spore sample, perhaps cryogenically; and establishing independent, self-maintaining occurrences of the species, either at historically known sites or elsewhere (Eberhart, Roos, and Towill 1991).

Land-Management Planning

Land protection must be followed by management decisions; few tracts should simply be left alone, without any planned management, due to such

factors as suboptimal size, edge effects, and risks of introductions of exotic organisms. The highest management priorities for a site often involve maintaining, or even increasing, the extent, abundance, and stability of its most
important rare plant species. Occasionally, conflicting needs of co-occurring
rare animal species add to management complexity, as in timing of controlled
burns. Relative importance of the species at a site can be determined first at a
global or range-wide scale, then at a national scale, then at a local scale such
as a state or ecoregion, with needs of globally rare species generally taking
precedence over needs of species merely locally rare. For example, in the
Linville Gorge area of the North Carolina mountains, controlled fires are
being used to remove shrubs of the state-rare but locally abundant sand-
myrtle (*Leiophyllum buxifolium*) that are overtopping plants of a globally
rare, federally listed local endemic, the mountain golden-heather (*Hudsonia
montana*).

Needs for Reintroductions and Population Restorations

Beyond management for the species populations already present, a site manager can consider such additional possibilities as *restoring* or enhancing a
local population or establishing additional nearby populations of species already at the site; *reintroducing* species historically known from the site but
now missing; or even *introducing* species appropriate to the region and habitat
but not particularly suspected to have occurred historically at the site. While
often controversial, any or all of these actions may be appropriate for some
species at some places. Maunder (1992), Gordon (1994), and White (Chapter
Three) provide further discussion.

A first step in making decisions about species restorations, reintroductions,
and introductions is to determine, to the extent practical, the spatial and temporal distribution of the species in question, especially in the general region
of the site being considered.

If the species is not present at the site but is known or strongly suspected to
have occurred there historically, then reintroduction may be appropriate.
One caution is that some plants can persist invisibly at a site for decades as
dormant seed, awaiting return of appropriate conditions for germination and
growth (Harper 1977; Grime 1979). Therefore, particularly for species of land-
scape-mosaic habitats (such as short-lived plants of fire systems), an attempt
to stimulate the seed bank might be made before a restoration from elsewhere
is attempted.

Even when there is no particular reason to suspect that the species did
occur at a site within a given temporal frame, an introduction within a
species' native range may be appropriate when significant to the conservation
of the species and not detrimental either to the other species present at the site

or to the integrity of the habitat itself there. Matching of such environmental factors as climate, microclimate, slope, aspect, substrate, soil, and vegetation should be considered to the extent appropriate for the species (see Fiedler and Laven, Chapter Seven). Generally, the rarer the species, the more carefully a potential introduction site should be examined.

On occasion, an introduction beyond the known historical range of a species may be deemed appropriate, particularly to establish additional populations of quite narrow endemics. Even for more common species, introductions outside the historic range may serve to mitigate possible losses due to anticipated climate change; sea-level rise; alterations to hydrologic regimes, regional pests or diseases; or other broadly based threats to a species (see Kutner and Morse, Chapter Two). However, such introductions should be considered experiments, not substitutes, for protection and management of the remaining natural occurrences of a species.

The following series of questions may help in determining the appropriateness of a particular reintroduction or restoration action in the broader context of the conservation of the species and habitats involved.

- First, does the proposed action *benefit the species* (for example, by increasing its sites within its appropriate range and habitat, by increasing its abundance at its better sites, or by increasing the number of places in which it reproduces successfully)?

- Second, does the action *benefit the site or place* (for example, by restoring a dominant or characteristic species of the habitat, by bringing back a rare species historically known there, or by helping restore an instance of a rare habitat or an ensemble of rare species)? On the other hand, might bringing this species to the site interfere with conservation of other significant species already there, disrupt habitat integrity, or make a different and more important reintroduction more difficult?

- Third, in the case of population restorations, does the action *benefit the population* of that species already at the site? For example, poorly conducted return of garden-grown material can bring in pests or diseases, and mixing of genotypes might (or might not) degrade local climate or ecological adaptations (see Guerrant, Chapter Eight).

- Finally, and most important in the long view, does the proposed action *promote the opportunity for future adaptation and evolution* in this species and in the broader phylogenetic line to which it belongs? One possibility is that relict populations in microclimatic refugia may retain characteristics lost by the species elsewhere and hence be important genetic reservoirs for the species' persistence, adaptation, and evolution if past conditions return. On the other hand, geographically or ecologically

marginal populations may be more likely than central populations to adapt or diverge into new taxa under novel conditions. In any case, great care should be taken to maintain the genetic distinctiveness of geographically or ecologically disjunct stands of a species.

Conclusion

Because of continuing loss, fragmentation, and degradation of existing populations, even in protected areas, reintroduction and population restoration are becoming important tools for rare-plant conservation, supplementing management of these species in their remaining natural sites. However, careful planning is necessary to assure effective actions and minimize interference with other values at the target sites. In general, the rarer the species, or the more ecologically unusual the target site, the greater the need for professional assistance in developing scientifically sound reintroduction or restoration plans.

REFERENCES

Angermeier, P. L., and J. R. Karr. 1994. "Biological Integrity Versus Biological Diversity as Policy Directives: Protecting Biotic Resources." *BioScience* 44: 690–697.

Barnes, B. V., and B. P. Dancik. 1985. "Characteristics and Origin of a New Birch Species, *Betula murrayana*, from Southeastern Michigan." *Canadian Journal of Botany* 63: 223–226.

Braun, E. L. 1961. *The Woody Plants of Ohio*. Columbus: Ohio State University Press.

Brooks, R. R. 1987. *Serpentine and Its Vegetation: A Multidisciplinary Approach*. Portland: Dioscorides Press.

Campbell, F. T., and S. E. Schlarbaum. 1994. *Fading Forests: North American Trees and the Threat of Exotic Pests*. New York: Natural Resources Defense Council.

Canfield, J. E. 1994. "Plant Conservation Blooms with Creation of Native Plant Conservation Committee." *Endangered Species Technical Bulletin* 19 (4): 9–10.

Clewell, A. F. 1985. *Guide to the Vascular Plants of the Florida Panhandle*. Tallahassee: Florida State University Press.

Coville, F. V. 1919. "The Threatened Extinction of the Box Huckleberry, *Gaylussacia brachycera*." *Science* 50: 30–34.

Cronquist, A. 1981. *An Integrated System of Classification of Flowering Plants*. New York: Columbia University Press.

Dahl, T. E., C. E. Johnson, and W. E. Frayer. 1991. *Wetlands: Status and Trends in the Conterminous United States, Mid-1970s to Mid-1980s*. Washington, D.C.: U.S. Fish and Wildlife Service.

Eberhart, S. A., E. E. Roos, and L. E. Towill. 1991. "Strategies for Long-term Management

of Germplasm Collections." In *Genetics and Conservation of Rare Plants*. Edited by
D. A. Falk and K. E. Holsinger. New York: Oxford University Press.

Falk, D. A. 1987. "Integrated Conservation Strategies for Endangered Plants." *Natural
Areas Journal* 7: 118–123.

Falk, D. A. 1991. "Joining Biological and Economic Models for Conserving Plant Genetic
Diversity." In *Genetics and Conservation of Rare Plants*. Edited by D. A. Falk and K.
E. Holsinger. New York: Oxford University Press.

Fiedler and Ahouse. 1992. "Hierarchies of Cause: Toward an Understanding of Rarity in
Plant Species." In *Conservation Biology: The Theory and Practice of Nature Conser-
vation, Preservation, and Management*. Edited by P. L. Fiedler and S. K. Jain. New
York and London: Chapman and Hall.

Franklin, J. F. 1977. "The Biosphere Reserve Program in the United States." *Science* 195:
262–267.

Gordon, D. R. 1994. "Translocation of Species into Conservation Areas: A Key for Natural
Resource Managers." *Natural Areas Journal* 14: 31–37.

Gottlieb, L. D. 1979. "The Origin of Phenotype in a Recently Evolved Species." In *Topics
in Plant Population Biology*. Edited by O. T. Solbrig et al. New York: Columbia University Press.

Grime, J. P. 1979. *Plant Strategies and Vegetation Processes*. Chichester: John Wiley
and Sons.

Hardin, J. W. 1977. "Vascular Plants." In *Endangered and Threatened Plants and Animals
of North Carolina*. Edited by J. E. Cooper, S. S. Robinson, and J. B. Funderburg.
Raleigh: North Carolina Museum of Natural History.

Harper, J. L. 1977. *Population Biology of Plants*. London: Academic Press.

Harris, L. D. 1984. *The Fragmented Forest: Island Biogeography Theory and the Preservation of Biotic Diversity*. Chicago and London: University of Chicago Press.

Hartesvelt, R. J., H. T. Harvey, H. S. Shellhammer, and R. E. Stecker. 1975. *The Giant Sequoia of the Sierra Nevada*. Washington, D.C.: National Park Service.

Holsinger, K. E. 1985. "A Phenetic Study of *Clarkia unguiculata* Lindley (Onagraceae)
and Its Relatives." *Systematic Botany* 10: 155–165.

Holsinger, K. E., and L. D. Gottlieb. 1991. "Conservation of Rare and Endangered Plants:
Principles and Prospects." In *Genetics and Conservation of Rare Plants*. Edited by D.
A. Falk and K. E. Holsinger. New York: Oxford University Press.

Hoose, P. M. 1981. *Building an Ark: Tools for the Preservation of Natural Diversity*. Washington, D.C.: Island Press.

Jenkins, R. E. 1985. "Information Methods: Why the Heritage Programs Work." *The Nature Conservancy News* 35 (6): 21–23.

Jenkins, R. E. 1989. "Long-term Conservation and Preserve Complexes." *The Nature Conservancy Magazine* 39 (1): 4–7.

Jensen, D. B. 1987. "Concepts of Preserve Design: What We Have Learned." In *Conservation and Management of Rare and Endangered Plants*. Edited by T. S. Elias. Sacramento: California Native Plant Society.

Jensen, D. B., M. S. Torn, and J. Harte. 1993. *In Our Own Hands: A Strategy for Conserving California's Biological Diversity*. Berkeley: University of California Press.

Kartesz, J. T. 1992. "Preliminary Counts for Native Vascular Plant Species of U.S. States
and Canadian Provinces." *Biodiversity Network News* [The Nature Conservancy] 5: 6.

Keener, C. S. 1983. "Distribution and Biohistory of the Endemic Flora of the Mid-Appalachian Shale Barrens." *Botanical Review* 49: 65–115.

Kruckeberg, A. R. 1984. *California Serpentines: Flora, Vegetation, Geology, Soils, and Management Problems*. Berkeley: University of California Press.

Kruckeberg, A. R., and D. Rabinowitz. 1985. "Biological Aspects of Endemism in Higher Plants." *Annual Review of Ecology and Systematics* 16: 447–479.

Marangio, M. S., and R. Morgan. 1987. "The Endangered Sandhills Plant Communities of Santa Cruz County." In *Conservation and Management of Rare and Endangered Plants*. Edited by T. S. Elias. Sacramento: California Native Plant Society.

Master, L. L. 1991. "Assessing Threats and Setting Priorities for Conservation." *Conservation Biology* 5: 599–563.

Maunder, M. 1992. "Plant Reintroduction: An Overview." *Biodiversity and Conservation* 1: 51–61.

McClintock, E. 1987. "The Displacement of Native Plants by Exotics." In *Conservation and Management of Rare and Endangered Plants*. Edited by T. S. Elias. Sacramento: California Native Plant Society.

McKnight, B. N., ed. 1991. *Biological Pollution: The Control and Impact of Invasive Exotic Species*. Indianapolis: Indiana Academy of Science.

Millar, C. I., and W. J. Libby. 1991. "Strategies for Conserving Clinal, Ecotypic, and Disjunct Population Diversity in Widespread Species." In *Genetics and Conservation of Rare Plants*. Edited by D. A. Falk and K. E. Holsinger. New York: Oxford University Press.

Morse, L. E. 1987. "Rare Plant Protection, Conservancy Style." *The Nature Conservancy Magazine* 37 (5): 10–15.

Morse, L. E., and M. S. Henifin, eds. 1981. *Rare Plant Conservation: Geographical Data Organization*. Bronx: New York Botanical Garden.

Morse, L. E., J. T. Kartesz, and L. S. Kutner. 1995. "Native Vascular Plants." In *Our Living Resources: A Report to the Nation on the Distribution, Abundance, and Health of U.S. Plants, Animals, and Ecosystems*. Edited by E. T. LaRoe, G. S. Ferris, C. E. Puckett, P. D. Doran, and M. J. Mac. Washington, D.C.: National Biological Survey, U.S. Department of the Interior.

Moseley, R., and C. Groves. 1990. *Rare, Threatened, and Endangered Plants and Animals of Idaho*. Boise: Idaho Department of Fish and Game.

National Fish and Wildlife Foundation. 1992. *FY 1993 Fisheries and Wildlife Assessment: U.S. Department of Agriculture Forest Service*. Washington, D.C.: National Fish and Wildlife Foundation.

National Heritage Data Center Network. 1993. *Perspectives on Species Imperilment*. Arlington, Va.: The Nature Conservancy.

National Research Council. 1993. *Setting Priorities for Land Conservation*. Washington, D.C.: National Academy Press.

Noss, R. F., and A. Y. Cooperrider. 1994. *Saving Nature's Legacy*. Washington, D.C. and Covelo, Calif.: Island Press.

Paris, C. A. 1991. "*Adiantum viridimontanum*, a New Maidenhair Fern in Eastern North America." *Rhodora* 93: 104–122.

Platt, W. J., and M. W. Schwartz. 1990. "Temperate Hardwood Forests." In *Ecosystems of*

Florida. Edited by R. L. Myers and J. J. Ewel, 194–229. Orlando: University of Central Florida Press.

Ranker, T. A., and A. M. Arft. 1994. "Allopolyploid Species and the U.S. Endangered Species Act." *Conservation Biology* 8: 895–897.

Russell, C., and L. E. Morse. 1992. "Plants." *Biodiversity Network News* [The Nature Conservancy] 5 (2): 4.

Sawhill, J. C. 1991. "Last Great Places: An Alliance for People and the Environment." *Nature Conservancy* 41 (3): 6–15.

Stein, B. S., L. L. Master, L. E. Morse, L. S. Kutner, and M. Morrison. 1995. "Status of U.S. Species: Setting Conservation Priorities. In *Our Living Resources: A Report to the Nation on the Distribution, Abundance, and Health of U.S. Plants, Animals, and Ecosystems*. Edited by E. T. LaRoe, G. S. Ferris, C. E. Puckett, P. D. Doran, and M. J. Mac. Washington, D.C.: National Biological Service, U.S. Department of the Interior.

Stibolt, V. M. 1981. "The Distribution of *Alnus maritima* Muhl. ex Nutt. (Betulaceae)." *Castanea* 46: 195–200.

Stolzenberg, W. 1992. "The Heritage Network: Detectives of Diversity." *Nature Conservancy* 42 (1): 22–27.

Strahan, J., and G. J. Wolley. 1987. "The Ring Mountain Restoration Plan." In *Conservation and Management of Rare and Endangered Plants*. Edited by T. A. Elias. Sacramento: California Native Plant Society.

Tessene, M. F. 1969. "Systematic and Ecological Studies on *Plantago cordata*." *Michigan Botanist* 8: 72–104.

Topik, C. 1994. "The National Forest System Rare Plant Conservation Program." *Endangered Species Technical Bulletin* 19 (4): 15–17.

U.S. Congress, Office of Technology Assessment. 1993. *Harmful Non-Indigenous Species in the United States*. Washington, D.C.: U.S. Government Printing Office.

U.S. Fish and Wildlife Service. 1993. "Plant Taxa for Listing As Endangered or Threatened Species: Notice of Review." *Federal Register* 58: 51143–51190, September 30, 1993.

U.S. Fish and Wildlife Service. 1994. "Box Score: Listings and Recovery Plans." *Endangered Species Technical Bulletin* 19 (4): 24.

von Droste zu Hulshoff, B., and W. P. Gregg, Jr. 1985. "Biosphere Reserves: Demonstrating the Value of Conservation in Sustaining Society." *Parks* 10 (3): 2–5.

Voss, E. G. 1972. *Michigan Flora*. Part 1, Gymnosperms and Monocots. Bloomfield Hills and Ann Arbor, Mich.: Cranbrook Institute of Science and University of Michigan Herbarium.

CHAPTER 2

Reintroduction in a Changing Climate

Lynn S. Kutner and Larry E. Morse

In the next few decades, climate change is likely to play an increasingly important role in the planning and management of conservation actions by affecting the survival of populations and species and altering the composition of natural communities. Because climate is a major factor that influences the distribution of plant species and vegetation patterns, the predicted changes in climate may have a substantial impact on the natural flora and vegetation. Accordingly, the design of reintroduction and restoration projects will probably be influenced by these considerations.

An analysis conducted by The Nature Conservancy on the potential effects of climate change on the native vascular flora of North America (Morse et al. 1993) provides a preliminary assessment of regional and floristic patterns of vulnerability to climate change. These results provide a framework for discussing some of the implications of climate change for the conservation and management of rare plant populations, including considerations that may influence plans for reintroduction and restoration projects.

Climate Change

Climate change has occurred naturally throughout the history of the earth, with substantial variations in the frequency and magnitude of changes occurring most notably during glacial cycles. Temperatures on global and local scales were both much warmer and much colder in the past than present-day averages (Pielou 1991; Peters 1992). In the next few centuries, however, climate is anticipated to change more rapidly than ever before due to anthropogenic influences.

Human activities, particularly the emission of carbon dioxide from burning fossil fuels, are altering the concentrations of "greenhouse" gases in the atmosphere (Schneider 1989). It is predicted that the equivalent concentration of carbon dioxide in the earth's atmosphere may double early in the next

century and change climate conditions. Models of climate change, as summarized in assessments by the Intergovernmental Panel on Climate Change (IPCC 1990, 1992), predict an increase in mean global temperature of approximately 1.5 to 4.5°C (2.7 to 8.1°F) in the next century at an approximate rate of 0.3°C (0.5°F) per decade.

While even a 3°C increase in mean global temperature would be a substantial change from the current climate, the rate of climate change may be the greatest determinant of the impact on biological diversity. At the end of the last major glaciation, the climate warmed over a period of several thousand years (Pielou 1991). Future climate change due to anthropogenic influences could occur many times faster than any past episode of global climate change (see IPCC 1990, 1992; Schneider, Mearns, and Gleick 1992).

Limitations of Climate Change Models

Despite a fairly general consensus on the magnitude of predicted increases in mean global temperature, there are differences between various climate change models' portrayals of future climate. General circulation models (GCMs) are used to represent physical processes and interactions on a large scale. Differences, however, in the treatment of climate variables cause discrepancies among GCMs. For example, in a comparison of GCMs by Dickinson (1986), there are variations among the precise predictions of future temperature change due to differing treatments of CO_2 increase, clouds, snow, and sea ice.

Climate change models are verified using several methods, including comparisons with current or ancient past conditions and isolation of individual components of a model to confirm a correspondence with actual climate conditions. Paleoecological evidence differs in some cases from estimates based on computer models of past climate conditions. The presence of fossil palms and cycads in Wyoming and Montana during the Eocene (50 million years ago) suggests that year-round temperatures there were above freezing, rather than the sub-zero Eocene temperatures predicted for this continental-interior region by some models (Kerr 1993).

Regional Changes in Climate

As suggested by global warming models, the quantity and timing of precipitation, rates of evaporation, and temperature extremes could vary significantly on regional or local levels. These changes are likely to be far greater in the interior of a continent than in coastal regions. Several models predict increased temperatures and decreased summer precipitation in mid-continental areas, which could increase wildfires (Schneider 1989). In contrast, coastal regions may experience less rapid temperature change due to the conductive abilities

of large bodies of water acting as a stabilizing influence on temperature (Ray et al. 1992). Increases in sea level are apt to be more important in coastal areas.

Regional predictions are especially difficult to formulate using models. Although high-resolution GCMs of the future that are capable of calculations over fifty years will have grids of approximately 100 kilometers, smaller-scale effects may still remain obscured. Variations in hills, coastlines, lakes, and soils and in storms and clouds could be imperceptible (Schneider et al. 1992), but these natural features have important influences on local climate. It is feasible that regional changes in temperature would be either greater or lesser than mean global changes (Schlesinger and Mitchell 1987, IPCC 1990). Using forecasts of regional changes in temperature, evaporation, and precipitation, the scenario developed by Schneider, Mearns, and Gleick (1992) indicates regional average changes of $-3°$ to $+10°C$ from the global average change.

Climate Change and Vegetation

Climate plays a significant role in determining the broad-scale distribution of plants. Temperature and precipitation and their seasonal patterns are some of the most important physical factors affecting the distribution of individual plant species, populations, and major communities (Brown and Gibson 1983). Patterns in such abiotic factors as fires and storms, regional winds, hydrologic regimes, and various microscale effects also influence plant distributions. On different scales, these factors are significantly affected by changes in the magnitude and patterns of regional temperature and precipitation (Schneider 1989; IPCC 1990, 1992).

The strong association between plant species distributions and climate suggests that rapid global climate changes would alter plant distributions, resulting in extensive reorganization of natural communities (Graham and Grimm 1990). These events, along with local extirpations of plant populations and species extinctions, would clearly have broad ecological impacts, including changes in ecosystem dynamics and alterations in habitat suitability for a broad variety of organisms. Climate change could thus present major challenges to the conservation of biological diversity at scales from genes to landscapes.

Climate and Disturbance Regimes
Climate change may alter the frequency and intensity of disturbances in addition to affecting the magnitude and seasonality of temperature and precipitation on global and local scales. Temperature extremes and changes in

disturbance regimes can have a greater influence than small shifts in the overall climate on plant species' survival at particular locations. This is shown by paleoecological studies (Davis 1989) that altered disturbance regimes can intensify the effects of climate change on plants and increase the total amount of vegetational change. More frequent droughts, fires, and pest and pathogen outbreaks are predicted to act in conjunction with climate change to significantly transform the landscape (Peters 1992). Some stress-tolerant species however could benefit from extreme climates if competitors are locally depleted or eliminated. An increased frequency of catastrophic events is likely to affect successional patterns and the influx of exotic species.

Fire patterns may be altered by increased temperatures, local droughts, and dramatic changes in precipitation patterns. Significant changes in the fire regime would affect all plant species, even those that have evolved mechanisms to resist fire or need periodic fires for regeneration, such as those species characteristic of the pitch pine forests of the New Jersey pine barrens or the chaparral vegetation in California. For example, the results of a study based on predictions from three GCMs indicate an approximate 46 percent increase in seasonal severity of Canadian wildfires if CO_2 doubles, with a similar increase in national burned area (Flannigan and VanWagner 1991). Although the study takes neither fire control nor ignition frequency into consideration, it does indicate a potentially important change in disturbance due to climate change.

Some climate change models (Emanuel 1988) predict an increase in sea surface temperature, which could spawn more frequent and more severe tropical storms. This phenomenon is liable to be especially important in coastal areas of the southeastern United States that already experience the devastating effects of hurricanes. If storms become more frequent and intense with global warming, the impacts could include high winds that uproot and break trees, and floods and avalanches that can reset stream and river ecosystems to early successional phases (Franklin et al. 1992). Together, predicted changes in climate and disturbance regimes would stress the already fragile balance established between native plants and their habitat and cause dramatic restructuring of existing ecosystems. In any area, some species would benefit, and others would not.

Another abiotic factor that could be affected by overall climate changes is regional wind patterns. In mountainous coastal areas such as the Pacific Northwest or Hawaii, winds bearing moisture-laden air are often the most significant determinant of local vegetation patterns. When the air reaches windward slopes, it rises and cools. Since cool air is capable of holding less moisture, the water is released as precipitation on the slopes. Areas on the leeward side of mountains are therefore in a "rain shadow" that maintains their drier

conditions. In Hawaii, some entire small islands fall in the rain shadows of larger islands (Stone and Stone 1989).

With larger changes in climate, the intensity of the winds may increase (Knox and Scheuring 1991) and/or the patterns themselves may shift. Stronger and more frequent storms would presumably enhance existing rain shadow conditions, creating even drier conditions on the leeward sides of such mountain ranges as the Cascades and the Sierra. Climate change could also produce alterations in the storm paths that are so influential in this region. In one study, Stine (1994) suggests that California's Sierra Nevada experienced severe drought conditions in about A.D. 1300, during the Medieval Warm Epoch. It appears that droughts in the region were both more severe and more persistent than they have been subsequently. Because there is similar evidence of extremely dry conditions in Patagonia during the same epoch, it is thought that drought conditions were caused by poleward shifts or contractions of the westerly winds in both hemispheres. Global climate change could again alter wind patterns and intensify rain shadow effects in California and similar areas, causing long and severe droughts.

Changes in Hydrology

Although climate changes would affect habitat suitability for all species, freshwater wetland ecosystems are likely to be the most vulnerable to changes in the amount and timing of precipitation and evaporation. Increased evapotranspiration would cause soils and habitats, particularly wetlands, to become drier. The hydrologic patterns of streams are also prone to substantial alterations if there are changes in the seasonal timing and amount of precipitation and the extent of runoff (Carpenter et al. 1992). Even minor fluctuations in water availability are known to affect habitat suitability for many wetland plant species.

In addition, numerous North American wetlands are associated with seacoasts and related estuaries. Development along most coasts has fragmented these habitats and in many places would prevent species from migrating upslope to slightly higher ground if sea levels rise significantly, as expected in most climate-change scenarios (Reid and Trexler 1991; Schneider 1989). If the sea level rises rapidly, landforms and vegetation may not have sufficient time to adjust. The most significant impact may be on rare species that grow only in estuaries or along coastal shores. These species include sensitive joint-vetch (*Aeschynomene virginica*), which currently occurs in Maryland, New Jersey, North Carolina, and Virginia, and 'ohai (*Sesbania tomentosa*), a native Hawaiian species that grows near shorelines on all of the islands. Such rare coastal species would be susceptible to losses of populations and are likely to become more rare.

Microclimates and Vegetation

In any but the most monotonous landscape, factors such as ground-level air temperatures, relative humidity, dew concentrations, exposure to winds, persistence of snow, length of frost-free growing season, and duration and intensity of sunlight vary considerably. The local microclimate variations due to these factors in almost any small area are greater than the differences in regional climate of places hundreds of kilometers apart (Wolfe, Wareham, and Scofield 1949). This heterogeneity creates microhabitats that often have unique species assemblages. During episodes of climate change, both warm and cold microclimatic refugia become floristically important, since individual plants experience these local conditions and not merely the regional climate.

Some microclimates are created nearly independently of regional climate and may be resilient to climate change. Cold-air talus slopes in Iowa (Frest 1984) and West Virginia (Hayden 1843; Core 1968) produce moist chilled air due to persisting underground ice deposits of apparently late-glacial age. Disjunct populations of once regionally abundant species such as twinflower (*Linnaea borealis*) remain in these microhabitats as ice-age relicts. Unless the ice deposits are depleted, the conditions at these sites could remain fairly constant despite substantial regional or global climate changes.

Representing warmer local microhabitat, the Appalachian shale barrens species currently occur in steep south-facing shale talus slopes (Core 1968). Species such as Kate's Mountain clover (*Trifolium virginicum*), shale-barren evening primrose (*Oenothera argillicola*), and white-haired leather flower (*Clematis albicoma*) may be resilient to substantial regional temperature increases and might even become more abundant in a warmer climate. At many sites, the barrens gradually grade into adjacent open woods, with a patchy transition dependent on local topography. With global warming, adjacent woods could become even more open and provide expansion space for shale-barren species even if the original sites become too warm or dry. Dispersal requirements would be well within the normal incremental dispersal capabilities of most species.

Migration

Anthropogenic climate change is anticipated to occur at rates more than five times faster than any changes since the last glacial maximum, including during the period of most rapid deglaciation (Overpeck, Bartlein, and Webb 1991). The response of many plant populations, species, and natural communities to rapid climate change may depend on the rates that they are able to migrate to more suitable areas. Each level of biological organization would

experience changes due to the differing abilities of individual species to respond to climate change.

Population Responses

Since there can be high levels of genetic variation among different populations of the same species, populations of some species are likely to have different climate tolerances and migration abilities. Genetic studies of ponderosa pine (*Pinus ponderosa*) indicate that there are at least four major ecotypes (Millar and Libby 1991) that may have slightly different responses to climate change. Harperella (*Ptilimnium nodosum*) is a rare semi-aquatic plant known from approximately fourteen sites from the mountains of West Virginia and Maryland to the coastal plain of Georgia (Maddox and Bartgis 1991). Recent studies of allozyme variation in the *P. nodosum* complex show that there are significant electrophoretic differences among three population groups that presumably have different climate tolerances (Kress, Maddox, and Ruesel 1994). These genetic differences indicate that species are not likely to react uniformly to episodes of climate change.

While some individual plants and populations may respond rapidly to changes, others may survive for several generations in place or persist as long-lived clones despite significant climate change. For example, individual trees of species such as giant sequoia (*Sequoiadendron giganteum*) and Great Basin bristlecone pine (*Pinus longaeva*) can live for several thousand years. This suggests that they may have germinated under climate conditions substantially different from the present. In addition, stands of certain self-incompatible clone-forming shrubs such as box huckleberry (*Gaylussacia brachycera*) appear to have spread vegetatively over thousands of years (Coville 1919; Wherry 1972), tolerating climate changes but not reproducing. These examples demonstrate that some species have a high degree of physiological tolerance to climate fluctuations and that there may be substantial differences among and within species' responses to climate change.

Species' Responses

Even if a species is vulnerable to extirpation in its present range, it has the potential to survive or even flourish during episodes of climate change if it can disperse rapidly to novel areas of more suitable climate and establish itself, compete, and persist there. At the peripheries of species ranges, species currently limited by intolerance of cold conditions may migrate to new sites as climatically marginal areas become more suitable. Species that are less vulnerable and more dispersible would be most likely to expand their ranges. One effect of these migrations may be the replacement of vulnerable, less dispersible rare species with resistant, dispersible, widespread species in particular regions.

The fossil pollen record provides evidence of migration rates for various tree species during the warming at the end of the last glacial period. These migration rates ranged from about 5 to 150 kilometers per century, depending on the species and local conditions (Shugart et al. 1986). Paleorecords indicate, for example, that maple and beech species migrated 10 to 20 kilometers per century, while oak and pine species migrated 30 to 40 kilometers per century (EPA 1989). Studies of eastern hemlock (*Tsuga canadensis*) fossil records show that hemlock, with its winged wind-dispersed seeds, migrated 20 to 25 kilometers per century during the Holocene (Davis 1990).

The natural histories of herbaceous forest species, such as trilliums (*Trillium*) and lady's-slipper orchids (*Cypripedium*), can be used to analyze possible climate change responses (Davis 1990). Many understory herbs produce few seeds, reproduce vegetatively, depend on particular soil conditions produced by the overstory trees, or are rare throughout their range. These characteristics would limit the ability of these species to migrate during periods of rapid climate change, possibly making understory herbs more dependent on human intervention to avoid losses in diversity and extinction (Davis 1990).

Various studies suggest that plant species survival during periods of rapid future climate changes would require shifts of plant species ranges of up to 500 kilometers within the next century, exceeding the known rates of migration for many plant species (Davis 1990; Davis and Zabinski 1992; Schwartz 1992). Species that cannot migrate rapidly enough to follow climate change and shifts of habitat ranges, or cannot otherwise persist or adapt, would thus be vulnerable to extinction.

Community Responses

Since species respond individually to climate change, migration rates will vary both within and among natural communities. It is unlikely that entire biological communities would move together and maintain their current compositions during episodes of substantial climate change (Graham and Grimm 1990). Instead, communities would be reconfigured according to the responses of their individual components. The fossil record provides evidence of time lags of decades or centuries in species migration (Davis 1989). These time lags and the ability of different species to tolerate changing and potentially unfavorable conditions will determine the degree of "ecological inertia" in a regional flora (Davis 1984; Pielou 1991). Wright (1984) and Webb and Bartlein (1992) discuss further the concept of ecological inertia and the degree that the current flora is in equilibrium with the climate.

Changes in community composition in response to climate change are documented in the fossil record through the disassociation and reassembly of groups of plant and animal taxa (Graham and Grimm 1990). The range shifts of hemlock and beech after the Wisconsin glaciation demonstrate this indi-

vidualistic response. About twelve thousand years ago these species had al-lopatric ranges in the southeastern United States. When the climate began to warm, hemlock and beech spread northward (Graham and Grimm 1990), forming part of a new species assemblage: the current northern hardwood forest. Conversely, three previously coexistent species of rodents—the prairie dog (*Cynomys ludovicianus*), eastern chipmunk (*Tamias striatus*), and northern bog lemming (*Synaptomys borealis*)—currently have disjunct distri-butions across North America (Graham 1988). These variations in species as-semblages display the ephemeral nature of former, existing, and future com-munity types. Davis (1981) shows that some modern forest communities developed within the last two thousand years. With climate change, differ-ences in migration rates and changes in resource competition and distur-bance regimes are likely to alter existing community types and create associa-tions with no past or present analogues (Davis 1989).

Preliminary Assessment of Climate Change Impacts on Plants

The Nature Conservancy conducted a preliminary assessment of the poten-tial effect of climate change on the vascular flora of North America (Morse et al. 1993; Morse et al. 1995). For this study, *climate envelopes*—as defined by the maximum and minimum mean annual temperature that each species ex-periences in its current distribution—were determined for the approximately fifteen thousand native vascular plant species in North America recognized in the checklist by Kartesz (1994).

Since most of the general circulation models suggest that doubled atmos-pheric CO_2 may cause an approximate 3°C (5.4°F) mean increase in global temperature, that estimate was used as the principal scenario for all analyses. As discussed previously, the nature and intensity of future climate change is likely to vary considerably from region to region, and current climate models differ in their regional predictions. Accordingly, uniform climate increase is an acceptable simplification for an initial analysis, allowing the development and application of a climate envelope technique for large numbers of indi-vidual species. The same biological and distribution data could be used to test the possible impacts on North America of different climate-change scenarios.

While many factors could be used in a climate envelope analysis, to esti-mate the viable climate envelopes for each species in the study area several simplifying assumptions were made about the relationships between plants and climate:

- Climate determines the range of plant species.

- Mean annual temperature adequately approximates climate.

- Species distribution appears to be in equilibrium with present climate.
- A species' current climate envelope is equivalent to its physiological tolerance of climate variation.
- The envelopes incorporate all the temperature variations within the region(s) inhabited by the species.

Together, these assumptions state that the current distribution of each species is greatly influenced by climate and that temperature adequately represents climate.

Clearly, each of the previous assumptions is not actually met for all native vascular plant species. Precipitation and soil moisture, for example, are more important determinants of range limits than temperature in some regions. Although climate change is predicted to cause both increases and decreases in moisture levels (Dickinson 1986), soil rather than temperature would remain the chief limiting factor in certain cases. Species restricted to specific soil types, such as the narrow endemics on serpentine soils in California (Kruckeberg 1984) are likely to continue to be distributed primarily according to unique edaphic characteristics. In other cases, such as those aforementioned, influences of disturbances, winds, or changes in hydrology may be more consequential than changes in temperature.

In addition, it is commonly accepted that current distributions of many species and communities are not in equilibrium with the present climate. Due to slow dispersal rates, competition, and persistence of existing plants in habitats that have become less favorable, vegetation types and individual species do not instantly respond to a changing climate. Current species distributions reflect in part climate fluctuations since the end of the last glacial period, approximately twelve thousand years ago.

Despite the limitations of the model discussed earlier, the use of simplified climate envelopes allows the initial identification of broad patterns of species vulnerability to climate change. The preliminary results presented here are part of an ongoing climate change analysis that aims to more accurately represent the relationship between plants and climate and incorporate regional and/or seasonal changes in climate patterns.

Examples of Climate Envelopes

Each species experiences a range of temperature, precipitation, and other climate parameters within its geographic range. The North American distributions of two species provide examples of actual climate envelopes and how they differ. Ponderosa pine (*Pinus ponderosa*), a tree species with a broad climate envelope, is common throughout the western mountains, extending from southern British Columbia into Mexico. Genetic studies, however, indicate that there are at least four major ecotypes (Millar and Libby 1991) that

may have unique temperature tolerances and thus slightly different climate envelopes. At the other extreme, the Harpers beauty (*Harperocallis flava*), a lily, has an extremely narrow climate envelope, occurring only at a few locations in the panhandle of Florida. The width of each species' climate envelope is used in this study to predict potential vulnerability to global warming.

Definition of Vulnerability

The reported geographical distribution of each species was matched with data from climate stations spanning the area of distribution. The width of each species' climate envelope was calculated from the difference between the maximum and minimum mean annual temperatures it experiences in its current distribution. Three slightly different methods were used to calculate climate envelopes to provide a range of vulnerability estimates. The width of each species' climate envelope was then compared to the predicted increase in mean annual temperature (the magnitude of climate change) to determine that species' vulnerability to climate change. If the width of the rangewide climate envelope of a species is narrower than the magnitude of climate change, then all populations are vulnerable to extirpation, and that species is considered vulnerable to extinction from climate change. If the width of the species' rangewide climate envelope is wider than the magnitude of climate change, some populations might be out of their climate envelope and thus vulnerable to extirpation, but the species as a whole is considered not vulnerable to climate change. Since the climate envelopes are based solely on maximum and minimum mean annual temperatures, the potential vulnerability of species to climate change ignores other climatic or ecological factors (such as availability of suitable soil types) and thus does not fully represent actual extinction risk.

Principal Findings

For the preliminary analysis, the North American climate was changed by uniformly increasing mean temperatures in 1°C increments up to an increase of 20°C above current mean annual temperatures. Many species were entirely out of their climate envelope in all scenarios of uniform temperature increase. With a mean global warming of 3°C, about 7 to 11 percent (1,060 to 1,670 species) of over 15,000 native plant species in North America were entirely out of their climate envelopes and thus considered vulnerable to extinction (Figure 2-1). About 15 to 20 percent of all species were entirely out of their climate envelopes in a 5°C climate change scenario. In the extreme case of a 20°C climate change, approximately 70 to 85 percent of all species were vulnerable. For comparison, only about 90 plant species in North America are believed to have gone extinct in the last two centuries (Russell and Morse 1992).

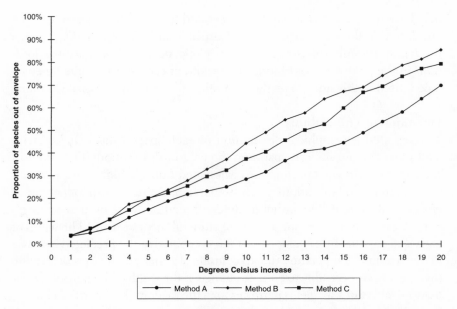

FIGURE 2-1.   The proportion of native vascular plant species that were entirely out of their climate envelopes as a function of the increase in temperature above mean annual temperature. Three methods (A, B, C) were used to determine climate envelopes to provide a range of results.

IMPACTS ON RARE SPECIES

In the data set analyzed, approximately forty-one hundred of the native North American vascular plant species are considered "rare" by The Nature Conservancy and the Natural Heritage Network. These species have been assigned the conservation status ranks G1, G2, or G3 (in descending order of rareness) and occur at fewer than one hundred sites or are comparably imperiled (Figure 2-2). The ranking system is discussed in depth by Morse (1987), Master (1991), and Stein et al. (1995). These rare plants constitute about 27 percent of the present flora but would be disproportionately affected by climate change. Approximately 10 to 18 percent of the rare species (ranked G1, G2, or G3) are considered vulnerable to a mean 3°C temperature increase (Figure 2-2). While only 1 to 2 percent of the common species (ranked G4 or G5) are vulnerable under these conditions, some species that are currently widespread could become rare. For example, parrot pitcher plant (*Sarracenia psittacina*) is ranked G4 (over one hundred populations) and is found in wet fields, bogs, and pine barrens in a fairly narrow range of climates near the Gulf coast from Louisiana to Georgia. This species, however, would be especially threatened by significant changes in local or regional hydrology, such as a lowered water table.

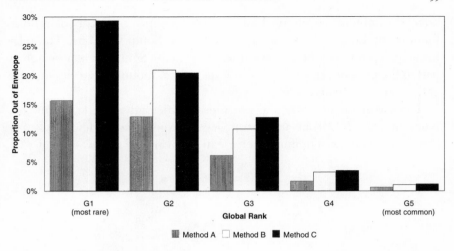

FIGURE 2-2.   The proportion of native vascular plant species in each global rank that were entirely out of their climate envelopes with a +3°C temperature change. Three methods were used to determine climate envelopes (A, B, C).

IMPACTS ON POPULATIONS AT SOUTHERN RANGE LIMITS

Because global warming is expected to push climatic zones poleward, areas in the Northern Hemisphere with many species at their southernmost range limits, such as those located at the southern ends of north-south mountain ranges, are likely to experience more species losses. The southern end of the higher Appalachians, in the area where Georgia, North Carolina, and South Carolina adjoin, is the southern range limit of about 150 species that do not extend downslope into the warmer climate of the adjacent Piedmont. Many of these species are already rare in states along their southern limits due to ecological stresses or limited habitat. The states with the highest proportions of southernmost range limits include Alabama, Arizona, California, Florida, Louisiana, and Texas. The results may be slightly inflated in the southwest, due to incomplete data on distributions into Mexico. Since areas with many species at their southernmost range limit are susceptible to the greatest losses from the local flora if the climate warms, this analysis could help identify regions that are likely to experience the greatest ecological impacts from climate change.

REGIONAL PATTERNS OF VULNERABILITY

Based on the uniform 3°C increase in mean annual temperature used for this analysis, there appear to be regional patterns to the proportion of species considered vulnerable to climate change in each state's or province's flora (Figure 2-3). The southeastern states have the highest percentage of species

out of their climate envelopes. Florida could lose up to 25 percent of its flora, followed by Louisiana, Alabama, Georgia, and South Carolina. The relatively high proportion of species vulnerability in the Southeast may be due in part to the presence in state floras of Appalachian mountain species that are at their southern range limits.

The Great Plains states and provinces, particularly Alberta, Manitoba, North and South Dakota, and Saskatchewan, experience the fewest losses to their floras based on a uniform increase in mean annual temperature of 3°C.

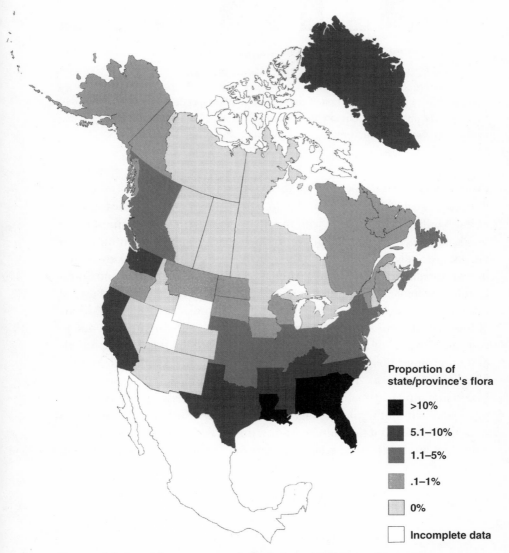

Proportion of state/province's flora

- \>10%
- 5.1–10%
- 1.1–5%
- .1–1%
- 0%
- Incomplete data

FIGURE 2-3.   The proportion of species that would be out of their climate envelope in each state or province with a +3°C temperature change.

As discussed above, however, global warming models suggest that the temperature and precipitation changes in the interior of the continent may be far greater than in coastal regions. Climate change impact analyses that incorporate regional changes in climate projected by GCMs will help refine our understanding of regional patterns of plant species' vulnerability to climate change.

POTENTIAL FOR DISPERSAL

A dispersal-ability scale was created for this analysis to assess the potential for different species to migrate. The scale is based on biological characteristics important to species mobility. Factors associated with increased species mobility include pollination by wind, partial (or complete) self-compatibility, dispersal of propagules by wind or birds, and short generation time. Characteristics such as dependence on specific pollinators (as with the yucca and yucca moth), dispersal by ants, or long generation time reduce the chances for successful rapid dispersal and establishment. Using these criteria, most of the species studied appear to have an intermediate dispersal potential.

The species in this analysis that would be vulnerable in a +3°C climate change appear to have characteristics that limit long-distance dispersal (Figure 2-4), suggesting that they may be forced to adapt in place to new conditions or be threatened by extinction. These vulnerable species include rare plants and narrow endemics, which often have restricted ranges, reduced seed sources, and possible dependency on specific microclimatic conditions for survival. In addition, many of these species are restricted to wetlands, unusual substrates, or other specialized habitats. Such plants would potentially have trouble migrating, regardless of their ability to disperse. For example,

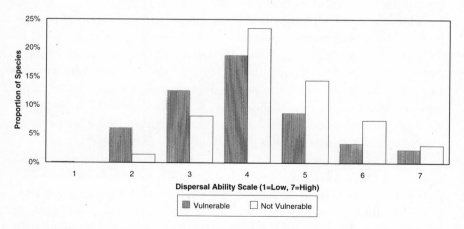

FIGURE 2-4.   The proportion of species on the dispersal-ability scale that were out of their climate envelopes (vulnerable) or in their climate envelopes (not vulnerable) with a +3°C temperature change. Full data were available for 8,668 species.

Boott's rattlesnake-root (*Prenanthes boottii*) and mountain avens (*Geum peckii*), endemic to alpine habitats in the northeastern United States, would be particularly sensitive to global warming. On the other hand, some very localized species, particularly distinctive substrate endemics, may have persisted at the same site through past climate changes. These species may have a greater tolerance to climate change than their present narrow distributions suggest.

Climate Change and Conservation Planning

Rapid climate change is likely to place novel demands and constraints on the conservation of rare plant species. The selection and design of projects such as population augmentations, reintroductions of species to historical sites, introductions of species to new sites, or community or habitat restorations could be greatly influenced by species' vulnerability to climate change. Site selection and genetic considerations will be the central issues in the design of these conservation plans. While materials for these projects may come from translocation of species among preserves or into new locations, *ex situ* propagation of critical species is likely to become increasingly important. If the climate changes as predicted, many native species could be further threatened by competition from weedy exotics. Finally, the uncertainties in climate change predictions and the highly fragmented nature of the current landscape indicate that microclimatic diversity and corridors will play a critical role in the design of future conservation actions, including preserves.

Reintroductions, Introductions, and Restorations

Strategies for the reintroduction of plant species and restoration of populations are becoming more important in the conservation of endangered or threatened species (Falk 1992; Maunder 1992). If the global climate changes rapidly in the next fifty to one hundred years, as projected by several general circulation models, then *ex situ* propagation of species of high interest and transplantation of plant populations to new areas are apt to become higher priorities. With higher temperatures, alterations in precipitation patterns, continued habitat fragmentation, and changes in disturbance regimes, endemic species or unique populations may be threatened with extinction in their present sites and/or unable to migrate to other suitable habitat.

While there is general agreement that reintroductions and *ex situ* propagation of critically rare species should be integrated with the conservation of existing populations, species, and communities, strategies that incorporate the possible effects of climate change are still being developed. At this point, most reintroduction and restoration projects should be considered experi-

mental supplements to *in situ* conservation. These projects, however, will provide excellent opportunities for ecological and genetic research if their design includes adequate controls and baseline data. Assessments of potential species vulnerability to climate change and regional patterns of losses, such as the study being conducted by The Nature Conservancy, could help set priorities for future conservation actions.

Climate change poses several difficult issues for the long-term success of reintroduction project design. Recognizing that it may be impossible to preserve current assemblages of species that presumably have individualistic responses to climate changes, it is not clear which species should become the highest priorities for intensive and expensive conservation efforts. It is challenging even to identify particular species, habitats, or regions that are most likely to be threatened by anthropogenic climate change. Rare species appear to be the most vulnerable to climate change due to their restricted distributions, narrow climate envelopes, limited dispersibility, and specialized habitat requirements, and current conservation and (re)introduction projects probably already include many of these species. Other species that would be negatively affected by climate change, based on temperature envelope and/or habitat limitations, should also be considered for intensive conservation efforts.

SITE SELECTION

Once priority species have been determined, the selection of sites for reintroduction projects will be controversial. In addition to meeting a variety of biological, physical, and logistical criteria (see Chapter 7), the sites selected could lie within the historical, the current, or the projected range of the species. There are advantages and disadvantages associated with each of these options.

The term historical (see Chapter 3) has been used to represent a variety of time scales, based on sources that range from herbarium specimens to paleobotanical records. Most frequently, historical distributions refer to those at the time of initial European settlement of North America. While the historical range, if well documented, can provide many options for reintroducing populations at sites that were once viable, this information may not be sufficient to address the potential effects of climate change. For example, if the historical sites are located at the southern limits of the species' range or are in areas that may be affected most by climate change, it would generally not be constructive to try to establish new populations.

The current range could provide the most detailed information on ecological and habitat requirements such as soils, aspect, temperature, and biotic interactions. For many rare species, however, the existing occurrences may not be abundant or large enough to plan reintroduction projects that incorporate

the potential effects of climate change. Numerous rare species have severely reduced ranges due to human impacts and/or habitat loss or are endemics that are restricted to specific habitats. These species are also the most likely to be vulnerable to climate change, based on their narrow climate envelopes. If existing sites are augmented with additional individuals of a species, those plants would presumably be vulnerable to climate change as well. Accordingly, nearby areas of similar climate, substrate, topography, soil, and vegetation should be considered for establishing additional experimental populations of species with critically limited distributions. If the original populations are extirpated during a period of rapidly changing climate, the experimental populations may survive and avert extinction of that species.

Finally, sites could be selected based on anticipation of future species' ranges if climatic regions move across the landscape. While the projected range of a species could encompass part or all of its historical or current distribution, it will probably include areas with no documentation of previous occurrences. In this case, conservation projects may actually involve the introduction of species to novel sites. The selection of sites based on projected range may help ensure that populations are established in locations that could become better climatically suited to that species' tolerances if the climate changes dramatically. If an introduction is planned, then the possible impacts of the newly introduced species on other rare species at the site and on the integrity of the site's existing natural communities should be considered. Introduced populations may provide at least some degree of insurance against climate change, but site selection would be highly speculative since climate change and impact assessment models cannot yet make accurate predictions at a local scale.

GENETIC CONSIDERATIONS

The design of reintroduction and restoration projects for success in a changing climate should include careful selection of the plant materials to be used. It is widely recognized that there can be considerable genetic variation within and among different plant populations, which affects physiological tolerances and responses to stochastic events (Huenneke 1991; Millar and Libby 1991). This variability could help species survive episodes of rapid climate change, but indiscriminate translocation or introduction of individuals could be detrimental to the species as a whole.

As discussed above, experimental populations could provide additional security for a species that is highly vulnerable to climate change. Natural ecotypic or clinal variation could help guide the selection of materials for these projects. For example, individuals that are currently adapted to a warmer climate could be translocated to slightly cooler northern or upslope sites in anticipation of climate change. If the proposed site has environmental condi-

tions that are significantly different, then the relocated individuals may have difficulty surviving. It may be preferable to establish populations at novel sites, at some distance from natural populations. Conservation geneticists have recommended that the natural distribution of ecotypic variation be maintained because random genetic additions often have a negative effect on the viability of the local population (Falk and Olwell 1992; Jensen et al. 1993). Translocated individuals that disrupt the existing populations' genetic structure by polluting coadapted gene complexes could reduce the species' ability to adapt to change.

Roles of Exotic Species

As certain species migrate into or out of given areas, the newly composed communities may be especially vulnerable to invasion by exotic weeds. Many weeds are able to expand into an area relatively quickly, posing serious threats to the existing species and overall biodiversity (Schwartz 1992). Invasive exotic species are often widespread, prolific, fast-growing annuals that are capable of colonizing disturbed habitats and are often favored by disturbances. For example, Australian paperbark (*Melaleuca quinquenervia*), an exotic species that thrives with disturbance, has already formed dense monotypic stands in the Florida Everglades, causing native species loss. Based on predicted climate change scenarios, this species is likely to spread (Peters 1992). Climate-induced changes could expose native plants to exotic competitors for the first time (Peters 1992), stressing the balance established between native plants and their habitats.

In addition, some highly dispersible native plants may move rapidly from their historic ranges in response to climate change and abundantly colonize new areas, away from their current competitors, pests, and diseases. Without these restraints, some species may become more abundant and threaten other native species at a site. Plants may also be deliberately moved outside of their known historic and/or current sites of natural occurrence as part of a species introduction project.

This movement of species across the landscape raises several important issues for the conservation of native plants (U.S. Congress 1993a). First, a reassessment of the concept of exotic species should be made. Currently, the term *exotic* is commonly used for species that have become established beyond their historic range through direct or indirect human intervention at some particular point in time. It is likely that this definition will become increasingly outdated and controversial with climate change as native species move to novel areas or are introduced to sites other than their known historical or current occurrences. Second, more intensive procedures may need to be taken to remove or inhibit movement of exotic species that are clearly recognized as harmful. Finally, under changing climate conditions, conservation

initiatives such as development of corridors and new protected areas or restoration of habitats and species could be important in protecting species from harmful weedy exotics (U.S. Congress 1993a).

SPECIES IN ANTHROPOGENICALLY FRAGMENTED HABITATS

The potentially rapid rates of warming, combined with habitat loss and fragmentation from human development, suggest that many species will not adjust as successfully to climate change as they have in the past. Most native plant species now exist in a highly fragmented landscape that further separates appropriate habitat patches, increasing the dependence of many species on relatively rare long-distance dispersal events. Furthermore, species often must disperse across hostile habitats, including roads, cities, suburbs, and farmland (Peters 1992). Finally, plants would need to establish themselves in landscapes where many of the open or disturbed areas have been colonized by aggressive weedy exotics, rather than in communities suitable for less aggressive native species. Human-altered landscape patches can thus constitute potentially insurmountable barriers to dispersal for many species. Habitat loss and fragmentation from human development could prevent many species from migrating to new suitable habitat and increase extinction rates (Peters 1992).

The negative effect of habitat fragmentation on the migratory ability of tree species is demonstrated by a simulation model based on Holocene migration rates that varies local population size and habitat availability (Schwartz 1992). The results, using two different dispersal models, predict that migration rates may be as low as 0.5 to 13 kilometers per century when only 30 percent of the habitat is available for colonization. In addition, rare species tend to migrate at slower rates, even with low levels of habitat fragmentation. Schwartz's model predicts that many tree species may be vulnerable to extinction because they could not migrate rapidly enough in an altered landscape to respond to future climate change, unless plants are translocated or corridors are established to link otherwise isolated fragments of natural habitats.

ROLE OF MICROCLIMATES

A continuing difference between the microhabitat and regional conditions may be sufficient to maintain species in place during episodes of climate change if those species are restricted to that microhabitat primarily because of competitive pressures from species that cannot tolerate the local conditions. If a species' distribution is limited by absolute tolerance of climatic conditions, however, then a cool refuge that warms by 3°C in parallel with regional temperature increases would become unsuitable for the heat-intolerant species it previously supported.

Microclimatic diversity may be increasingly significant in the design of conservation projects, including intentional translocation of rare species and preserves. The potential for microclimates to maintain a suitable local climate during regional climate changes should be considered in planning and implementating these projects, because local climatic differences created refugia that helped protect species during past episodes of climate change. In addition, a variety of climates in close proximity to each other may allow natural species migration to more suitable sites if the regional climate changes. It should be noted, however, that sites with regionally rare microclimates are likely to contain rare species or communities that might be affected by introductions of additional species.

Preserve Selection and Design

Since reintroduction and restoration projects should be performed in coordination with other conservation actions, climate change will also influence the selection and design of preserves. The potential for rapid anthropogenic climate change in the next century received little attention, if any, during the selection and design of present national and state parks, wilderness areas, wildlife refuges, and privately managed nature preserves such as those of The Nature Conservancy. Many such preserves were established to protect particular species or vegetation types, with the unstated assumption that appropriate habitat conditions could be maintained indefinitely. In a changing climate, it is likely that the local habitat would be altered, species would retreat from southern range limits or low elevations, and future distributions would be further fragmented. As portrayed by Peters (1992), these effects could result in reserves that are no longer capable of maintaining the original communities or species for which they were designed.

As it is anticipated that northward and upward species migrations will become more important during periods of changing climate, corridors that link isolated protected areas should be included in conservation plans. Such corridors could provide protected zones for migration of flora and fauna that may in turn reduce inbreeding among small populations and maintain species diversity. A long, flat, east-to-west corridor, however, linking two protected areas that have a history of exotic species invasions may not provide a protected migration route. The gradient and orientation may not permit elevational or south-to-north migration. Furthermore, the corridor could offer easy passage of harmful exotic species into an area that would otherwise be fairly insulated (Simberloff and Cox 1987). Although corridors can be a valuable conservation strategy (U.S. Congress 1993b), their effectiveness and design will need to be carefully evaluated to take a variety of factors, including climate change, into consideration.

In the coming years, preserves and reintroductions with the greatest

probability of successfully maintaining rare species and biological diversity will probably include microclimate and microhabitat variety. Conservation sites with heterogeneous topography would possibly be more resilient to climate change since they offer more opportunities for upslope and southern-to-northern aspect shifts in species' populations (Weiss and Murphy 1990; Peters 1992). Locations with few weedy exotics that include diverse habitats and geologically unique areas with different soil types would also be critical to include in preserves. These areas may provide additional locations to which vulnerable plants might disperse or be introduced through human intervention.

Conclusion

History illustrates that species react and migrate as individuals in response to climate change and not as intact community types. Therefore, the unique response of individual species needs to be focused upon when incorporating climate change into future restoration and reintroduction projects. Ultimately these species will constitute future community types, however they are assembled. Differences between historical and current distributions, uncertainties in projected range, and variations among climatic tolerances will determine the selection of sites for future reintroduction and restoration work. If there are several available options for relocation projects, however, it may be optimal to include a few sites that are slightly poleward, upslope, or less exposed than the existing or historical distribution, including areas with topographic and microclimatic variety. This would provide species that are potentially vulnerable to climate change in their current distributions an extra margin for survival. These actions, however, should be performed cautiously, with consideration for potential ecological or genetic impacts.

A significant number of rare plant species that have been the focus of intensive conservation efforts could be further threatened by anthropogenic climate change. There has been a considerable investment in the preservation of many critically threatened or endangered species, including the creation and management of preserves, *ex situ* cultivation, and reintroductions. An early warning that some of these species may be vulnerable to conditions under a rapidly changing climate might allow new conservation strategies to be developed and implemented before a crisis occurs. This could improve the prospects of success for the protection of rare plants while minimizing costs.

ACKNOWLEDGMENTS
We wish to acknowledge Ellen Lippincott and Kristen Spangler for their assistance in preparing this manuscript and Brandy Clymire for preparing the

graphics. The climate change impacts analysis is supported in part by the Electric Power Research Institute.

REFERENCES

Brown, J. H., and A. C. Gibson. 1983. *Biogeography*. St. Louis: The C. V. Mosby Company.

Carpenter, S. R., S. G. Fisher, N. B. Grimm, and J. F. Kitchell. 1992. "Global Change and Freshwater Ecosystems." *Annual Review of Ecology and Systematics* 23: 119–140.

Core, E. L. 1968. "The Botany of Ice Mountain, West Virginia." *Castanea* 33: 345–348.

Coville, F. V. 1919. "The Threatened Extinction of the Box Huckleberry, *Gaylussacia brachycera*." *Science* 50: 30–34.

Davis, M. B. 1981. "Quaternary History and the Stability of Forest Communities." In *Forest Succession: Concepts and Application*. Edited by D. C. West, H. H. Shugart, and D. B. Botkin. New York: Springer-Verlag.

———. 1984. "Climatic Instability, Time Lags, and Community Disequilibrium." In *Community Ecology*. Edited by J. Diamond and T. J. Case. New York: Harper and Row.

———. 1989. "Insights from Paleoecology on Global Change." *Ecological Society of America Bulletin* 70 (4): 222–228.

———. 1990. "Climatic Change and the Survival of Forest Species." In *The Earth in Transition: Patterns and Processes of Biotic Impoverishment*. Edited by G. M. Woodwill. Cambridge: Cambridge University Press.

Davis, M. B., and C. Zabinski. 1992. "Changes in Geographical Range Resulting from Greenhouse Warming Effects on Biodiversity in Forests." In *Global Warming and Biological Diversity*. Edited by R. L. Peters and T. L. Lovejoy. New Haven and London: Yale University Press.

Dickinson, R. E. 1986. "How Will Climate Change?" In *The Greenhouse Effect, Climatic Change, and Ecosystems*. Edited by B. Bolin, B. R. Döös, J. Jäger, and R. A. Warrick. Chichester and New York: John Wiley and Sons.

Emanuel, K. A. 1988. "Toward a General Theory of Hurricanes." *American Scientist* 76 (4): 371.

EPA. 1989. "The Potential Effects of Global Climate Change on the United States." Washington, D.C.: United States Environmental Protection Agency.

Falk, D. A. 1992. "From Conservation Biology to Conservation Practice: Strategies for Protecting Plant Diversity." In *Conservation Biology: The Theory and Practice of Nature Conservation, Preservation, and Management*. Edited by P. L. Fiedler and S. K. Jain. New York and London: Chapman and Hall.

Falk, D. A., and P. Olwell. 1992. "Scientific and Policy Considerations in Restoration and Reintroduction of Endangered Species." *Rhodora* 94 (879): 287–315.

Flannigan, M. D., and C. E. Van Wagner. 1991. "Climate Change and Wildfire in Canada." *Canadian Journal of Forest Research* 21 (1): 66–72.

Franklin, J. F., F. J. Swanson, M. E. Harmon, D. A. Perry, T. A. Spies, V. H. Dale, A. McKee, W. K. Ferrell, J. E. Means, S . V. Gregory, J. D. Lattin, T. D. Schowalter, and D. Larsen. 1992. "Effects of Global Climatic Change on Forests in Northwestern North America." In *Global Warming and Biological Diversity*. Edited by R. L. Peters and T. L. Lovejoy. New Haven and London: Yale University Press.

Frest, T. J. 1984. *National Recovery Plan for the Iowa Pleistocene Snail* (Discus macclintocki). Rockville, Md.: Fish and Wildlife Reference Service.

Graham, R. W. 1988. "The Role of Climatic Change in the Design of Biological Reserves: The Paleoecological Perspective for Conservation Biology." *Conservation Biology* 2 (4): 391–394.

Graham, R. W., and E. C. Grimm. 1990. "Effects of Global Climate Change on the Patterns of Terrestrial Biological Communities." *Trends in Ecology and Evolution* 5 (9): 289–292.

Hayden, C. B. 1843. "On the Ice Mountain of Hampshire County, Virginia." *American Journal of Science and Arts* 45: 78–83.

Huenneke, L. F. 1991. "Ecological Implications of Genetic Variation in Plant Populations." In *Genetics and Conservation of Rare Plants*. Edited by D. A. Falk and K. E. Holsinger. New York: Oxford University Press.

IPCC (Intergovernmental Panel on Climate Change). 1990. *Climate Change: The IPCC Scientific Assessment*. Geneva: World Meteorological Organisation (WMO) and United Nations Environment Programme (UNEP).

IPCC (Intergovernmental Panel on Climate Change). 1992. *1992 IPCC Supplement*. Geneva: World Meteorological Organisation (WMO) and United Nations Environment Programme (UNEP).

Jensen, D. B., M. S. Torn, and J. Harte. 1993. *In Our Own Hands: A Strategy for Conserving California's Biological Diversity*. Berkeley: University of California Press.

Kartesz, J. T. 1994. *A Synonymized Checklist of the Vascular Flora of the United States, Canada, and Greenland*. Portland, Oreg.: Timber Press.

Kerr, R. 1993. "Fossils Tell of Mild Winters in an Ancient Hothouse." *Science* 261: 682.

Knox, J. B., and A. F. Scheuring. 1991. *Global Climate Change and California: Potential Impacts and Responses*. Berkeley and Los Angeles: University of California Press.

Kress, W. J., G. D. Maddox, and C. S. Ruesel. 1994. "Genetic Variation and Protection Priorities in *Ptilimnium nodosum* (Apiaceae), an Endangered Plant of the Eastern United States." *Conservation Biology* 8 (1): 271–276.

Kruckeberg, A. R. 1984. *California Serpentines: Flora, Vegetation, Geology, Soils, and Management Problems*. Berkeley and Los Angeles: University of California Press.

Maddox, G. D., and R. L. Bartgis. 1991. "Recovery Plan for Harperella (*Ptilimnium nodosum*)." U.S. Fish and Wildlife Service.

Master, L. L. 1991. "Assessing Threats and Setting Priorities for Conservation." *Conservation Biology* 5: 559–563.

Maunder, M. 1992. "Plant Reintroduction: An Overview." *Biodiversity and Conservation* 1: 51–61.

Millar, C. I., and W. J. Libby. 1991. "Strategies for Conserving Clinal, Ecotypic, and Disjunct Population Diversity in Widespread Species." In *Genetics and Conservation of Rare Plants*. Edited by D. A. Falk and K. E. Holsinger. New York: Oxford University Press.

Morse, L. E. 1987. "Rare Plant Protection, Conservancy Style." *The Nature Conservancy Magazine* 37 (5): 10–15.

Morse, L. E., L. S. Kutner, G. D. Maddox, J. T. Kartesz, L. L. Honey, C. M. Thurman, and S. J. Chaplin. 1993. "The Potential Effects of Climate Change on the Native Vas-

cular Flora of North America: A Preliminary Climate-Envelopes Analysis." Report TR-103330. Palo Alto, Calif.: Electric Power Research Institute.

Morse, L. E., L. S. Kutner, and J. T. Kartesz. 1995. "Potential Impacts of Climate Change on North American Flora." In *Our Living Resources: A Report to the Nation on the Distribution, Abundance, and Health of U.S. Plants, Animals, and Ecosystems.* Edited by E. T. LaRoe, G. S. Farris, C. E. Plunkett, P. D. Doran, and M. J. Mac. Washington, D.C.: U.S. Department of the Interior, National Biological Service.

Overpeck, J. T., P. J. Bartlein, and T. Webb, III. 1991. "Potential Magnitude of Future Vegetation Change in Eastern North America: Comparison with the Past." *Science* 254: 692–695.

Peters, R. L. 1992. "Conservation of Biological Diversity in the Face of Climate Change." In *Global Warming and Biological Diversity.* Edited by R. L. Peters and T. L. Lovejoy. New Haven and London: Yale University Press.

Pielou, E. C. 1991. *After the Ice Age: The Return of Life to Glaciated North America.* Chicago and London: University of Chicago Press.

Ray, G. C., B. P. Hayden, A. J. Bulger, and M. G. McCormick-Ray. 1992. "Effects of Global Warming on the Biodiversity of Coastal-Marine Zones." In *Global Warming and Biological Diversity.* Edited by R. L. Peters and T. L. Lovejoy. New Haven and London: Yale University Press.

Reid, W. V., and M. C. Trexler. 1991. *Drowning the National Heritage: Climate Change and U.S. Coastal Biodiversity.* Washington, D.C.: World Resources Institute.

Russell, C., and L. E. Morse. 1992. "Plants." *Biodiversity Network News* [The Nature Conservancy] 5 (2): 4.

Schlesinger, M. E., and J. F. B. Mitchell. 1987. "Climate Model Simulations of the Equilibrium Climatic Response to Increased Carbon Dioxide." *Reviews of Geophysics* 25 (4): 760–798.

Schneider, S. H. 1989. "The Changing Climate." *Scientific American* 290 (9): 70–79.

Schneider, S. H., L. Mearns, and P. H. Gleick. 1992. "Climate-Change Scenarios for Impact Assessment." In *Global Warming and Biological Diversity.* Edited by R. L. Peters and T. L. Lovejoy. New Haven and London: Yale University Press.

Schwartz, M. W. 1992. "Modelling Effects of Habitat Fragmentation on the Ability of Trees to Respond to Climatic Warming." *Biodiversity and Conservation* 2: 51–61.

Shugart, H. H., M. Y. Antonovsky, P. G. Jarvis, and A. P. Sandford. 1986. "CO_2, Climatic Change and Forest Ecosystems." In *The Greenhouse Effect, Climatic Change, and Ecosystems.* Edited by B. Bolin, B. R. Döös, J. Jäger, and R. A. Warrick. Chichester and New York: John Wiley and Sons.

Simberloff, D., and J. Cox. 1987. "Consequences and Costs of Conservation Corridors." *Conservation Biology* 1 (1): 63–70.

Stein, B. A., L. L. Master, L. E. Morse, L. S. Kutner, and M. Morrison. 1995. "Status of U.S. Species: Setting Conservation Priorities." In *Our Living Resources: A Report to the Nation on the Distribution, Abundance, and Health of U.S. Plants, Animals, and Ecosystems.* Edited by E. T. LaRoe, G. S. Farris, C. E. Puckett, P. D. Doran, and M. J. Mac. Washington, D.C.: U.S. Department of the Interior, National Biological Service.

Stine, S. 1994. "Extreme and Persistent Drought in California and Patagonia During Mediaeval Time." *Nature* 369: 546–549.

Stone, C. P., and D. B. Stone. 1989. *Conservation Biology in Hawai`i*. Honolulu: University of Hawaii Press.

U.S. Congress. Office of Technology Assessment. 1993a. *Harmful Non-Indigenous Species in the United States, OTA-F-565*. Washington, D.C.: U.S. Government Printing Office.

———. Office of Technology Assesment. 1993b. *Preparing for an Uncertain Climate-Volume II*, OTA-O-568. Washington, D.C.: U.S. Government Printing Office.

Webb, T., III, and P. J. Bartlein. 1992. "Global Changes During the Last 3 Million Years: Climatic Controls and Biotic Responses." *Annual Review of Ecology and Systematics* 23: 141–173.

Weiss, S. B., and D. D. Murphy. 1990. "Warm Slopes and Cool—Topographic Criteria in Conservation Planning." *Endangered Species UPDATE* 10–11: 6.

Wherry, E. T. 1972. "Box-Huckleberry As the Oldest Living Protoplasm. *Castanea* 37: 94–95.

Wolfe, J. N., R. T. Wareham, and H. T. Scofield. 1949. "Microclimates and Macroclimate of Neotoma, a Small Valley in Central Ohio." *Ohio Biological Survey, Bulletin* 41.

Wright, H. E., Jr. 1984. "Sensitivity and Response Time of Natural Systems to Climatic Change in the Late Quaternary." *Quaternary Science Reviews* 3: 91–131.

Spatial and Biological Scales in Reintroduction

Peter S. White

———————————➤•◦•◄———————————

One of the implicit assumptions of the first conservationists was that nature's "balance" would produce persistence in natural populations. A corollary was that nature needed protection only from destruction or overuse. Today, biological diversity is threatened not only by habitat loss and by the exploitation and purposeful eradication of species but also by the changes within surviving natural areas that result from many factors: habitat fragmentation, changes in the physical and chemical environment, invasions of exotic species, and human alterations of natural processes (such as changes in fire and hydrologic regimes). Threats such as air pollution and exotic species invasions affect even large wilderness areas such as Everglades National Park (Davis and Ogden 1994); these threats do not respect the boundaries of sites protected for *in situ* populations (White and Bratton 1980; Bratton and White 1980). Certainly, the first job of conservation is legal protection, but this no longer guarantees the survival of genes, species, or ecosystems. In addition, as pristine areas become rare, we will have only human-affected lands to bolster the conservation of biological diversity. Restoration becomes the only solution in these cases. (For a discussion of the restoration of Central American dry forests, see Janzen 1988.)

Our scientific understanding has also changed. We now accept that stasis in populations either is absent or exists only at particular spatial and temporal scales of observation. We have learned that even large wilderness areas can experience substantial natural change (such as the influence of natural disturbances and the fluctuating populations of herbivores), and, if any dynamic equilibrium occurs, it is likely to be at large spatial scales as the consequence of independent, smaller-scale dynamics. Global change stands as the ultimate challenge to conservation of wild areas *in situ* because it threatens to reset all species ranges—and thus to reset the relation between particular species and particular sites.

The changes in the amount of nature left, in our scientific understanding, and in the threats to natural areas shape our conservation choices. For some places (such as Hawaii and parts of the California landscape), conservationists must now use reintroduction and even introduction as the only choice for some plant species (see Howald, Chapter Thirteen). Conservation of some endemic species in Hawaii will require that wildland gardens be established at a series of elevations and habitats (see Mehrhoff, Chapter Five; see also Center for Plant Conservation 1992). In areas where wild habitats are extant, by contrast, reintroduction is viewed as a last resort, used only after habitat protection and management have failed. Where habitats remain intact, introduction is disparaged as potentially changing the natural distribution of rarity, obscuring the original biogeography, potentially causing outbreeding depression if there are nearby natural populations, and changing the role of isolation in evolution. The amount and rate of global environmental change become critical issues—they will cause reintroduction, introduction, and restoration to be used on a grand scale rather than a restricted one (see Kuntner and Morse, Chapter 2). Introduction will be particularly important if the rate of environmental change exceeds the rate at which species can migrate to appropriate environments; the ability of species to migrate under changing environments is further diminished because remnant natural areas are often isolated in a sea of developed lands.

In using reintroduction and restoration to prevent the erosion of biodiversity, we often focus on single species, populations, and sites. This narrow spatial focus is often accompanied by a narrow temporal one. Understandably, the top priority is short-term survival. Further, we usually lack the information to address the multiple-species context of reintroduction. Nonetheless, spatial and biological context will affect the success of reintroduction, introduction, and restoration, whether the surrounding matrix is heavily disturbed (a relatively straightforward situation) or fairly intact (a more subtle situation). The influence of surrounding lands on a population or habitat of interest must decrease as distance increases, but the distance at which the surroundings become irrelevant to conservation success will vary with particular circumstances and is, in any case, likely to be unknown a priori. When we design restoration projects for heavily disturbed habitats, we must also examine scale issues: The size and isolation of sites influence their ultimate species richness and composition.

As our editors in the Introduction note, interest in scale effects that derive from the size and isolation of habitats has a long history in conservation biology in the form of island biogeography and preserve design (Shafer 1990) and the species-area relation (Palmer and White 1994). Spatial scale plays a major role in studies of landscapes (Gardner et al. 1989) and metapopulations (Hanski and Gilpin 1991). At the ecosystem level, Magnuson et al. (1991) created a strong image for the importance of spatial and temporal scale. They

suggest that we need to use long-term research to make visible the "invisible present" (that is, to put present observations in a longer temporal context) and spatially extensive research to make visible the "invisible place" (that is, establish the landscape context for intensive study sites). The potential for future environmental change heightens the importance of understanding behavior of populations and communities at larger spatial scales. The ability of species to shift with environmental change will most certainly depend on spatial context. The potential increase in the resilience of biodiversity with spatial scale is a strong argument for regional conservation planning (Noss 1993).

Concerns with temporal and spatial context lead to some general questions: How is the survival of biological diversity affected by processes at larger landscape and regional scales, and which issues should conservation biologists address at these scales to increase the sustainability of biodiversity? These questions lead to a corollary question about the efficiency of our conservation actions: Can planning and management at larger spatial scales decrease the cost of conservation management at the site scale by increasing the resilience of biological diversity? Even if integrity is not attainable at any spatial scale, answers to these questions will focus attention on management actions needed to maintain biological diversity in the face of low natural resilience of populations and communities.

In this chapter I explore principles for the integration of site-based reintroduction and restoration with larger-scale landscape and regional conservation. The first section describes influences of the spatial context of sites. Since species composition and species richness are a function of spatially constrained processes, I consider scale issues in establishing the reference state or goal of restoration. I then turn to a discussion of scale in the composition and structure of individual populations, clustered populations, communities, and landscapes. This will be followed by a description of general implications of these ideas for use of reintroduction and restoration in conservation strategies. I use examples throughout from the literature and from my own experience in the southern Appalachians. I assume that our goal is that rare species survive rather than be made common (Harper 1981, 1992); this assumption focuses the discussion on the survival of a semblance of natural biological diversity and the processes that support and provide resilience for that diversity, including processes responsible for the ubiquitous inequality in the natural abundance of organisms.

The Spatial Context of Sites

The factors that are a function of spatial context can be physical or biological—or, like natural disturbances, they can be produced by an interaction of the two. The surrounding matrix may influence the temperature, humidity,

hydrology, wind speed, and fire regime of a conservation area. It may also determine whether the conservation area is invaded by exotic species; coevolved species such as pollinators, seed dispersers, and mycorrhizal fungi; and native herbivores (whether through direct habitat effects on herbivore populations or indirect effects through changes in predator populations). The effects of spatial context may be positive or negative. An example of a positive effect is when a pollinator for a rare plant has larvae that are supported by plants in habitats adjacent to the rare plant's habitat. An example of a negative effect is the eastern United States' white-tailed deer problem, in which a native herbivore exists in large populations because of expanded habitat and loss of predators. Changes in spatial context can lead to transient effects. One interesting result from a forest fragmentation experiment was a short-term increase in species richness and density of some groups of animals as populations displaced from destroyed habitat invaded remnant patches (Bierregaard et al. 1992). However, these increases were temporary, and species loss became the dominant process after fragmentation.

The effect of spatial context involves two broad kinds of mechanisms. First, the surrounding matrix of a particular population or habitat can produce immediate and measurable changes (such as changes in hydrology, fire regime, or herbivory) in the physical or biological environment. These factors can cause relatively quick and straightforward structural and compositional change in populations and communities, although inertia may lengthen the time scale of this change. Species generally have unique niche characteristics, and change in the physical and biological environment alters demographic parameters and competition. The second broad mechanism is that effects of spatial context may result purely from the spatial configuration of habitats. For example, an increase in isolation and a decrease in habitat size will affect population dynamics, gene flow, and genetic diversity, even if there are no changes in the locally measured physical or biological environment. Even in wholly natural landscapes, the size and isolation of habitats influence natural processes such as gene flow and dispersal, pathogen movement, and the spread of natural disturbances among habitat patches. Such spatially mediated processes have been important in shaping how biological diversity is distributed (White and Nekola 1992). Although both environment and spatial configuration of habitats can have independent effects on species presence, sites are often affected by both directional change in the environment (such as loss of habitat quality) and habitat fragmentation.

Human-caused fragmentation of habitats imposes a new spatial configuration on populations. The loss of diversity through "species relaxation" after habitat fragmentation is predicted to occur as a function of the size and isolation of conservation areas, even if no other influence of the matrix and no other environmental change is present (Shafer 1990). This species loss occurs

because of an increase in extinction rate (caused by a reduction in population sizes) and a decrease in immigration rate (caused by an increase in distance between populations). A major issue is whether the species lost from fragmented areas or those absent from small natural habitats are a predictable or random subset of the species pool, a part of the general question concerning whether species lists are nested subsets as a function of area (Patterson 1987).

When changes in the physical and biological environment cause population loss, conservation managers must look beyond site boundaries in an attempt to better manage or restore the relevant processes. When those processes cannot be managed beyond site boundaries, management within the site (through such methods as local alteration of hydrology, exclosures to eliminate herbivory, and hand pollination) may be able to stabilize populations. An example of the local management of sensitive populations is the Virginia chainfern (*Woodwardia virginiana*) in Great Smoky Mountains National Park. This species, known from only a single small population, was probably eliminated from the park by deer browsing or wild boar rooting (Table 3-1); the inability to control the effects of these animals in the landscape as a whole necessitates fencing the most sensitive areas, and reintroducing the extirpated species. When population loss is caused by fragmentation effects on extinction and immigration, conservation managers can address the possibility of increasing local population sizes by creating new sites and populations within dispersal distances; moving seeds, cuttings, spores, or pollen to nearby sites; and creating *ex situ* collections for augmentation and reintroduction.

Scale and Restoration

The contrast between the effects of environmental change and spatial configuration on species composition is also important in setting goals for restoration (White and Nekola 1992). Restoration requires the definition of the reference state that we are trying to achieve, whether that reference state is determined by composition, structure, species richness, or ecological processes. We might initially assume that restoring the environment (in the form of restored hydrology and soils and appropriate disturbance regime), coupled with species availability, will produce the composition of interest. But there are many problems with this assumption, one of which is understanding all dimensions of the species-environment relationship well enough to establish the reference state for a particular case. Let us assume for the moment, however, that we do understand the species-environment relationship well enough to establish a restoration target. An additional problem is that nature does not always produce a tight coupling of site environment and species

TABLE 3-1. Population and Community Structure as Function of Scale, with Examples from the Rare Plants of Great Smoky Mountains National Park, North Carolina and Tennessee.

Level of Organization/Description of Structure

Populations
    Populations all-sized and/or all-aged at some spatial scale at one point in time
        Sizes and ages interspersed in space
            Example: Rugel's ragwort (*Cacalia rugelia*)
            An endemic of high-elevation spruce-fir and northern hardwood forest.
        Sizes and ages not interspersed
            Reproduction found in gaps adjacent to older individuals
                Example: Kentucky yellowwood (*Cladrastis kentukea*)
                A scarce tree reproducing mostly by vegetative reproduction within mesic coves.
            Reproduction found on sites disjunct from older individuals
                Example: American red raspberry (*Rubus idaeus*)
                A successional species dominating after large patchwise disturbance in high-elevation forests.
    Populations not all-sized or all-aged at any spatial scale at one point in time
        Reproduction occurs after death of older stems on patches occupied by these stems
            Example: Synandra (*Synandra hispidula*)
            A biennial mint of rocky limestone woods.
        Reproduction by seed generally absent
            Populations of mature plants seemingly stable
                Example: Mountain Avens (*Geum radiatum*)
                An endemic of crevices in high-elevation rock outcrops.
            Populations of mature plants seemingly in decline (relictual species)
                Example: Three-forked rush (*Juncus trifidus*)
                A disjunct from alpine areas northward; possibly extirpated in the park.
Clustered populations
    Metapopulations
        Example: Virginia meadowsweet (*Spiraea virginiana*)
        A species of rocky streamsides colonizing newly disturbed sites after flood scour.
    Multiple independent populations
        Example: Carey's saxifrage (*Saxifraga careyana*)
        A high-elevation endemic, widely scattered on wet rocks. There are several examples of the vulnerability of extirpation for single populations:
            Three-forked rush (*Juncus trifidus*), Virginia chainfern (*Woodwardia virginiana*), and Pursh's rattlebox (*Crotalaria purshii*) are presumed lost from the park's flora over the last several decades. They were all known from single populations.
Communities
    Equilibrium at the within-community scale
        Gap phase dynamics in mesic hardwoods and hemlock-hardwoods.
            Example: Fraser's sedge (*Cymophyllus fraseriana*)
            This species occurs in scattered populations in mesic coves.
    Equilibrium at the landscape scale
        Fire disturbance on dry ridges
            Example: Mountain pieris (*Pieris floribunda*)
            This species is prominent in pine stands that owe their origin to past burning. They may also benefit from added light after southern pine beetle outbreaks. Long-term succession to oak forest occurs in the absence of fire and will reduce pieris populations.
    No equilibrium at any scale
        Grassy balds
            Example: Azalea hybrids (*Rhododendron spp*)
            Succession is eliminating most of the grassy bald habitat in the park, with the exception of two managed balds. Grassy balds are 150-year-old anthropogenic communities that harbor some rare plant populations.

composition, for reasons having to do purely with the spatial configuration of sites.

Let us assume that we define a particular habitat type for restoration. The definition may include either homogeneous areas or gradients of continuous change (such as from the center to the edge of a wetland). Given this designation of a habitat of interest, the influence of spatial configuration on species richness and configuration will result from the size and isolation of the habitat. The ecological contrast between the habitat of interest and its surroundings is also important (it will determine whether species can disperse through this matrix), but let us assume that this variable is subsumed under isolation. The greater the ecological contrast between habitat and surrounding matrix, the more effectively isolated the habitat will be, all else being equal.

The species-environment relation shows that the size of a particular site determines the number of species it can support (see Palmer and White, 1994). Regardless of how many species we start with, this suggests that the number of species remaining will ultimately be a function of habitat size. If we start with all possible species, the species that persist on a particular small site may be a random subset (if extinctions are random) or a biased subset of the initial species pool (if extinctions are biased toward particular extinction-prone or area-sensitive species). If extinction-prone species are lost from all sites, then a series of small sites will have similar compositions, and the sum of their individual species lists will not add many, if any, to the species list from any particular site. Species richness will be held primarily within rather than among sites. If extinctions are random, however, then a series of small sites with the same environment may have different compositions, and the sum of their species lists will include many more species than can persist on any one site. In this case, species richness is held among rather than within sites.

The need to assess within-site and among-site contributions to species richness underlies Simberloff and Abele's important observation (1976) that the theory of island biogeography was neutral with regard to preserve design. Conservation strategy is partly a function of how biological diversity is distributed as a function of spatial scale. This problem is more tractable if we formally separate *scale* into its components: *grain* (a unit of observation, such as a quadrat) and *extent* (the distance over which observations are made) (Palmer and White, 1994). A similar problem occurs in designing sampling strategies for *ex situ* germplasm collections: We must understand how much genetic diversity is held within versus among populations (Falk and Holsinger 1991).

Isolation can also be important. The more isolated a series of sites are from one another, the less immigration will average out the composition among

them. Thus we might expect that the similarity between sites will, on average, decrease with the distance between the sites, a phenomenon termed the distance decay of similarity (Palmer 1992; Nekola and White, submitted manuscript). There are many reasons to expect that similarity will decrease with distance. On average, the farther apart two sites are, the less likely they are to be similar in temperature, rainfall, soils, history, or species availability. The chance of encountering a significant barrier to dispersal also increases with distance. Differences in spatial configuration may control the rate of similarity loss with distance. In spruce-fir forests, similarity declined more quickly for fragmented versus continuous habitats, for larger- versus smaller-seeded and spore-dispersed species, and for herbs versus trees (Nekola and White, submitted manuscript). The distance-related decline of similarity adds to the between-site component of biological diversity and suggests that we should not use a single compositional state for all restorations of a particular habitat. The potential importance of spatial configuration was suggested by a study of the number of rare species present in the southern Appalachians. Richness of rare species was most strongly correlated with an index of topographic complexity (Miller and White 1986; Miller, Bratton, and White 1987) and the number of rare species occurrences increased at a steep rate with mountaintop area and the maximum elevation reached (White and Miller 1988).

Thus, both habitat size and isolation can affect the overlap in species presence among sites; in extreme cases (where there is low similarity among sites), more species will be held across a series of sites than can be held within any one of the sites individually. In these cases, species composition is not entirely predictable from site environment. Even if we artificially bring all possible species to all possible sites, and even if the environment is restored, we may not be able to retain all of the biodiversity at that single site. The goal of restoration in such a case must address larger spatial scales than that of a single site; otherwise, we will not be able to maintain the original species richness of a landscape. By similar argument, multiple populations of a species sometimes hold more genetic diversity than single populations (see Falk and Holsinger 1991), and the truncation of range and loss of sites may represent the loss of genetic diversity (Perring 1974).

Even when we correlate environment with compositional patterns to determine reference states for restoration, scale effects may be important. Reed et al. (1993) reported that the variables best correlated with compositional variation in communities changed with the scale of the observations, with scale represented by a series of quadrats from 0.0125 m^2 to 256 m^2. In a study of rare plant distributions, Wiser (1993) found that the most important environmental correlates of population distribution were different at two scales of observation (1 m^2 and 10 m^2).

The Search for Stable Structure and Composition: Populations, Clustered Populations, Communities, and Landscapes

Whether the threat to biological diversity is caused by environmental change or habitat fragmentation per se, and whether or not human management must replace natural processes, the larger issue is whether we can achieve roughly stable population structures and community compositions at some spatial and temporal scale. I now turn to the issue of stability through a consideration of structure and composition as a function of scale. *Stability* will be used in the sense of "unchanging" at the scale of observation; at some level our goal is the sustainability of biological diversity. Understanding processes often begins with examining the population and community composition and structure. Further, monitoring of structure and composition plays a key role in assessing conservation success. The decline of populations and directional change in composition often is used to trigger a change in conservation management.

Populations

Zedler and Goff (1973) presented an analysis of the effect of sampling scale in the population structures of two tree species. For the shade-tolerant species, young and old individuals occurred within the same stand, but for the shade-intolerant early successional species, young and old individuals occurred on disjunct patches. Nonetheless, both populations could be fit by a reverse J–shaped distribution that describes constant proportional mortality and thus a "self-reproducing" population (Figure 3-1). The difference was in the scale at which the stable population structure occurred. Within one community, the early successional species underwent directional change leading eventually to extirpation on that site; at larger scales, the data suggest persistence because at that scale new sites for the establishment of new individuals were dependably created by disturbance.

Such data on population structure suggest regional persistence of both species, but a cautionary note must accompany any interpretation of spatial data. The data were collected at one point in time; to conclude that stability is present, we must assume that the pattern (the population structure) is in equilibrium with the ongoing processes of recruitment and mortality. (In Zedler and Goff's case, the critical process is the disturbance regime that creates new sites for establishment of the shade-intolerant species.) If spatial structure does reflect stationary (nonvarying) processes, and if there consequently is a dynamic equilibrium, spatial analysis can be substituted for long-term observations. This is analogous to the "space-for-time substitution" in studies of succession (Pickett 1989). For populations, a reverse J spatial structure suggests

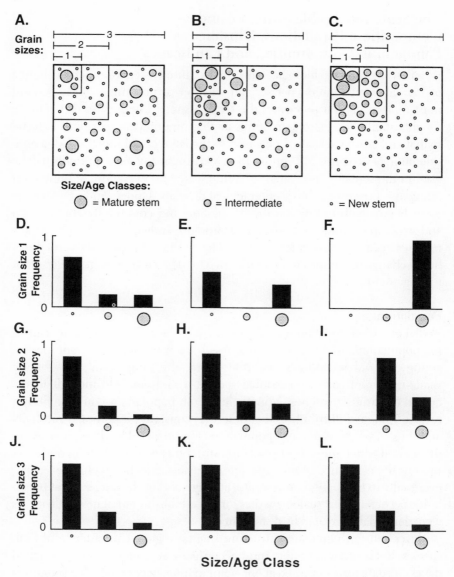

FIGURE 3-1. Population structure and spatial scale. Three populations (A, B, and C) differing in the spatial distribution of young and old stems are sampled at three grain sizes (1, 2, and 3). In all populations there are sixty-four new stems, sixteen intermediate stems, and four mature stems. In Population A, the size/age classes are interspersed. In Population B, mature stems are clustered. In Population C, each size class is spatially segregated from the others, as might occur if reproduction occurs in patches after disturbance.

constant recruitment and proportional mortality. However, the assumption that spatial structure reflects temporal dynamics is a dangerous one to make for rare species and must be tested with investigations of demographic processes and with monitoring data.

If the spatial structure reported by Zedler and Goff (1973) does reflect equilibrium, it is an example of a form of coexistence that has been reported for a wide variety of taxa (Denslow 1985). Species that are superior competitors for scarce resources form stable populations at small scales. Species that are poor competitors for limited resources but strong colonizers of newly opened sites form locally unstable populations but may have stable population structures at a larger spatial scale if disturbances create appropriate new sites with regularity and impose limits to spatial dispersion. In other words, the new sites must open up when propagules are available from older patches and must be within the species' dispersal distance. A species that forms a long, persistent seed bank may be less sensitive to the frequency and dispersion of new sites. Processes that allow periodic reproduction of fugitive species will often result in patchwise successional change (discussed later), but the changes need not be at the community scale. Because adult individuals take up space and use resources, mortality often imposes a patchiness on reproduction, even if it occurs within a population at relatively small spatial scales. At least at the scale of the individual, reproduction is apt to be patchy (White and Pickett 1985).

The tradeoff between competitive ability and colonizing ability produces two extremes for the spatial structure of plant populations. At one extreme, younger individuals or individuals at an earlier stage of development are interspersed with adult reproductive plants. A stable size or age structure, if present, occurs at small spatial scales. At the other extreme, younger and earlier-stage plants are found only on patches that are disjunct from adult populations. If a stable age or size structure occurs, it will occur only on large spatial scales (as shown earlier in Figure 3-1). This perspective forces us to ask questions about the distribution in time and space of younger and smaller stems relative to older, reproductive stems. These questions must be asked even if colonizing and competitive ability do not underlie the pattern.

For some rare species, reproduction occurs within existing populations and younger individuals are interspersed with adults. For other species, reproduction occurs on patches scattered within the site, adjacent to the adult plants. For yet other species, reproduction occurs on the same site the adults dominate but only after the death of the adults in the patch (recolonization may then come from seed shed before adult death, from the seed bank, or from other populations). Finally, there are species that must colonize new sites that are so distant from the adults that these sites are not usually considered part of the same community.

In some cases, the existing adults may saturate all the available microsites

at particular times of observation, and the population will be characterized most of the time by an absence of young individuals. For example, mountain avens (*Geum radiatum*) in the southern Appalachians may, over time, saturate all the available rock crevices—or to the degree that is possible in the face of competition from other species. Opportunities for the establishment of new seedlings will be very rare. For this reason, a population that lacks younger individuals at a particular time is not necessarily in decline.

Furbish lousewort (*Pedicularis furbishiae*) is an example of a rare plant that colonizes newly disturbed patches (Menges 1990; Menges and Gawler 1986). This species does not reproduce on sites where older individuals dominate. Such species will require management at a different spatial scale than species that do reproduce at the site of the adult population. The persistence of such species as Furbish lousewort is dependent on a source of colonists and therefore on the presence of clustered populations, a subject that will be discussed in more detail later.

If a population does not establish sufficient numbers of new individuals on sites where its adults occur, even immediately after adult mortality opens space in the population, that particular population or patch will become extirpated as the adults die. The loss of individual populations must be balanced by the initiation of new populations if the species is to persist at a larger spatial scale. If the new sites become open within the reproductive lifespan of the adults, and if they are located within appropriate dispersal distances, then our conservation focus will be to maintain or restore natural processes that open new sites for colonization and the system of interacting populations that provide the dispersing individuals for colonization.

A species whose young and adult individuals do not co-occur may not, however, fit this model of larger-scale persistence. New sites may be opened at such a low frequency that there are no surviving reproductive adults when the openings occur. The openings may occur at too great a spatial distance for dispersal to be dependable, a problem that is likely to be made worse by human alteration of the matrix surrounding nature reserves and habitat fragmentation (see Primack, Chapter Nine). A particular nature reserve may be smaller than the size of a typical population, and interreserve distances may be larger than typical dispersal distances. In these cases the reserve may be pathologically isolated, and managers will have to maintain critical life stages and genetic diversity through intensive means. If reproduction typically occurs at a spatial scale that cannot be restored or maintained in nature, then we must attempt to retain all individual populations. We may have to create an *ex situ* collection of genetic material to prevent the loss of genetic diversity or the fortuitous loss of a critical life-history stage.

An extreme case for populations that do not reproduce on sites of adult occurrence is that of the relictual population. This would occur, for example, if

adults persist during climatic warming, but new individuals could only establish farther to the north or on cooler microsites. The adults thus represent inertia in the system—they are relictual from a defunct climate and may be under physiological stress or in population decline (see Ledig, Chapter Eleven). Habitat fragmentation is likely to hinder the migration of such species; and conservationists will then have to determine the appropriateness of new sites for establishment and artificially disperse species there. Threats to such relictual populations may require us to move the species to new sites outside the original range. This will challenge conservation managers to address some difficult scientific questions: How do we choose appropriate new sites for establishment? How do we ensure that pollinators, dispersers, and other support species are present? How, given decade-scale environmental variation and the broad fundamental niches of species, can we demonstrate that the population really is relictual on a particular site?

In sum, we can expect that recruitment in rare plant populations can be either continuous or periodic and can occur either on the same sites as adults, adjacent to adults, or disjunct from the adults. As a consequence of these observations, we must distinguish between species that reproduce on site from those that depend on the colonization of new sites, and we must document the scale-dependence of population structure and the processes that lead to successful reproduction. We must also determine if recruitment is continual or periodic and identify the sites and processes that lead to recruitment. Even if a particular population shows abundant reproduction at a relatively small, within-population scale, there are additional questions that must be addressed before we determine the appropriate conservation-management regime. The most basic questions are, Are the processes that lead to the spatial structure stationary, so that there is a dynamic equilibrium? Does the spatial structure reflect continual reproduction at the appropriate scale? Given year-to-year variation in the environment and longer-term environmental trends, population monitoring will be required before we can confirm overall population trends deduced from spatial pattern. Uncertainty about population trends underscores the need for *ex situ* collections as a last resort, failsafe, backup for *in situ* management.

Clustered Populations: Metapopulations and Independent Populations

Several to many populations of one species are often clustered in space, although the scale at which this clustering occurs varies tremendously. The existence of such systems of multiple populations can often enhance but may sometimes lessen the probability of species survival. Individual populations within a cluster of populations may interact forming a *metapopulation*: a system of semi-independent individual populations (Hanski and Gilpin 1991). Even if individual populations are fully independent and thus do not

form a metapopulation, there are important implications for species persistence. I describe each of these cases in turn. It is important to note, however, that the degree of interaction between populations is likely to show continuous variation. Further, interaction might be zero in some years and high in others and may change with decade-scale environmental change. Finally, clustered populations can be independent with regard to some parameters (such as seed dispersal) but interactive with regard to others (such as pathogen spread) (Figure 3-2). Thus, true metapopulations and clustered but independent populations should be considered the ends of a multidimensional continuum.

METAPOPULATIONS

Metapopulations are clustered, noncontiguous populations that interact through gene flow and dispersal (Hanski and Gilpin 1991). To produce a behavior that is different from that of the individual populations, this interaction must be positive (zero interaction means that populations are independent) yet lower than the interaction among individuals within a single population (even if distribution among individuals is patchy, a high level of interaction essentially defines a single population). Hanski and Gilpin defined three scales for species distributions: the *population scale*, over which individuals or their propagules freely move; the *metapopulation scale*, the scale at which individuals must cross an interpatch matrix that has lower usefulness or higher mortality rate than the patch of origin; and the *scale of the geographic range*, over which no individual ever moves.

One of the first questions to be addressed in the metapopulation literature was, Under which circumstances would a metapopulation produce stable persistence, even if local populations were unstable? In a slightly different form, this is the central question in the application of the metapopulation concept to plant conservation: When does a metapopulation support persistence of a rare species and lessen extinction risk?

The models that were constructed to answer this question started with a series of simplifying assumptions:

1. All patches that could be occupied by the species had the same (high) habitat quality.

2. All patches were the same size.

3. Habitat patches were equally spaced.

4. Within-patch population dynamics could be ignored (once occupied, the patch would support a population size equal to the carrying capacity of the patch).

The simplest metapopulation model, parallel to the logistic model for popu-

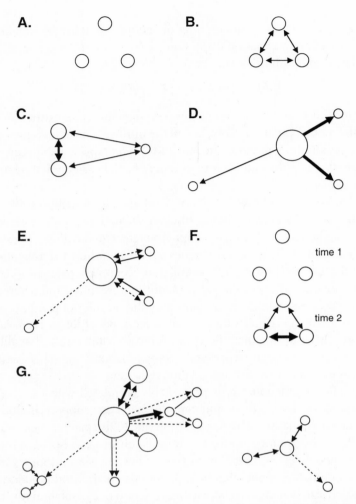

FIGURE 3-2. Patterns of clustered populations (see also Harrison 1991). The arrows indicate interpopulation interactions:

A. Multiple independent populations.

B. Simplest metapopulation structure, in which patches are equal sized and have equal probabilities of interacting.

C. Interaction is a function of distance between populations (stronger interactions at close distance, weaker ones at far distances).

D. Core-satellite population structure (the satellite populations are so far and small that they do not affect the core population and the arrows are unidirectional—the island biogeography model in its simplest form).

E. Core population and nearby populations interact in two ways, but a small distant population is affected by only one kind of interaction (solid arrows indicate gene flow, dashed arrows indicate disease spread).

F. Populations interact in some years but not others.

G. A combination of interactions diagramed in A through F.

lation growth, stated that the number of occupied patches (populations) (p) was a function of population extinction rate (e) time (t), and migration rate, (m) (Hanski and Gilpin 1991):

$$dp/dt = (m - e)\, p\, [1 - p/(1 - e/m)].$$

In this formulation, the difference between migration and extinction rate ($m - e$) is the metapopulation equivalent of the intrinsic rate of increase in the logistic equation for population growth, and $1 - e/m$ represents the carrying capacity, in this case the equilibrium number of patches occupied (Hanski and Gilpin 1991).

Under these conditions, the ratio of intrapatch extinction rate to interpatch migration rate (e/m) defines whether the metapopulation confers persistence on the species. When the migration rate is greater than the extinction rate, the number of occupied patches increases; when the migration rate equals the extinction rate, the number of occupied patches is constant (though the occupancy of the patches continually shifts); and when the migration rate is less than the extinction rate, the number of patches declines. This simple model suggests that human effects, which often increase patch isolation (thus lowering migration) and decrease patch size and quality (thus increasing extinction rate) will destabilize populations whose persistence was originally dependent on metapopulation dynamics.

Population extinction rate is presumably correlated with the size of the patch and/or the size of the population the patch can support (including the effects of patch quality). The immigration rate to the patch is determined by its isolation (the distance among patches). Patch size and isolation thus have joint consequences on species persistence. (See Hanski 1991 for a version of the metapopulation model that incorporates patch area and isolation.) For a given patch size (which correlates with extinction rate), isolation must be less than a threshold value to allow population persistence. For a given degree of isolation (which correlates with immigration rate), patch size must be greater than a threshold value to allow population persistence. If patch size and isolation vary independently, they can have compensatory effects on species presence (Hanski 1991). The worst case for population persistence occurs when isolation is high and patch size is small. Carter and Prince (1991) and Hanski (1991) suggested that at the margin of a species range, habitat patch size decreases and isolation increases, leading to low probability of population occurrence. The actual edge of a species geographic range might be truncated because of the spatial configuration (the size and isolation) of appropriate habitats. Isolated and small patches of favorable environments may generally be unoccupied toward the periphery of a species' range, a phenomenon that will increase the rarity of some species and ensure that available habitats are not all occupied at any one time.

Nisbet and Gurney (1982) and Hanski (1991) proposed that, for a given number of patches occupied (p) a metapopulation persistence would increase as a function of the inverse of the square root of the number of patches (or the islands of appropriate habitat), (H). Thus, for long persistence, the percentage of patches occupied must exceed a threshold:

$$p > 3H^{-0.5}$$

The number of patches required for a given length of persistence will decrease as the patch occupancy increases. For example, a metapopulation occupying only 10 percent of nine hundred available sites at any one time would have the same persistence time as a metapopulation occupying 50 percent of thirty-six sites.

While the mean extinction rate is important (a high extinction rate on patches must be compensated by a high immigration rate for populations to persist), variance in extinction rate also plays a role. Even if the extinction rate is less than immigration rate, high variance signifies a chance that each population will become extinct during the same time period. This was termed regional stochasticity by Hanski (1991) (Table 3-2) and is analogous to environmental stochasticity within populations. While such an effect might be produced by chance alone (that is, by sampling from an extinction rate with variance, all populations might have a high extinction probability in the same year), extinction rates could be positively correlated and change synchronously among populations because they are affected by the same environmental or human factors. Correlations in environmental factors across

TABLE 3-2. Stochasticity in Populations and Metapopulations (Hanski 1991; Menges 1990).

Kind of stochasticity	
Population	
Demographic	Chance events of births and deaths that are not correlated among individuals or with environment
Environmental	Variation in the physical or biological environment that affects all individuals in the population, although different life stages may be affected differently
Metapopulation	
Immigration-extinction	Chance events of initiation and extinction of local populations within a metapopulation that are not correlated among local populations
Regional	Variation in the physical or biological environment that affects all local populations within a metapopulation

populations will tend to destabilize a metapopulation; persistence of a meta-population is promoted by asynchronous dynamics of the individual popula-tions (Hanski 1991). With regard to forces that would synchronize the separate populations, this case borders on that of truly independent populations; inde-pendence is desirable in environmental stochasticity yet undesirable for re-colonization after local extinction.

Of course, species possess differences that will cause them to react differ-ently to the same spatial arrangement of populations. For example, Harrison (1991) suggested that species with small seeds will have lower establishment rates than species with large seeds, leading to a lower importance of metapop-ulation dynamics in regional persistence. However, species with smaller seeds might also have higher seed crops and greater dispersal distances and might experience higher establishment rates after natural disturbance when resources are high and competition low (thus eliminating the need for in-vestment in seed energy stores and allowing small seed sizes to evolve). It might be expected that if rare species have poor dispersal, low reproductive rates, or high sensitivity to small habitat size, and/or if they are limited by ap-propriate habitat, they are unlikely to have been dependent on metapopula-tion dynamics prior to human impact. However, one case of metapopulation dynamics we have discussed previously is the Furbish lousewort. The initia-tion of new populations occurs only after erosion creates open sites. Colo-nization of these new sites is, of course, dependent on the presence of older populations of reproductive individuals in the landscape. The populations on the older sites are doomed to local extinction.

The kind of patchiness assumed in the simplest metapopulation models rarely holds in nature (see Figure 3-2, presented earlier). More often, there is a hierarchical arrangement of patches that differ in size, quality, and/or isola-tion. A common situation is to have a core area in which a population never suffers extinction and is surrounded by satellite populations, as with a main-land surrounded by islands. Such a hierarchy of populations may also involve spatial autocorrelation in environmental factors, so that there is a gradient of decreasing patch quality from center to periphery. Distance may correlate with the degree of departure from the ideal environment, leading to higher extinction rates at the periphery regardless of patch size.

If a large central population results in a high immigration rate to sur-rounding patches, individuals will be present on those patches even if extinc-tion rate is high. This has been called the rescue effect (Brown and Kodrick-Brown 1977). When the individuals established are adjacent to the reproductive individuals and serve to expand the species distribution on an environmental or spatial gradient onto marginal sites, the phenomenon has been called mass effect (Shmida and Wilson 1985). If no reproduction occurs on the peripheral patches, and mortality rate is very high, the phenomenon has been described as source-sink. Obviously, in such cases, peripheral popu-

lations will not have stable age or size structures. Hanski and Gilpin (1991) have summarized the variation in patch size and structure by noting that island biogeography, within its mainland-island model, stands at one end of a continuum for metapopulations, with the original metapopulation model (with its equal sizes, qualities, and spacing of habitat patches) at the other (see Figure 3-2).

There are also potential negative effects of interaction between populations. For example, organisms will increase as distance decreases the immigration of diseas. Hess (1994) used a simulation model to compare the persistence of a series of isolated independent populations with the same number of populations linked in a metapopulation. The metapopulation system usually persisted the longest, but if disease-caused mortality was low enough that infected individuals could spread the disease before they succumbed but high enough to reduce individual populations to a point that demographic and environmental stochasticity took over, isolated populations persisted longer. He went on to cite numerous examples of devastating disease spread among clustered populations and cautioned against overemphasizing the positive effects of connectivity.

Another potential negative effect of metapopulation dynamics concerns the loss of genetic diversity. Gilpin (1988) has argued that, for a given number of individuals, the effective population size is much smaller for a metapopulation than for one population of the same total size. Individual small populations experience drift, and the founder effect comes into play when patches are colonized. If there is a maximum number of individuals available for restoration, Gilpin's work argues against subdivision into separate populations as part of a metapopulation.

Multiple Independent Populations

Distance between populations may counter spatial autocorrelation in environmental fluctuation and thus produce uncorrelated exposures to catastrophe. Multiple populations will buffer extinction risks when the risks of catastrophic loss are independent. Guerrant (1992) has argued that when populations are truly independent, and the chance of any one population lasting one hundred years is 0.5, there is a 75 percent chance of at least one of two surviving and an almost 95 percent chance of one of four independent populations surviving. However, if endangerment for all populations were driven by the same external cause, or if all populations were subject to the same human errors in their establishment or management, there would be no benefit of increasing the number of populations. However, it is interesting to note that suspected extirpations of species from Great Smoky Mountains National Park all involve species known from single populations (see Table 3-1); species represented by single populations are clearly vulnerable.

The central idea of metapopulation persistence is that instability in local populations is compensated by immigration from nearby populations. If metapopulations are an important mechanism of species persistence, then we must minimize the distances between populations and increase the connections between habitats and the quality of the intervening matrix for the species in question. These factors will increase dispersal rate. However, as the degree of synchrony of local population dynamics increases, the importance of metapopulation dynamics for persistence decreases; human effects, including management that is driven by the same lack of biological understanding, may tend to synchronize population fluctuations among remnant habitat patches. As isolation of populations increases, regional stochasticity (the regional correlation in mortality rates and extinction rates) will decrease, leading to increased persistence, a phenomenon that finds complete expression when populations are truly independent. Isolation will decrease immigration rates for disease organisms but will decrease immigration rates of the target species and thus prevent reestablishment after local extinction.

Quinn and Hastings (1987) addressed the relative influence of population size and subdivision (multiple independent populations) on extinction risk through reliability theory. Reliability theory predicts the distribution of failure times for systems of redundant subunits whose individual failure times are known. This analysis allows us to compare the benefits of increasing the reliability of subunits (in our context, this would be the lowering of extinction risk within individual populations) versus increasing the redundancy of subunits (in our context, increasing the number of independent populations, without increasing their "reliability"). Quinn and Hastings (1987, 1988) concluded that if environmental stochasticity is the cause of extinction, subdivision decreases extinction risk, whereas, when demographic stochasticity is important, subdivision increases extinction risk. They argued that environmental stochasticity was generally more important than demographic stochasticity and that, therefore, subdivision was called for. Gilpin (1988) criticized this conclusion on several grounds. First, he noted that in reliability theory it does not matter whether all or only a few of the subunits are functioning at any one time—they function in parallel, and as long as one subunit is functioning, the system has not failed. For conservation, he argued, the number of subunits (independent populations) in each time interval will control the genetic diversity that survives. It makes a large difference whether, for example, one of sixteen or fifteen of sixteen populations survive. Second, Gilpin argued that populations may not be truly independent but tend to be influenced by the same environmental shock. Gilpin supported decreasing extinction risk through increased population size and decreased subdivision. Experiments by these authors (Forney and Gilpin 1991; Quinn et al. 1989) and others have supported one or the other sides of this argument, presum-

ably because both population size and multiple independent populations can lessen extinction risk in various circumstances.

Given these conflicting trends, we cannot expect a priori a solution that will fit all cases, unless we are allowed the luxury of large populations and many populations, some of which are independent and some of which are within dispersal distances. While human fragmentation of natural areas will generally increase extinction rates and decrease immigration rates, not all rare species were dependent on metapopulation dynamics before human effects. The negative and positive effects of isolation should be investigated in populations. The processes that operated before human impacts will often be one reference point for designing conservation strategies for the future.

Communities and Landscapes

The analysis of pattern and process in plant communities (Watt 1947) seeks to link observations of state variables (such as numbers, sizes, spatial distribution of individuals within communities) with understanding of ongoing dynamic processes (such as growth, mortality, disturbance, reproductive rate, dispersal, and colonization). Watt attempted to shift research in plant ecology away from emphasis on static community pattern and toward investigations of dynamic processes that shape communities, a theme subsequently taken up by many plant ecologists (White 1979). Whereas Watt emphasized the influence of process on pattern, pattern can have a correspondingly important influence on process (Turner 1989). A simple example is the way that pollination success depends on the spacing of flowering individuals. The field of landscape ecology addresses the interplay of pattern and process with explicit reference to the spatial configuration of landscapes.

Watt's examples involved processes intrinsic to the plant communities studied that produced a cyclically changing structure and composition on any one patch in the community. These changes occurred on a small scale and thus would rarely be termed successions, although the changes were clearly analogous to successional change. Some researchers described the cycles as minisuccessions (White 1979). In Watt's examples, the sites of successful establishment of younger individuals were spatially displaced from the living reproductive adults. Indeed, it was mortality of the adults (whether caused by external factors or not) that allowed for successful reproduction on the sites where reproductive adults had previously dominated. Watt's view was subsequently extended to larger scales, with small-scale, within-community processes such as individual mortality grading into large-scale, natural disturbances that initiated succession (White 1979).

Whether at the within-community or larger scale, if the distribution of patches (that is, the kinds of patches and the frequency distribution of these patches) is stable as a function of mortality rate or disturbance regime, a

dynamic equilibrium holds. This implies that the elements of the pattern, the communities and species, are persistent (though they may be fleeting on any one site), which is often precisely what we are trying to accomplish in conservation management. If there is no patch dynamic equilibrium at any scale, there will be directional change in the pattern, possibly resulting in loss of some of the elements. If we are restoring or managing a successional community, species that characterize the community may not be reproducing within the community.

One of the first questions for any community or landscape in conservation is whether the patterns and processes are at equilibrium. As for populations, equilibrium (and persistence) will often be present at larger scales but not at smaller ones. The relation between the size of the patch and size of the landscape in which it occurs will often be important (Pickett and White 1985; Shugart 1984). If we sample small patches consisting of single successional states, we will conclude that the individual patches are undergoing directional change. On the other hand, if we sample across a larger area so that we encompass patches of different ages, there may be a stable frequency distribution of patch types. Local patchwise dynamics can produce a dynamic equilibrium at larger spatial scales. Some communities and landscapes may not be at equilibrium at any scale; others may have been originally in equilibrium, but the remnant conservation areas may not be large enough to encompass the full dynamic mosaic (White 1987). Factors that tend to produce equilibrium are stable disturbance regimes, small patch size relative to landscape size, and a feedback between patch state and probability of disturbance (such as susceptibility to wind, fire, or insect increases through succession or time since last disturbance; White and Pickett 1985; Table 3-3). Whereas it can be relatively straightforward to evaluate the latter two factors, it is harder to document the stability of disturbance regimes, given the long time scales involved and the annual- to century-scale environmental variation that imposes different frequencies and intensities of disturbance on whatever successional mosaic is present.

Romme (1982), Romme and Knight (1982), and Turner et al. (1994) examined population, community, and ecosystem consequences for fire regime in Yellowstone National Park. Romme and Knight showed that the size of natural fires was large relative to the size of the watershed, with the result that the watershed as a whole saw dramatic changes in the habitats present over time (Figure 3-3). They suggested that such watershed-scale changes (specifically the successional development of stands after the most recent large fires some 150 to 200 years ago) lay behind contemporary declines in mountain bluebird populations (through change of habitat structure) and in native fish populations (through loss of aquatic productivity due to decreases in nutrient additions from the landscape). They suggested that a dynamic equilibrium and

TABLE 3-3. Factors leading to patchy dynamic equilibrium (Shugart 1984; White and Pickett 1985).

Pattern/process	Equilibrium	Disequilibrium
Patch size relative to landscape area	Small	Large
Frequency of disturbances for landscape as a whole	High	Low
Biotic feedback: Correlation of probability of disturbance to time since last disturbance	Pos. Corr.	Uncorrelated
Disturbance regime/Physical environment	Stable	Changing
Population age/size structure across landscape	All-aged	Even-aged patches
Composition/structure	Stable	Unpredictable

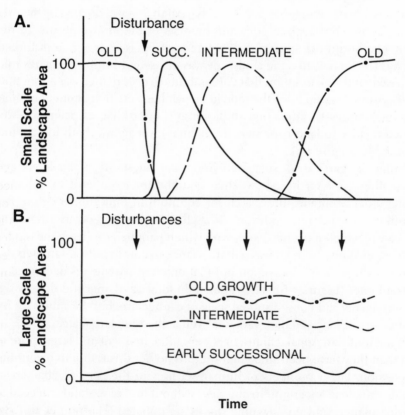

FIGURE 3-3. Percentage of landscape area in various successional states at two spatial scales (see also Romme and Knight 1982). Dramatic successional change occurs at small scales. In a small landscape (A), the effect is wide swings in the habitats present. In a large area (B), if disturbances are independent in the landscape (for example, if the probability of disturbance is controlled by successional time since last disturbance), the landscape as a whole shows more constant representation of various habitats.

stability of populations for many species would only occur at a much larger spatial scale—beyond the watershed they studied—for the fire regime that prevailed (large fires reoccurring at long intervals). The Yellowstone fires of 1988 showed that if a fire-generated equilibrium occurs, it does so at a scale larger than the national park itself (Christensen et al. 1989). Turner et al. (1994) concluded, based on simulation studies, that the 1988 fires placed Yellowstone in a region of stable landscapes with high variance and that landscape composition and structure was affected much more by the largest observed disturbances, rather than the mean disturbance sizes. They also argued that the chances of a dramatic shift in landscape dynamics would decrease as the size of the protected area increased.

Shugart (1984) developed a prediction for patch dynamic equilibrium based on simulation modeling. He determined that, if the size of the landscape was fifty or more times the size of the patch that was created by the prevailing disturbance regime, then the biomass of the landscape was in dynamic equilibrium (Figure 3-4). If the ratio of landscape area to disturbed area was less than 50 to 1, the landscape was in disequilibrium. Shugart's rule was based on the assumption that vulnerability to disturbance was a function of successional age and that the individual patches were independent in their dynamics. Shugart's simulation studies also showed that, as patch size decreased relative to landscape size, the habitat diversity was both higher and more stable (Figure 3-5).

While Shugart's work suggested important questions at the landscape scale, other research has shown that equilibrium is difficult to define—among other problems, the search for equilibrium forces us to define the quantitative bounds for equilibrium. Whether equilibrium occurs (as well as the scale at which it occurs) varies with which parameter is being measured. Busing and White (1993) showed that variance in community parameters decreased with scale of observation but that different parameters decreased at different rates (Figure 3-6). The variance in total basal area and density decreased steeply at an observational scale of 0.4 hectare, but the values for individual species showed high variance even at the 1-hectare scale.

In the southern Appalachians, fires were more frequent and larger prior to 1930 than they have been since that time. Whether this is due to recent fire suppression, larger numbers of human-set fires in the past, or changes in overall climate is uncertain. In any case, without fire, pine stands succeed to oak dominance, and no new pine stands are initiated (Harmon et al. 1983; White 1987). The number of lightning ignitions observed over the last several decades does not seem high enough to account for the pine dominance now observed, but fire sizes may be much smaller now due to fire suppression. Because the landscape is on a net successional trajectory, a number of pine forest species are declining. It will probably be necessary to allow lightning

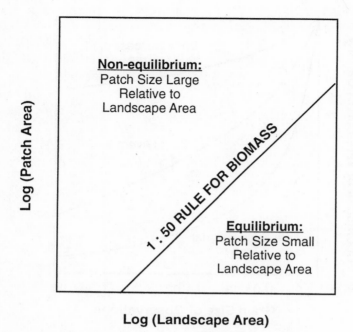

FIGURE 3-4. Shugart's 1 to 50 rule for biomass equilibrium in a landscape affected by patchwise disturbance.

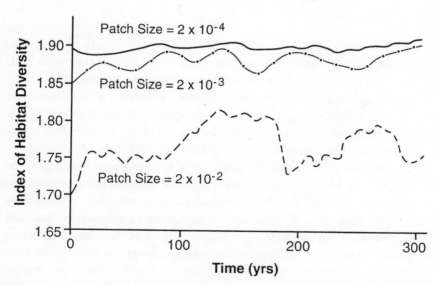

FIGURE 3-5. Habitat diversity as a function of the patch size relative to the landscape area (from Shugart 1984). The smaller the patch size, the higher and more stable the habitat diversity across the landscape.

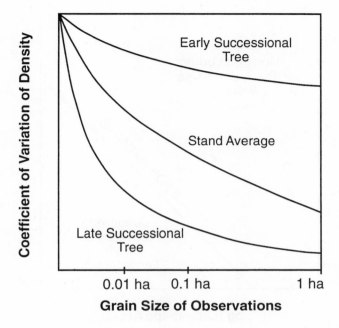

Grain Size of Observations

FIGURE 3-6. Variation in community parameters as a function of the spatial scale (here, grain or quadrat size) of the observations (after Busing and White 1993). The coefficient of variation decreases as a function of observation scale, but late successional trees, which have abundant shade-tolerant reproduction, have a coefficient of variation that decreases more quickly than early successional trees, which reproduce on widely scattered and relatively large patches.

fires to burn and/or to use management fire to reestablish a larger scale equilibrium of this system and the species that occur within it (Table 3-1).

In general, then, stability in patch dynamics within and among communities, like the stability of population structure, can be expected to vary with spatial scale. Scales at which pattern and process are in equilibrium are likely to vary from one community to another and from one landscape to another. For the conservationist, the principle challenge is determining at which scale stability occurs for key processes, whether the scale is practical for conservation management, and how to manage processes that are not stable at the scale of the conservation area. If community structure and composition and landscape pattern are in dynamic equilibrium, conservation strategy essentially becomes synonymous with preserving a large enough area for equilibrium and protecting the integrity of natural processes. On the other hand, if equilibrium does not occur at any scale, we will have to determine whether equilibrium can be induced through restoration of natural processes or whether ongoing management, including reintroduction of individuals and

populations, will be necessary to retain all important elements of the landscape mosaic. Since we are unlikely to know if a patch dynamic equilibrium applies, monitoring of community composition and structure and analysis of natural processes are critical.

While patch dynamic equilibrium may be an outcome of natural processes, the size of particular conservation areas may not be large enough to encompass this equilibrium. When conservation areas are small relative to the patch size produced by natural dynamics, there is a danger that particular successional stages will dominate the entire nature reserve at a particular point in time (Pickett and Thompson 1978). Species dependent on other successional stages will be lost from the system, with the result that the reserve will both lose biological diversity in the short-term and lack the species to respond to future disturbances. In essence, the reserve is too small for the natural patch dynamics of the system. There is also a danger that in our zeal to reintroduce natural processes (such as fire) to remnant natural areas we will not take scale effects into account. For example, if we burn an entire, relatively small natural area, we may remove fire-sensitive species or life stages that would have persisted in a larger burned landscape because of spatial variation in fire effects, even if fire is a natural process. Beyond the issue of patch dynamics and scale, of course, disturbances such as fire vary in intensity. We must determine not only the scale of burned patches relative to the conservation area but also the conditions (season, weather, and fuel conditions) under which to burn. While fires may be a natural process, not all fires are equal. It is likely that too much homogeneity (whether in a uniform intensity or season of burn, or even the uniform absence of fire) will simplify a system biologically. Essentially, we will have less freedom to manage by natural process in a landscape dominated by habitat fragmentation, and we will have to be more careful to manage for spatial variability and temporal dynamics per se.

Individual species differ in requirements for establishment, growth, and reproduction. This presents a challenge: The management regime necessary for one species may be detrimental to another. To maintain all of the species all of the time obviously requires environmental variation in both space and time (that is, the protection of all parts of an environmental gradient and the presence of all successional stages). That species differ in their requirements, with the stochastic nature of colonizations and extinctions, means that we should ideally have enough protected area (whether in one block or among reserves) for replication of both gradients and disturbances. In any case, monotony of environmental factors and disturbance regimes is to be avoided at all costs because they will tend to simplify natural systems. The reverse of this statement is the intermediate disturbance hypothesis (Connell 1978) and the role of disturbance in coexistence (Huston 1979). Given the stochastic nature of population recruitment and dispersal, redundancy will also support the

survival of biological diversity. This suggests an important role for reintroduction in biological management.

Approaching Reintroduction at the Landscape and Regional Scales

Reintroduction and restoration both imply a "putting back" to some original or natural condition. However, *original* is not the same as *natural*, and neither should be equated a priori with *stable*. For example, a conflict between restoration of a historic state versus restoration of natural processes underlies a recent controversy over fire management in Sequoia-Kings Canyon National Park (Bonnickson and Stone 1985). Some researchers argued that natural fires should be allowed to burn; others argued that the system first had to be restored to its 1880s structure and composition. The latter view holds that, since fire suppression had altered forest structure, fire behavior would no longer be natural, regardless of ignition source. Recent evidence has suggested that fires often swept into the park from lower-elevation chaparral lands and that the park therefore may not be able to achieve a patch dynamic equilibrium (N. Christensen, personal communication, August 1988).

Let us assume that a single rare plant species has disappeared from a community and that the site is now protected and managed for conservation. Let us also assume that we have an adequate sample of the original gene pool stored off-site. Finally, let us assume that the threats that caused the extirpation in the first place have been eliminated. The putting back of an original gene pool into a site of original occurrence and into a community with restored function is reintroduction in its narrow sense. We assume that the biological context—the friends and enemies of our target species—is intact.

Reintroduction to a site of original occurrence may represent the most frequent goal of reintroduction, but there are problems with this simple formula. In addition to assuming that we have documentation that the population occurred on the original site, the formula implies we know the original condition (original gene pool, original habitat) and that the original condition was in fact supportive of a sustained population of the particular species. While the narrow case of the "natural-original" condition may sometimes be applicable and may often be the motivation of reintroduction, the material presented previously and the following discussion suggests that we cannot restrict ourselves to this situation. First, there will be problems in defining the original site as a reference point or using it for reintroduction. Further, the degree of threat (including loss of the original habitat and changes to the spatial context of remnant areas) will influence our options; moreover, climate change

has the potential to drastically reset the "natural" and "original" distribution of environments. Moreover, as our editors note in the introduction, concepts such as "original" and "historic range" have little meaning except if stated explicitly with regard to biological, spatial, and temporal scale. In the text that follows, I discuss problems with the definition and use of the "original" site as a criterion for reintroduction and then describe the implications of imperfect knowledge.

The Original-Natural Site

To what degree of precision should the original site be defined? In one of the few quantitative statements, the Nature Conservancy Council of Britain used < 1,000 m as the definition of an original site (Birkinshaw 1991). However, distance is obviously a crude measure of the variables of interest. Knowledge of the appropriate habitat, original patch dynamics of the population, and gene flow distances are all important. A definition based on distance would preclude the re-creation of interacting subpopulations if original sites had been destroyed.

The National Park Service has a conservative policy on introductions. They prohibit, for example, the release of a native fish above a waterfall that represents a natural barrier to its dispersal (M. Ruggerio, personal communication, May 1990). Obviously, release of the fish above the barrier would change the natural role of isolation in community composition and in evolution; the fish is, technically, an exotic species above the barrier. This leads to a "natural" and narrow definition of sites for reintroduction. It implies a strict reliance on natural process. An increase in range (such as above a barrier in a stream) must be the result of natural processes, not human artifact. A management reintroduction would be limited to unambiguous cases of human-caused loss of populations. Natural extirpations or failures to disperse or establish would be viewed as producing desirable absences of species from some sites on which they could occur. These would be desirable because they would be natural; however, since nature had many absences explained by barriers to dispersal rather than by unfavorable environmental conditions, the natural process of evolution would be influenced by the pattern of absences.

Reintroduction requires temporal scaling as well. How recently must we document the original presence on the site? Many conservationists would probably be comfortable with tens of years but might extend to larger temporal scales (such as hundreds of years) if a past extirpation were known to be the result of human action. The importance of time in the definition of an original occurrence is illustrated by a recent National Park Service action to reject the reintroduction of a Mexican desert tortoise to Big Bend National Park because it was uncertain if the species had occurred within the park

more recently than several thousand years ago (M. Ruggerio, personal communication, May 1990). Natural shifts in species distributions would be expected over that time span in response to climate change.

As argued previously, spatial and temporal scales also become important in populations that fluctuate naturally in abundance on any one site. For example, some species occur naturally on unstable sites (such as the Furbish lousewort, Menges 1990). Older patches of reproductive plants are doomed to extirpation, with new sites being continually colonized. The best sites for the establishment of new individuals may not be near or immediately adjacent to adult plants (for example, many shade-intolerant species must disperse to newly open sites). Disturbance-generated dynamics suggest the importance of larger functional scales than that measured by the distribution of existing populations (Pickett and White 1985). The importance of interacting patches—for example, in metapopulations—also argues against an overly narrow definition of the site of original occurrence.

Where patches of different successional states are maintained through the operation of disturbance, and where a particular rare species is dependent on the early successional patches, successional change on any one site will result in species decline. Conservationists can manage against succession on that one site or manage in a larger context for the continual production of new sites for colonization. Where conservationists opt to manage for the presence of rare plants within a larger spatial context (a watershed or a landscape) without regard to specific sites, "original site" becomes a less meaningful criterion for reintroduction. As with the reintroduction of mobile animal species, presence becomes a property of the larger system rather than the specific site.

In the extreme, if metapopulation dynamics holds and/or a species depends on colonizing sites that are disjunct from sites of mature plants, then individual sites will ultimately be as fleeting as the individuals themselves. In this sense, we cannot manage for populations to occupy a particular site forever, although our goal (persistence of the species or its descendants) remains intact at the larger spatial scale.

The Problem of Uncertainty

While some conservationists may be committed to reintroduction based on original conditions in a philosophical sense, we may not have sufficient knowledge to describe that state. We may be uncertain about the range, environmental conditions, threats, and gene pool of an "original" occurrence. We may, for example, in the face of uncertainty about the gene pool, start with as much genetic diversity as we can. Faced with uncertainty about the

species–environment relation, we may decide to plant individuals along an environmental gradient within a site.

At greater spatial scales uncertainty may also occur. It is often assumed that historic occurrences define the best habitats for reintroductions. However, this assumes an equilibrium between species distribution and environment. There is an asymmetry in the evidence: Presence tells us that the species can live in the habitat (but not how beneficial that habitat is for the species relative to other habitats); absence does not distinguish between an inappropriate habitat, a failure to disperse to that habitat, or a recent stochastic extinction unrelated to the species–environment match. An important research question is how to use the extant range (with information on extirpated populations) and experimental findings to define the envelope of possible sites for reintroduction. Depending on the type of rarity, some rare species might be expected not to saturate their potential habitats (they do not fill the envelope defined by their physiological and competitive abilities)—their distributions and performance might be difficult to predict from site conditions and they may exhibit high population variance in nature. In general, the small size and isolation of some rare plant habitats means that the species–area relation, random extinctions, and dispersal limits will cause absences that are not predictably from biological or environmental factors. The simplest form of the theory of island biogeography predicts the number of species as a function of the size and isolation of islands suggests that there will be a constant number of species on an island, but that there will also be an ongoing turnover of species. The theory does not predict which species are present at any one time. Historic events, such as the elimination of a species from one side of a natural barrier and its failure to recolonize thereafter, will also tend to produce distributions that are not predictable from the environment.

The southern Appalachians have many examples of rare species that do not saturate the seemingly appropriate habitats. Rugel's ragwort (*Cacalia rugelia*) is a narrow endemic of the Great Smoky Mountains and yet is abundant in spruce-fir forests and in bordering northern hardwood forests. There is nothing environmentally unique about its sites—the explanation probably lies in its inability to disperse across the major river valleys south and east of the Smokies. Species richness patterns in the high-elevations also suggest historic effects, including extinction during the warmest postglacial times (White and Miller 1988; Wiser and White, forthcoming). Wiser (1993) modeled the distributions of several rare species of high-elevation cliffs in the southern Appalachians. She found that models based on existing populations suggested that the blue ridge goldenrod (*Solidago spithamaea*) was found on only a narrow subset of possible sites. Further, she found that most variables deemed important in the models changed with the spatial scale of the data

used in the models (1 m² versus 100 m²) and that primary variables for community composition differed from those important for individual rare species.

A Continuum for Reintroduction and Introduction

Using site and gene pool for illustration, our actions can range from the purely natural (known original gene pool on known original site) to highly artificial (manipulated gene pool on a new site outside the natural range). At one extreme, human manipulation will alter natural patterns and processes if applied to remnant natural areas. We may lose our ability to study natural patterns, the role of isolation, and the importance of absences on ecological and evolutionary processes. If humans play a dominant role in the distribution of species and genes, the original biogeographic pattern will be lost and unavailable for study in the future. At the other extreme, the most hands-off, "natural" options will doom species to extinction if rapid environmental change occurs. We should proceed stepwise, from the conservative actions to the more manipulative actions, judging the necessity of action for the survival of species and gauging impacts of introduction to intact natural areas and intact gene pools.

In essence, this discussion suggests that there is no one right answer; rather, our decisions are contingent on context. Three critical elements of the context are degree of endangerment to a particular element of biological diversity, degree of environmental change (which will potentially reset all species distributions), and the kind of natural processes that lead to species persistence. In other words, our reintroduction plans need "if–then" clauses (Gordon 1994). When we are dealing with natural areas that are protected and managed for conservation, that have high ecological integrity, and that are not affected by climate change, the reintroduction of a species extirpated because of human activity prior to conservation of the habitat is a straightforward action and one at the conservative end of the scale. When we are dealing with a degraded habitat and changed natural process, more drastic human activities and multiple species introductions may be called for (Janzen 1988).

Conclusion

The poet Robert Frost wrote that he wanted the following epitaph on his gravestone: "He had a lover's quarrel with the world." A conservationist's encounter with reintroduction, and more broadly with restoration, brings Frost's poem to mind because these strategies, as positive as they are, become appropriate only when the conservation of wild, intact nature has failed. Further, we should have a healthy skepticism about our ability to restore nature given

our inadequate understanding of natural processes, species interactions, and ecosystem function (Diamond 1987; Fahselt 1988). At the population level, reintroduction challenges us to create self-sustaining populations that do not require ongoing human intervention. A wild population will often have a better chance of survival than a newly introduced population, especially if we have corrected any problems with site environment or management. Thus, reintroduction cannot be seen as having equal standing with *in situ* conservation (Falk and Olwell 1992; New England Plant Conservation Program 1992). Finally, reintroduction and restoration may, through our own ignorance of nature, result in increased artificiality and diminished natural processes in surviving conservation areas (Diamond 1987). In essence, these strategies force us to address the question, What should our human role in nature be?

Regardless of our answers to this philosophical question, however, reintroduction and restoration have become necessary to conserving biological diversity because of the pace of habitat loss and degradation (Falk and Olwell 1992). These strategies are already critical for the most endangered species and ecosystems (Janzen 1988; Center for Plant Conservation 1992). Their importance is increasing. As the last large, relatively pristine areas become either protected or developed, only degraded lands will remain, ensuring an important role for restoration. Further, the loss of populations usually precedes the outright loss of species, so that reintroducting or establishing new populations will be necessary to reestablish the natural role of multiple, semi-independent, populations in lessening extinction risk and in maintaining genetic variation among populations. Perring (1974) illustrated this phenomenon by analyzing 10-kilometer grid cell records of British vascular plants. The number of grid cells lost (thus providing a minimum estimate for loss of populations) was very dramatic, even though few species became extinct. Humans often make ecosystems and landscapes monotonous; reintroduction seeks to counter this trend and diversify our simplified landscapes.

Given the uncertainties associated with reintroduction, we must ask, At which time scale do we judge the success of reintroduction? Certainly the period should depend on the life span and reproductive characteristics of the species in question. Once we get to a ten-year evaluation period we have effectively merged our time scale with that of environmental fluctuation. The only conclusion we can come to is that, for critically endangered species, reintroduction functions like a long-term experiment and should be part of a species survival plan and ongoing monitoring. Our subsequent actions should be based on a periodic evaluation of the condition of our population. We must design these experiments in such a way that even failures will increase our understanding. A further conclusion is that when the reintroduction arises from an *ex situ* germplasm collection, we should always keep a backup collection of stored material.

Reintroductions can be experiments in the narrow sense of the term as well. We may want to establish some of the initial parameters as variables if we cannot, as we often cannot, predict which values of those parameters will optimize establishment, reproduction, and survival. For example, if we do not know the precise microsite requirements of a species, we can plant individuals across a local gradient of conditions. Alternative microsites may even perform differently in different years, and individuals within those microsites may have different reproduction in different years, thus buffering the population against extinction.

When we use reintroduction, introduction, and restoration, we must assess how we can use spatial scale to increase our chances of success and to reduce our costs of ongoing management. At the population level, this will often mean investigating how the age or size structure of populations changes with scale of observation, combined with assessment of natural processes responsible for reproduction and local population dynamics. We must investigate the spatial scale at which reproduction and the establishment of new individuals or new populations occurs. If natural processes produce a stable age or size structure at some spatial scale, and that scale is encompassed within conservation areas, our job is reduced to maintaining natural processes and monitoring populations. If natural processes do not produce a stable structure or if they operate only at a scale larger than the area managed for conservation, then direct management will be required to prevent population loss and/or the loss of critical life-history stages.

Similar questions about community composition and structure must be asked: At which scale do natural processes that initiate succession operate? How does this scale compare to the area protected for conservation? As with populations, if the conservation area is small, the natural dynamics may result in the entire conservation area becoming dominated by a single successional stage, with a resulting loss of populations and species dependent on other successional stages. If we cannot integrate the site with conservation management on surrounding lands, we will have to manage for these stages if we are to retain biological diversity.

For species originally dependent on metapopulation dynamics, we must assess human effects on within-patch extinction rate and between-patch migration rate. If human impacts have increased extinction rate, emphasis should be placed on increasing patch and intrapatch population size. If human impacts have decreased migration rates, then emphasis should be placed on decreasing interpatch distances or artificially increasing dispersal. Persistence time can also be increased by increasing the number of patches and/or increasing the percentage of the patches occupied at equilibrium. In general, decreasing the synchrony of extinction and causes of mortality among patches will increase persistence time. However, not all, and perhaps

not many, rare species were originally dependent on metapopulation dynamics. In fact, subdivision of a population may accelerate the loss of genetic diversity. There may still be good reasons to create new populations. In general, persistence may increase with the number of multiple independent populations; however, if we create new populations, we must address habitat quality, habitat management, and the introduction of coevolved species (such as mycorrhizal fungi, pollinators, and dispersers).

Given the natural dynamics and possible global environmental change, it is probably unwise, in the long term, to tie reintroduction and restoration to a narrow definition of original conditions. The presence of some species should be seen in the context of the ecosystems in which they occur, rather than in terms of a site narrowly defined in space and time. Such ecosystems may experience substantial local fluctuation. Because of this, because redundancy of population will generally increase species persistence, and because of environmental change, conservation planning must ultimately address landscape and regional scales.

REFERENCES

Bierregaard, R. O., T. E. Lovejoy, V. Kapos, A. A. dos Santos, and R. W. Hutchings. 1992. The biological dynamics of tropical rain forest fragments: A prospective comparison of fragments and continuous forest. *BioScience* 42:859–866.

Birkinshaw, C. R. 1991. Guidance notes for translocating plants as part of recovery plans. Nature Conservancy Council. CSD Report 1225.

Bonnicksen, T. M., and E. C. Stone. 1985. Restoring naturalness to national parks. *Environ. Manag.* 9:479–486.

Bratton, S. P., and P. S. White. 1980. Rare plant management: After preservation what? *Rhodora* 32:49–75.

Brown, J. H., and A. Kodrick-Brown. 1977. Turnover rates in insular biogeography: Effect of immigration on extinction. *Ecology* 58:445–449.

Busing, R. T., and P. S. White. 1993. Effects of area on old-growth forest attributes: implications of the equilibrium landscape concept. *Landscape Ecol.* 8:119–126.

Carter, R. N., and S. D. Prince. 1991. Epidemic models used to explain biogeographical distribution limits. *Nature* 293:644–645.

Center for Plant Conservation. 1992. A strategy for conserving the endangered plants of Hawaii. St. Louis: Center for Plant Conservation, Missouri Botanical Garden.

Christensen, N. L., J. K. Agee, P. F. Brussard, J. Hughes, D. H. Knight, G. W. Minshall, J. M. Peek, S. J. Pyne, F. J. Swanson, J. W. Thomas, S. Wells, S. E. Williams, and H. A. Wright. 1989. Interpreting the Yellowstone fires of 1988. *BioScience* 39:678–685.

Connell, J. H. 1978. Diversity in tropical rain forests and coral reefs. *Science* 199:1302–1310.

Davis, S. M., and J. C. Ogden. 1994. *Everglades: The ecosystem and its restoration.* Delray Beach, Fla.: St. Lucie Press.

Denslow, J. S. 1985. Disturbance-mediated coexistence of species. In *The ecology of*

natural disturbance and patch dynamics, eds. S. T. A. Pickett and P. S. White. New York: Academic Press.

Diamond, J. 1987. Reflections on goals and on the relationship between theory and practice. In *Restoration ecology,* eds. W. R. Jordan, M. E. Gilpen, and J. D. Aber. Cambridge: Cambridge University Press.

Fahselt, D. 1988. The dangers of transplantation as a conservation technique. *Natural Areas Journal* 8:238–244.

Falk, D. A., and K. E. Holsinger. 1991. *Genetics and conservation of rare plants.* New York: Oxford University Press.

Falk, D. A., and P. Olwell. 1992. Scientific and policy considerations in restoration and reintroduction of endangered species. *Rhodora* 94:287–315.

Forney, K. A., and M. E. Gilpin. 1991. The genetic effective size of a metapopulation. *Biol. J. Linn. Soc.* 42:165–175.

Gardner, R. H., R. V. O'Neill, M. G. Turner, and V. H. Dale. 1989. Quantifying scale-dependent effects of animal movement with simple percolation models. *Landscape Ecology* 3:217–227.

Gilpin, M. E. 1988. A comment on Quinn and Hastings: Extinction in subdivided habitats. *Conserv. Biol.* 2:290–292.

Gordon, D. 1994. Translocation of species into conservation areas: A key for natural resource managers. *Nat. Areas J.* 14:31–37.

Guerrant, E. O., Jr. 1992. Genetic and demographic considerations in the sampling and reintroduction of rare plants. In *Conservation biology,* eds. P. L. Fiedler and J. K. Jain. New York: Chapman and Hall.

Hanski, I. 1991. Single-species metapopulation dynamics: Concepts, models, and observations. *Biol. J. Linn. Soc.* 42:17–38.

Hanski, I., and M. Gilpin. 1991. Metapopulation dynamics: brief history and conceptual domain. *Biol. J. Linn. Soc.* 42:3–16.

Harmon, M. E., S. P. Bratton, and P. S. White. 1983. Disturbance and vegetation response in relation to environmental gradients in the Great Smoky Mountains. *Vegetation* 55:129–139.

Harper, J. L. 1981. The meanings of rarity. In *The biological aspects of rare plant conservation,* ed. H. Synge. New York: John Wiley and Sons.

Harper, J. L. 1992. Foreword. In *Conservation biology,* eds. P. L. Fiedler and J. K. Jain. New York: Chapman and Hall.

Harrison, S. 1991. Local extinction in a metapopulation context: An empirical evaluation. *Biol. J. Linn. Soc.* 42:73–88.

Hess, G. R. 1994. Conservation corridors and contagious disease: A cautionary note. *Conserv. Biol.* 8:256–262.

Huston, M. 1979. A general hypothesis of species diversity. *Am. Nat.* 113:81–101.

Janzen, D. J. 1988. Tropical ecological and biocultural restoration. *Science* 239:243–244.

Magnuson, J. J., T. K. Kratz, T. M. Frost, C. J. Bowser, B. J. Benson, and R. Nero. 1991. Expanding the temporal and spatial scales of ecological research and comparison of divergent ecosystems: Roles for LTER in the United States. In *Long-term ecological research: An international perspective,* ed. P. G. Risser. New York: John Wiley and Sons.

Menges, E. S. 1990. Population viability analysis for a rare plant. *Conserv. Biol.* 4:52–62.

Menges, E. S., and S. C. Gawler. 1986. Four-year changes in population size of the en-

demic Furbish's lousewort: Implications for endangerment and management. *Nat. Areas. J.* 6:6–17.

Miller, R. I., and P. S. White. 1986. Considerations for preserve design based on the distribution of rare plants in Great Smoky Mountains National Park, USA. *Environ. Manag.* 6:119–124.

Miller, R. I., S. P. Bratton, and P. S. White. 1987. A regional strategy for reserve design and placement based on an analysis of rare and endangered species distribution patterns. *Biol. Conserv.* 39:255–268.

Nekola, J. C., and P. S. White. Submitted manuscript. The distance decay of similarity in biogeography and ecology.

New England Plant Conservation Program. 1992. *Wild Flower Notes* 7(1). Framingham, Mass.: New England Wild Flower Society.

Nisbet, R. M., and W. S. C. Gurney. 1982. *Modeling fluctuating populations.* New York: John Wiley.

Noss, R. F. 1993. A conservation plan for the Oregon Coast Range: Some preliminary suggestions. *Nat. Areas J.* 13:276–290.

Palmer, M. W. 1992. The coexistence of species in fractal landscapes. *Am. Natur.* 139:375–397.

Palmer, M. W., and P. S. White. 1994. Scale dependence and the species–area relationship. *American Naturalist* 144:717–740.

Patterson, B. D. 1987. The principle of nested subsets and its implications for biological conservation. *Conserv. Biol.* 1:323–334.

Perring, F. H. 1974. Changes in our native vascular plant flora. In *The changing flora and fauna of Britain*, ed. D. L. Hawksworth. New York: Academic Press.

Pickett, S. T. A. 1989. Space-for-time substitution as an alternative to long-term studies. In *Long-term studies in ecology: Approaches and alternatives*, ed. G. E. Likens. New York: Springer-Verlag.

Pickett, S. T. A., and J. N. Thompson. 1978. Patch dynamics and the design of nature reserves. *Biol. Conserv.* 13:27–37.

Pickett, S. T. A., and P. S. White. 1985. *The ecology of natural disturbance and patch dynamics.* New York: Academic Press.

Quinn, J. F., and A. Hastings. 1987. Extinction in subdivided habitats. *Conserv. Biol.* 1:198–208.

Quinn, J. F., and A. Hastings. 1988. Extinction in subdivided habitats: Reply to Gilpin. *Conserv. Biol.* 2:293–296.

Quinn, J. F., C. L. Wolin, and M. L. Judge. 1989. An experimental anaysis of patch size, habitat subdivision, and extinction in a marine intertidal snail. *Conserv. Biol.* 3:242–251.

Reed, R. A., R. K. Peet, M. W. Palmer, and P. S. White. 1993. Scale dependence of vegetation–environment correlations: A case study of a North Carolina piedmont woodland. *J. Veg. Sci.* 4:329–340.

Romme, W. H. 1982. Fire and landscape diversity in supalpine forests of Yellowstone National Park. *Ecol. Monogr.* 52:199–221.

Romme, W. H., and D. H. Knight. 1982. Landscape diversity: the concept applied to Yellowstone Park. *BioScience* 32:664–670.

Shafer, C. 1990. Nature reserves: Island theory and conservation practice. Washington, D.C.: Smithsonian Institute Press.

Shmida, A., and M. V. Wilson. 1985. Biological determinants of species diversity. *J. Biogeogr.* 12:1–20.

Shugart, H. H. 1984. *A theory of forest dynamics*. New York: Springer-Verlag.

Simberloff, D. S., and L. G. Abele. 1976. Island biogeography theory and conservation practice. *Science* 191:285–286.

Turner, M. G. 1989. Landscape ecology: the effect of pattern on process. *Annu. Rev. Ecol. Syst.* 20:171–197.

Turner, M. G., W. H. Romme, and R. H. Gardner. 1994. Landscape disturbance models and the long-term dynamics of natural areas. *Nat. Areas J.* 14:3–11.

Watt, A. S. 1947. Pattern and process in the plant community. *J. Ecol.* 35:1–22.

White, P. S. 1979. Pattern, process, and natural disturbance in vegetation. *Bot. Rev.* 45:229–299.

White, P. S. 1987. Natural disturbance, patch dynamics, and landscape pattern in natural areas. *Nat. Areas J.* 7:14–22.

White, P. S., and S. P. Bratton. 1980. After preservation: The philosophical and practical problems of change. *Biol. Conserv.* 18:241–255.

White, P. S., and R. I. Miller. 1988. Topographic models of vascular plant richness in the southern Appalachian high peaks. *J. Ecol.* 76:192–199.

White, P. S., and J. C. Nekola. 1992. Biological diversity in an ecological context. In *Air pollution effects on biodiversity*, eds. J. R. Barker and D. T. Tingey. New York: Van Nostrand Reinhold.

White, P. S., and S. T. A. Pickett. 1985. Natural disturbance and patch dynamics: An introduction. In *The ecology of natural disturbance and patch dynamics*, eds. S. T. A. Pickett and P. S. White. New York: Academic Press.

Wiser, S. K. 1993. Vegetation of high-elevation rock outcrops of the southern Appalachians. Ph.D. diss., University of North Carolina, Chapel Hill.

Wiser, S. K., and P. S. White. Forthcoming. High-elevation outcrops and barrens of the southern Appalachian mountains. In *The phytogeography and ecology of rock cliffs, barrens, and glades in eastern North America*, ed. J. M. Baskin.

Zedler, P., and F. G. Goff. 1973. Size-association analysis of forest successional trends in Wisconsin. *Ecol. Monogr.* 43:79–94.

CHAPTER 4

The Regulatory and Policy Context

Charles B. McDonald

The U.S. Fish and Wildlife Service (FWS) is the principal federal agency responsible for implementing the Endangered Species Act (ESA) of 1973, as amended (Table 4-1). In that role the FWS enacts regulations and establishes policies intended to carry out the purposes of the ESA in a consistent and responsible way. Because plant introductions, reintroductions, and population augmentations are only a small part of the overall FWS endangered species program, they are the subject of relatively few FWS regulations and policies. Nevertheless, the few existing regulations and policies concerning these activities provide a framework for carrying out introductions, reintroductions, and population augmentations when needed to support the purposes of the ESA.

Regulations have the force of law and are binding on all who come under the jurisdiction of the issuing government—in this case the U.S. government. The FWS has the authority to create regulations that implement the purposes of the ESA. These and all other federal regulations are established through a process called rulemaking. In this process, a proposed regulation is published in the *Federal Register*,[1] and public comment is invited. After an appropriate comment period, all comments are considered, and either a final regulation is published in its original or altered form, or the proposal is withdrawn. Final regulations and withdrawals are also published in the *Federal Register*. Final regulations amend the Code of Federal Regulations[2] and have the force of law.

Adding species to the federal list of endangered and threatened wildlife and plants and establishing procedures that direct the conservation of species under the ESA are rulemaking processes (Table 4-2). (The lists are found in the Code of Federal Regulations at Title 50, Part 17, Sections 11 and 12, which may be abbreviated 50 CFR 17.11 and 17.12.)

Policies, unlike regulations, are only binding on the establishing group, so technically only the FWS is required to follow FWS policies. Policies may be

issued in any number of forms (memoranda, director's orders, handbooks, guidelines, procedures manuals, and so on).

U.S. Fish and Wildlife Service Regulations Relating to Endangered Plant Introductions, Reintroductions, and Population Augmentations

FWS regulations relating to endangered and threatened plant introductions, reintroductions, or population augmentations are intended to interpret and implement the ESA prohibitions against certain activities with endangered and threatened species. As will become evident from this discussion, the ESA gives the FWS only limited regulatory authority over individuals who wish to conduct introductions, reintroductions, or population augmentations with endangered and threatened plants.

TABLE 4-1.   An index to some topics in the Endangered Species Act of 1973, as amended (U.S. Congress 1988).

Sections and subsections of the act	Subjects covered
2	Findings, purposes, and policies
2(b)	Purposes of the Act
3	Definitions
4	Determination of species as endangered or threatened
4(a)(1)	Factors for determining endangered or threatened status
4(a)(3)	Requirement to designate critical habitat
4(b)(3)	Petitions to list, delist, or revise critical habitat
4(b)(7)	Emergency listing
4(d)	Authority to create special rules for threatened species
4(f)	Recovery plans
4(f)(4)	Requirement for public review of draft recovery plans
5	Land acquisition
6	Cooperation with the states
6(c)(2)	Cooperative agreements for states with plant-conservation programs
7	Interagency cooperation
8	International cooperation
9	Prohibited acts
9(a)(2)	Prohibited acts for plants
10	Exceptions
10(a)(1)(A)	Permits for research or conservation activities
10(j)	Experimental populations
11	Penalties and enforcement

Prohibitions and Permits Under the ESA

Prohibited activities with endangered and threatened plants are described in section 9(a)2 of the ESA and in the Code of Federal Regulations (50 CFR 17.61 and 17.71). It is unlawful under the ESA for any person subject to the jurisdiction of the United States to import or export endangered plants from the United States or engage in commercial interstate transport or sale. It is further prohibited to damage, destroy, or remove endangered plants from areas under federal jurisdiction or damage, destroy, or remove them from any area in knowing violation of a state law or regulation, including state criminal-trespass law. The ESA contains prohibitions only for endangered plants but states that the prohibitions may be extended to threatened plants through regulation. Regulations describing prohibited activities for threatened plants are

TABLE 4-2. An index to some ESA regulations in Title 50 of the Code of Federal Regulations.

Part	Subject covered
17.12	List of endangered and threatened plants
17.61	Prohibitions for endangered plants
17.61(c)(4)	Exceptions to endangered plant–collecting prohibitions for employees or agents of state conservation agencies with a Section 6 cooperative agreement
17.62	Permit procedures for endangered plants
17.71	Prohibitions for threatened plants
17.71(b)	Exceptions to threatened plant–collecting prohibitions for employees or agents of state conservation agencies with a Section 6 cooperative agreement
17.72	Permit procedures for threatened plants
17.81	Listing experimental populations
17.86	Special rules for plant experimental populations (reserved but not yet used)
17.96	Critical habitats for plant species—legal boundaries, constituent elements, and maps
23.23	List of species protected under the Convention on International Trade in Endangered Species of Wild Fauna and Flora (CITES)
81	Procedures for cooperation with the states under Section 6
402	Procedures for federal interagency cooperation under Section 7
424	Procedures for revising the endangered and threatened species lists and designating critical habitat under Section 4
424.11	Factors for listing, delisting, or reclassifying species
424.12	Criteria for designating critical habitat
424.14	Petitions
424.15	Authority to publish lists of candidate species in the *Federal Register*
424.17	Time limits and required actions for procedures under Section 4
424.20	Emergency rules

given in the Code of Federal Regulations at 50 CFR 17.71. The regulations are the same as for endangered plants except the regulations have not been updated to include the 1988 ESA amendments, which make it a violation to maliciously damage or destroy endangered plants on areas under federal jurisdiction or to damage, destroy, or remove them from any area in knowing violation of a state law or regulation, including state criminal trespass law.

The FWS may issue permits to undertake certain activities that are prohibited under the ESA or federal regulation. The authority for granting permits to undertake conservation activities that would otherwise be prohibited is provided at Section 10(a)(1)(A) of the ESA.

Permits have two principal purposes. First, they let the FWS set terms and conditions under which a permit is valid. The FWS, for instance, could stipulate in a permit to collect an endangered plant from federal land strictly for the purpose of a taxonomic study or, if an endangered plant is being collected for an introduction project, the FWS could describe in the permit the methods, locality, and other details of the project. The second purpose of permits is to provide a record-keeping mechanism for activities with various species. Such record keeping might, for instance, be used to determine the commercial demand for certain artificially propagated cacti, which might in turn help indicate the likelihood of illegal field collecting.

Permits are issued only to allow otherwise prohibited activities and a careful reading of the prohibited activities for plants indicates several circumstances where no permit is required. No permit is required, for instance, to collect endangered plants on private land as long as no state law is violated. No permit is required simply to possess, cultivate, or propagate endangered plants as long as no interstate commercial activity is involved. (These two provisions differ from the similar provisions for endangered animals. For animals, taking is prohibited anywhere within the United States [private lands included], and endangered animals cannot be possessed [either dead or alive] without a permit, except for those animals possessed prior to their being added to the federal endangered species list.)

Under the present set of ESA prohibitions, endangered plant permits cannot function as a record-keeping mechanism for tracking plant propagation and introduction, reintroduction, or augmentation activities because of the large number of instances when no permit is required. Further, permits cannot be used to direct introduction, reintroduction, or augmentation projects when the original propagation stock comes from private lands, because no permit is required to collect or possess these plants. Thus the ESA establishes no mechanism for the FWS to function as a national record-keeper or coordinator for endangered and threatened plant introduction, reintroduction, or augmentation projects. Agencies, groups, and individuals conducting such projects are encouraged to coordinate their work with the FWS and

other interested parties, but presently, in many instances there is no legal requirement to do so.

Experimental Population Designations Under the ESA

The ESA and FWS policy allow introduction of endangered species into unoccupied habitats. However, many proposals to do so have been fervently resisted because the FWS could not assure other federal agencies, state and local governments, and private landowners that transplanted populations would not limit their future land-management options. Introduced or reintroduced plants or animals have full protection under the ESA, including the taking prohibitions of Section 9 and the federal interagency consultation requirements of Section 7. To help alleviate the resistance to introductions, the ESA was amended in 1982 to include the possible designation of an introduced population as "experimental" (Section 10(j)).

"Experimental populations" must be wholly separate geographically from nonexperimental populations of the same species and are to be designated either essential or nonessential. An "essential experimental population" is one whose loss would likely appreciably reduce the likelihood of the species' survival in the wild. All other experimental populations are to be designated nonessential. Congress expected that most experimental populations would be designated nonessential.

The FWS's intention to establish an experimental population must be formally announced through publication of proposed and final regulations in the *Federal Register*. An experimental population proposal must identify the boundaries of the experimental population area; indicate whether the population is essential; describe management restrictions, protective measures, or other management concerns for that population; and describe the periodic review process for evaluating the introduction's success and its effect on the species' conservation and recovery (50 CFR 17.81).

Most experimental population designations also contain "special rules" that provide greater management flexibility than the prohibitions of the ESA would ordinarily allow. For instance, a "special rule" might remove ESA protection from individuals of the endangered species that leave the experimental population area, or fishermen might be allowed to catch and release endangered fish of an experimental population without violating the ESA's taking provisions.

When considering whether to designate an introduced population as experimental, the FWS must first determine if such designation is needed. If no local or other opposition exists to introducing or reintroducing a population, experimental population designation is unnecessary. Experimental populations have been designated for the red wolf, southern sea otter, black-footed ferret, Colorado squawfish, and several other animals, but none have been

designated for endangered or threatened plants, despite a number of introduction projects. These plant projects have, in general, encountered little public opposition or even public attention. There are myriad reasons for this: (1) most plant introductions are done within a relatively small area, often only a few acres; (2) plants usually stay within the introduction area; and (3) endangered plants are not protected on private land, so no ESA violation would occur if a cooperating landowner or subsequent landowner destroyed an introduced population. Despite present lack of use, the experimental population designation remains available as a management tool for plant introductions.

FWS Policies Relating to Endangered Plant Introductions, Reintroductions, and Population Augmentations

The policies discussed here apply only to species that are federally listed as endangered or threatened. They do not apply to species that are candidates for federal listing or to other rare or restricted species, although the principles involved can apply to these species as well.

The Historical Range Policy

The first FWS policy relating to introductions was enunciated in an agency memorandum dated June 25, 1981 (U.S. Fish and Wildlife Service 1981). The policy states, "Endangered and Threatened species will not be relocated or transplanted outside their historical range without specific case-by-case approval from the Director." *Historical range* is the "known general distribution of the species or subspecies as reported in the current scientific literature" (50 CFR 17.11 and 17.12). This policy is intended to help carry out a major purpose of the ESA, which is "to provide a means whereby the ecosystems upon which endangered species and threatened species depend may be conserved" (ESA, Section 2(b). Under this purpose, the goal is to rehabilitate ecosystems so they can support endangered species, rather than simply to move endangered species away from ecosystems that are imperiled.

In addition to conforming to the purposes of the ESA, there are biological reasons for not introducing species outside their known historical range. Two reasons were enunciated in an FWS memorandum dated July 9, 1982 (U.S. Fish and Wildlife Service 1982):

> 1. *Doubtful Survival of Transplanted Populations.* The historical (natural) range limits of a species are determined by the interaction of physical and biotic factors in its environment, including such influences as extreme temperature minima, competition with other species, susceptibility to disease under varying habitat conditions,

precise substrate composition, and so forth. These interactions may be subtle and may occur only sporadically or cyclically at long intervals. When a given species is absent from what superficially appears to be suitable habitat near its historical (natural) range (i.e. within limits of dispersal for the species), it may generally be assumed that its absence reflects some natural quality of the habitat that precludes the species' long-term survival. Biological information is often lacking as to species' microenvironmental requirements. Transplants into habitats resembling but outside endangered and threatened species' historic[al] range are thus unlikely to hold much potential for the species' survival over the long run, although initial establishment may be possible.

2. *Potential Alteration of Gene Pools.* The occurrence of a species (or subspecies or distinct population) in its present form is the product of a long evolutionary process involving close adaptation to particular habitat conditions. Introduction of representatives of a species into nonhistorical range inevitably subjects them to new selection pressures and may result in significant genetic change, so that eventually the protected transplanted population, if it survives, may not in fact be the same organism we were attempting to conserve. Even more drastic would be the introduction of a listed species (or subspecies or distinct population) into the range of a closely related taxon with which hybridization could occur. In this case we would run the risk of significantly altering the gene pools of both taxa.

This policy and the just-stated operational assumptions of the FWS have guided the agency in conflicts over development of occupied endangered species habitat. In instances of conflict, the usual preferred solution put forth by developers is to simply translocate species to "safe" habitat and allow development to proceed. Such proposals might even be accompanied by proposals to "enhance" the species through propagation and introduction of more individuals than would be moved in the first place. When possible, the FWS has resisted such proposals because attempts at this kind of "conservation" have usually failed due to inadequate biological understanding and also because they involve the exchange of known suitable habitat for habitat of unknown quality.

Adherence to the policy of not introducing endangered species outside their historical range is extended by regulation to any state agency that has a cooperative agreement with the FWS under Section 6 of the ESA (50 CFR 17.61(c)(4)(iii)). By this regulation, qualified employees or agents of state

conservation agencies may collect endangered plants from areas under federal jurisdiction without a permit, provided the collecting is not anticipated to introduce the species into an area beyond its historical range.

The Captive Propagation Policy

The policy titled "Captive Propagation/Artificial Propagation of Native Threatened and Endangered Species" is found in the Fish and Wildlife Service Manual, which contains the standing and continuing directives of the FWS (U.S. Fish and Wildlife Service 1993b). The policy pertains to propagation programs for producing individuals for research, establishing and maintaining refugia populations, eventual introduction or reintroduction into the wild, or augmentation of existing populations. Animal propagation programs provided the initial impetus for this policy because of the significant resource commitments that such programs often require. However, most of the guidance in the policy is equally applicable to plant propagation programs. The policy reads as follows:

> Captive propagation of animals and artificial propagation of plants are recognized in certain situations as essential tools for the conservation and recovery of species, subspecies, or populations. The Service has used this tool to enhance the recovery of several species and successfully return them to the wild. However, to ensure prudent use of limited funds, the long-term resource benefits must be critically assessed and evaluated relative to alternative conservation measures and other recovery priorities nationwide. Therefore, it is the policy of the U.S. Fish and Wildlife Service (Service) that captive propagation or artificial propagation of native threatened and endangered species, subspecies, or populations:
>
> 1. Will be conducted in accordance with the regulations implementing the Endangered Species Act, the Animal Welfare Act, and the Departmental and Service procedures relative to the National Environmental Policy Act;
>
> 2. Will be based on the specific recommendations of recovery strategies identified through approved recovery plans and in accordance with the recovery *task* priority system. The recovery plan should clearly identify: (a) the role of propagation as a recovery strategy, (b) the role of Service facilities in propagation efforts as appropriate, (c) the role of Service cooperators and partners in recovery strategies involving propagation (e.g., Center for Plant Conservation, American Association of Zoological Parks and Aquaria), (d) the estimated cost of propagation efforts, including an analysis

of expected capital and operational expenditures, (e) estimate of the number of individuals (FTEs) and training which will be required for implementing propagation/maintenance, and (f) an estimate of task duration;

3. Will be used as a recovery strategy only when other measures employed to maintain or improve a species', subspecies', or population's status in the wild have failed, are determined to be likely to fail, or would be insufficient to ensure/achieve full recovery. Every effort should be made to accomplish conservation measures that enable a species, subspecies, or population to recover naturally in the wild, with or without human manipulation (e.g., translocation), prior to contemplating captive/artificial propagation for reintroduction or augmentation. *Propagation programs will not be employed in lieu of habitat conservation or other measures that would stimulate natural recovery in the wild.* Propagation programs intended for reintroduction/augmentation should be closely coordinated with habitat management efforts, and both propagation programs and habitat conservation efforts should be periodically reviewed;

4. Will be implemented only after appropriate consideration of the potential effect on wild populations of the removal of individuals for propagation purposes (e.g., following a population viability analysis and/or risk assessment in the instance of severely depleted populations). In those instances where individuals propagated in captivity are to be introduced to suitable habitat or are to be used to augment an existing population, consideration of the potential effects of such introductions on the receiving population and other resident species will be evaluated;

5. Will be based on sound genetic principles to preserve the genetic variability and integrity of wild populations of the species, subspecies, or population involved. Intercrosses will not be considered for use in propagation programs unless absolutely necessary to preserve unique genetic material of species critically close to extinction. Use of intercross individuals for species conservation will require written justification and Director's approval. Propagation should not be initiated without the completion of an approved genetics management plan. Such a plan should be comparable to existing standards (e.g., American Association of Zoological Parks and Aquaria, Species Survival Program, Center for Plant

Conservation guidelines) and insure [sic] that the genetic makeup of propagated individuals is similar to that of wild populations. The genetic management plan will include all necessary consultations and permits, including those required by States. The genetic management plan, in addition to other elements (e.g., maintenance of genetic variability), should specifically address the issue of disposal of individuals found to be unfit for introduction to the wild, unfit to serve as broodstock, surplus to the needs of research,[3] or surplus to the recovery needs for the species (e.g., to preclude genetic swamping). Exceptions to these general guidelines may be granted at the Director's discretion when the species in question has an ephemeral or very short (1–2 year) life span which necessitates propagation for the purposes of maintenance in refugia or for purposes of required research;

6. Will be conducted in a manner to produce individuals that are behaviorally and physiologically suitable for release to the wild;

7. Will be conducted in a manner that minimizes potential introduction or spread of diseases and parasites of concern into captive or wild environments;

8. Will be conducted in a manner which will prevent the escape of captive stock outside their historic[al] range;

9. Will, when feasible, be conducted at more than one location in order to reduce the potential for catastrophic loss at a single facility;

10. Will be coordinated, as appropriate, with organizations and investigators both within and outside the Service. The Service will make efforts to cooperate with other Federal, State, Tribal, and local governments;

11. Will be conducted in cooperation with the Wildlife Conservation Management committee of the American Association of Zoological Parks and Aquaria (AAZPA) in maintaining studbooks and registration of animals with the Species Survival Program (SSP) and the International Union for the Conservation of Nature's (IUCN's) International Species Information System (ISIS) as appropriate. Plant propagation programs will be coordinated with the Center for Plant Conservation or other appropriate groups or investigators;

12. The policy and guidelines contained herein will be subject to exceptions on a species-by-species or case-by-case basis only when biologically supported and approved by the Director.

Three major points are emphasized in this policy. The first and most important is that the FWS must use considerable restraint in employing propagation for conservation of endangered and threatened species. Propagation for introduction should be a final rather than an initial option and is totally inappropriate if done in lieu of protecting the habitat needed to support existing or introduced populations. The final two points provide guidance on how to assure a high likelihood of success for propagation projects determined to be essential. These points involve the requirements (1) to carefully plan and coordinate projects prior to initiation and (2) to employ sound genetic and biological principles in their execution.

Recovery Plans as Policy Documents

The ESA (Section 4(f)) requires the FWS to develop plans (referred to as recovery plans) for the conservation and survival of endangered and threatened species. Recovery plans represent the official position of the FWS on the goals for achieving species' recovery and the tasks required for reaching those goals. Recovery plans are to be updated regularly and revised as needed to reflect new findings and changes in species' status. A public comment period is required prior to approval of any new or revised recovery plan (ESA, Section 4(f)(4)).

Recovery plans are written in three parts. The first part provides background information discussing, to the extent known, the species' taxonomy, distribution, abundance, ecology, threats, and past conservation efforts. The second part describes the objectives and criteria for achieving recovery and provides an outline of tasks to accomplish the objectives. The third part gives a schedule for implementing the recovery tasks; it assigns priorities, duration, costs, and responsible parties for the recovery tasks. Recovery plans identify all parties the FWS anticipates will be involved in recovery. Parties other than FWS are identified to aid planning and to help those parties justify seeking and expending funds for recovery tasks.

Introduction, reintroduction, and population augmentation tasks are identified in species' recovery plans if anticipated as necessary to prevent extinction or accomplish full recovery. As indicated in Items 2 and 3 of the previous policy, propagation for introduction, reintroduction, or population augmentation is to be initiated only if it has been identified as necessary in an approved recovery plan and only if other higher-priority recovery tasks to protect the species' habitat have failed or appear likely to fail.

Approximately 25 percent of the FWS plant-recovery plans identify introduction or reintroduction as a needed part of the recovery program, thus indicating the importance the FWS places on this recovery activity for many species (Falk and Olwell 1992). Although the need for introductions or reintroductions is identified in plans, the details of these programs are seldom described. It is anticipated that the contents of this book will provide much of the information needed for FWS personnel and others to plan and execute successful introduction, reintroduction, and population augmentation programs.

Summary

1. The ESA allows individuals to possess endangered and threatened plants and collect them from private lands without a permit. Therefore, permits to undertake prohibited activities with endangered and threatened plants do not provide the FWS with a reliable mechanism for tracking or supervising non-federal endangered-plant introduction projects.

2. The possibility of designating experimental status for introduced or reintroduced populations of endangered or threatened species was included in the ESA to reduce public resistance to endangered species introductions. Thus far no introduced plant populations have been given experimental status because the few such projects have lacked public controversy.

3. FWS policy requires that the agency introduce endangered and threatened species only within their historical ranges and only after habitat conservation efforts have failed, appear likely to fail, or would be insufficient to accomplish full recovery.

4. An FWS policy establishes the biological and procedural framework for the propagation of endangered species for introduction, reintroduction, and population augmentation projects. The policy emphasizes that such projects should be initiated only after habitat conservation actions alone have been deemed inadequate to recover the species. The policy also emphasizes the need for full consideration of genetic and biological principles in conducting propagation projects.

5. Introduction, reintroduction, or population augmentation tasks are described in FWS recovery plans when they are determined to be needed to conserve a species or accomplish its full recovery. Few recovery plans, however, describe specific details for implementing these tasks.

ENDNOTES

1. The *Federal Register* is a daily publication intended to inform the public of actions of the executive branch (including regulations) that affect them. All federal agencies publish in the *Federal Register* proposed rules, final rules, and notices that cover the programs under their authority. The Office of the Federal Register is an agency of the National Archives and Records Administration.

2. The Code of Federal Regulations is a codification of the general and permanent rules published in the *Federal Register* by the executive departments and agencies of the federal government. The code is divided into fifty titles that represent broad areas subject to federal regulation. Each title is divided into chapters that usually bear the name of the issuing agency. Each chapter is further subdivided into parts covering specific regulatory areas. Regulations created by the U.S. Fish and Wildlife Service are found at Title 50: Wildlife and Fisheries, Chapters I and IV (1993a).

3. "Captive propagation research and production for introduction purposes generally should not be conducted simultaneously, though it may be desirable in some instances. The primary objective of captive propagation research is to conduct studies which will provide for future propagation success in establishing and maintaining refugia populations and producing individuals for release into the wild."

REFERENCES

Falk, D. A., and P. Olwell. 1992. Scientific and policy considerations in restoration and reintroduction of endangered species. *Rhodora* 94:287–315.

U.S. Congress. 1988. Endangered Species Act of 1973, as amended through the 100th Congress. Washington, D.C.: U.S. Government Printing Office.

U.S. Fish and Wildlife Service. 1981. Memorandum. Subject: Reintroduction of endangered and threatened species. Washington, D.C.: U.S. Department of Interior, Fish and Wildlife Service, Office of Endangered Species.

————. 1982. Memorandum. Subject: Fish and Wildlife Service policy on transplanting listed endangered and threatened species. Washington, D.C.: U.S. Department of Interior, Fish and Wildlife Service, Office of Endangered Species.

————. 1993a. Code of Federal Regulations, Title 50: Wildlife and Fisheries, Chapters I and IV. Washington, D.C.: Office of the Federal Register, National Archives and Records Administration.

————. 1993b. Captive propagation/artificial propagation of native threatened and endangered species. *Fish and Wildlife Service Manual, Part 733* Washington, D.C.: U.S. Department of Interior, Fish and Wildlife Service, Division of Endangered Species.

FOCUS
Reintroducing Endangered Hawaiian Plants

Loyal A. Mehrhoff

⟶⟶◦⟵⟵

Hawaii has been justifiably called the extinction and endangerment capital of the United States. Almost one-third of the entire flora is either extinct or in danger of becoming so (Mehrhoff 1993). With so many rare and endangered species, it is no wonder that there have been numerous attempts to reintroduce rare Hawaiian plants. These attempts raise many questions about the role that reintroduction can and should play in conservation programs. Are reintroduction efforts successful in dealing with endangered species? Have we learned anything from these attempts? Are there any general principles concerning the transplantation of endangered plants? In this chapter I look at Hawaii's experience and try to answer these questions. In many respects, the reintroduction of Hawaiian plants, with their extremely localized distributions, high endemism, high habitat specificity, short seed viability, small population sizes, and large number of threats present unprecedented conservation challenges and difficulties. I hope that by developing an understanding of Hawaii's successes and failures land managers may gain important insights into conservation techniques of value to other areas. While some generalities from Hawaii's experiences may apply to temperate continental situations, many will be most applicable to oceanic islands or subtropical areas undergoing catastrophic ecosystems collapses resulting from large-scale land conversions and invasions of significant alien species (such as tropical dry forests).

Conservationists who undertake reintroduction efforts in Hawaii are confronted not just with technical problems of how to transplant rare species or how to control threats but also with philosophical questions such as the definition and measurement of success, the use of augmentation versus reintroduction, the role that translocations associated with mitigation projects

should play in conservation, and the question of justification (if any) for introducing a species outside of its known historical range. (Several of these important issues are further addressed in other chapters by Pavlik [Chapter Six], Berg [Chapter Twelve], and White [Chapter Three].)

Hawaiian Biology

In order to understand the role plant transplantations have played in Hawaiian conservation, one has to understand both the uniqueness of Hawaii and the magnitude of Hawaii's extinction crisis. The Hawaiian archipelago is one of the most isolated places on the face of the earth. Over 3,200 kilometers (2,000 miles) separate these tropical volcanic islands from the nearest continent. The biological consequences of this isolation have been profound (Carlquist 1980; Wagner, Herbst, and Sohmer 1990). Isolation directly resulted in low immigration rates and may have played a key factor in the development of Hawaii's many spectacular examples of adaptive radiation (Carlquist 1980). These two factors—low immigration and extensive speciation from a few colonists—probably led to the islands' exceptionally high level of endemism (Table 5-1). It is thought that Hawaii was at one time home to as many as ten thousand endemic species of plants, birds, and invertebrates (Gagne 1988). Nearly 90 percent of all flowering plants native to Hawaii are found nowhere else on Earth, making Hawaii truly unique from a biological perspective (Sohmer 1990). This high degree of endemism at the archipelago level occurs on finer scales as well. Many species are restricted to a single island or even a portion of an island. Another important characteristic of the Hawaiian flora is that there are several conspicuous groups of organisms that either have failed altogether to become established or are severely underrepresented. Before human contact, there were no land mammals, (except insectivorous bats), no land reptiles, no amphibians, no ants, no termites, no social bees, no mosquitoes, no earthworms, no coconut trees, no obligatory epiphytic flowering plants, and only three native orchid species. In fact, the state of Alaska has almost ten times as many native orchids as does Hawaii (twenty-nine in Alaska versus three in Hawaii).

Human Impacts on Hawaii

With the arrival of Polynesians sixteen-hundred years ago (Kirch 1985), Hawaii began to lose its unique biological diversity. Humans cleared land for agriculture (Kirch 1985), hunted Hawaii's large flightless birds (James and Olsen 1991), and brought with them many alien species, which also altered

TABLE 5-1. The status of the Hawaiian flowering plants. Percentages shown indicate the percentage of the total native flora (1131 species and subspecies) that falls into a specific category (such as endemic, extinct, or endangered) (Wagner, Herbst, and Sohmer 1990; Center for Plant Conservation 1992; Mehrhoff 1993; and unpublished data).

	Number of species
Total native species	1131
Endemic species	1032 (91%)
Extinct species	101 (9%)
Endangered species*	358 (32%)
Naturalized alien species	856

*Species having fewer than one thousand individuals or less than five extant populations and threatened by factors (such as alien ungulates, specific diseases/insects, or habitat conversion) known to eliminate entire populations.

the Hawaiian ecosystems (Smith 1985; Cuddihy and Stone 1990). The most notable of these aliens were the Polynesian rat, domestic pigs, and dogs. While the Hawaiians extensively modified lowland forest habitats and probably caused the extinction of many endemic birds (Olsen and James 1990), it was European colonization and the alien species accompanying this colonization that wrought the most destruction. Some species, like cockroaches, Norway rats, earthworms, and ants were probably simple tagalongs. Still others were intentional introductions. Early sailing ships released cattle, sheep, and goats. Two other species, axis deer and mouflon sheep, were brought in later for sport hunting. Alien herbivores have been particularly damaging, because they both directly feed on endemic plants unadapted to mammalian herbivory and because these aliens can convert native ecosystems into alien ecosystems by promoting conditions that favor other alien species (Stone 1985; Smith 1985; Vitousek et al. 1987; Cuddihy and Stone 1990). In short, mammalian herbivores are incompatible with the survival of native Hawaiian ecosystems. Given time, native areas not protected from alien goats, pigs, deer, and sheep will eventually degrade to the point that the mature biota is lost.

The presence of alien pests in Hawaii has not gone unnoticed, but sometimes the cure has been as bad as the problem. For example, mongooses were brought in to control alien rats, but the mongoose has since had as profound an effect on endemic birds as the rat populations it was brought in to control. Similarly, the alien predatory snail (*Euglandina rosea*) was brought in from Florida in an attempt to control another alien snail (*Achatina fulica*) that was thought to be an agricultural pest. Unfortunately, this predatory snail has since contributed to the decline or extinction of many endemic snails (Solem 1990).

Today, over eleven thousand alien plant species have been introduced into

Hawaii (G. Staples, personal communication 1994), with over 850 of these species becoming naturalized (Wagner, Herbst, and Sohmer 1990; Mehrhoff unpublished 1995). Almost ninety alien plants are considered significant threats to native ecosystems and endemic species (Smith 1985).

In total, these human-related activities have resulted in the loss of approximately 70 percent of Hawaii's native ecosystems (Gagne 1988). Habitat loss has not been distributed randomly, most of the destruction has been concentrated in lowland areas, which harbored unique dry forests and species-rich mesic forests (Figure 5-1). As the habitat for Hawaii's flora has vanished, so have the plants comprising these ecosystems (Figure 5-2).

Conservation Efforts

With 9 percent of its flowering plants extinct (84 species) and 23 percent (227 species) at risk of extinction (fewer than one thousand individuals or fewer than five populations), Hawaii truly is the extinction and endangerment capital of the United States. Unfortunately, efforts to stop this trend have been complicated by several important factors.

First, many species now have so few individuals remaining that they are extremely difficult to preserve. Five percent of the entire flora (47 species) consists of species that have fewer than ten plants remaining in the wild (Table 5-2). How does one manage a species that has only one or two individuals?

Second, Hawaii is poorly known from an ecological perspective. Because Hawaii is isolated from areas with large numbers of ecologists and because there are relatively few local ecologists, there have been only a handful of studies on reproductive biology, population dynamics, and restoration techniques. This lack of information means that many transplantation efforts have been little more than educated guesses.

Third, several ecosystems are on the verge of complete collapse. For example, cattle, goats, and human activities have converted so much lowland dry forest that the ecosystem is almost gone, and most of the remaining fragmented forests are not regenerating. Introduced avian malaria and pox have decimated endemic Hawaiian honeycreepers and left most of the forests below 600 meters (2,000 feet) in elevation virtually without native birds. The loss of bird pollinators and seed dispersers can only be guessed at, but it is likely to be a significant factor in the decline of at least some of the more than fifty extinct or endangered plants found in this habitat. Perhaps equally important in the erosion of plant diversity in dry forests is the impact of alien weed species, such as fountain grass (*Pennisetum setaceum*). This species was originally introduced into Hawaii as an ornamental and has since spread over vast areas. Fountain grass not only competes with native species for light,

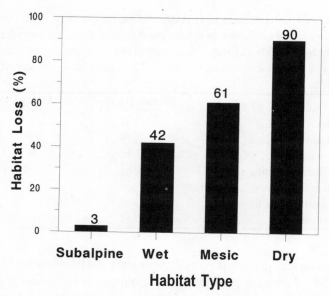

FIGURE 5-1.   Estimated habitat loss since the arrival of humans (circa 400 A.D.) among Hawaiian habitat types (Hawaii Department of Land and Natural Resources 1991).

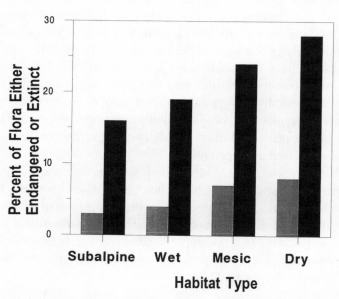

FIGURE 5-2.   Extinction and endangerment of Hawaiian flowering plants among different habitats. Note that when these data are compared to habitat loss in Fig. 5-1, there is a trend showing increasing extinction and increasing endangerment as habitat loss increases. (Wagner et al, 1990; Center for Plant Conservation 1992; Mehrhoff 1993; unpublished data).

TABLE 5-2.   The magnitude of Hawaii's endangerment crisis. The number of
species in each category is shown, as is the cumulative total and the cumulative percentage of the total number of native flowering plants (1131 species and subspecies)
(Wagner, Herbst, and Sohmer 1990; Center for Plant Conservation 1992; Mehrhoff
1993; unpublished data).

	Number of species	Cumulative total	
Extinct species	101	101	(9%)
Outplanted individuals only	4	105	(9%)
One plant remaining	12	117	(10%)
Two to ten plants remaining	81	198	(18%)
Eleven to one hundred plants remaining	86	284	(25%)
One hundred-one to one thousand plants remaining	151	435	(38%)

water, nutrients, and space but alters these ecosystems by promoting hot, frequent wildfires. Fountain grass readily recolonizes after fires and it does so
much more rapidly than native species. Surveys of dryland forests before and
after fires indicate that approximately 90 percent of the native shrub and tree
layers are irreversibly lost after a single fire, and 99 percent are lost after a
second fire (Takeuchi 1990). Unfortunately, systems such as the dryland forest
will continue to collapse even if all development and human activities were
to stop today. By releasing and fostering the establishment of alien weeds,
pigs, goats, and insects, we have already set in motion all of the factors necessary to eliminate large segments of the Hawaiian biota. Hawaii is one place
where we can sit back, do no additional harm, and still watch ten thousand
life forms found nowhere else slide into extinction. We can not fix Hawaii's
extinction crisis simply by stopping further development, revoking grazing
leases, or making it illegal for people to destroy endangered species. Hawaii's
ecosystem and the plants and animals that depend upon them will require intensive and costly human intervention.

While habitat protection is costly in both time and money, it can also be
very effective. This is particularly true when the situation involves a limited
number of threatening factors. For example, the elimination of feral pigs and
goats from areas of Haleakala National Park resulted in a fifteen-fold increase
in Haleakala silversword (*Argyroxiphium sandwicense* var. *macrocephalum*) to
over sixty thousand plants (Loope and Medeiros 1990) and significant recovery of high-elevation bogs.

Transplantations of Endangered Species

The practice of reintroducing Hawaiian plants back into the wild started in
the early 1900s when the territorial government became alarmed at the wide-

spread deforestation associated with cattle grazing and agricultural development. Between 1910 and 1960, foresters transplanted 78 species of native plants and 948 species of alien plants onto public lands (Skolmen 1979). Of these, 13 species are now considered to be endangered. Some of these species, such as uhiuhi (*Caesalpinia kavaiensis*) and kauila (*Colubrina oppositifolia*) were planted in large numbers, while other attempts consisted of only one or two plants. To my knowledge, none of these transplanted populations of endangered plants is alive today. Nonetheless, these were ambitious, if not very successful, efforts.

These and other outplantings of common, rare, and endangered plants were undertaken for a number of reasons: to reforest denuded watersheds; to re-establish valuable timber trees that happened also to be endangered; to satisfy research purposes; to provide mitigation or compensation for development projects; and to fulfill conservation goals such as reestablishing viable wild populations. Several projects simply were last-ditch efforts undertaken in the hope that a few individual plants of an otherwise doomed species might somehow survive. In at least one instance, a conservation group was accused of conducting unauthorized transplantation of endangered plants to stop a proposed construction project (Trask 1992). While this accusation appears to be unfounded, it does illustrate one potential problem with the transplantion of endangered species.

In total, I have been able to document that 103 species of native Hawaiian plants have been transplanted into wild or semi-wild situations. Thirty-five of those species (Table 5-3) are at risk of extinction. These attempts have been carried out by many people and organizations: State or Territorial foresters, the National Park Service, the U.S. Fish and Wildlife Service, university researchers, conservation organizations, botanical gardens, private individuals, and development corporations.

Examples

Four transplantation programs involving federally listed endangered species have been particularly intense. Two of these (*Achyranthes splendens* var. *rotundata* and *Chamaesyce skottsbergii* var. *skottsbergii*) involved transplantations as part of mitigation for industrial development. One of the remaining projects was aimed at augmenting a declining population of the highly endangered *Argyroxiphium sandwicense* ssp. *sandwicense*, the Mauna Kea silversword. The last of these intensive efforts attempted to re-establish populations of a species that had been previously extirpated from the wild (*Kokia cookei*).

TABLE 5-3   Endangered Hawaiian plants used in transplantation efforts.

Taxon	Outcome*	Purpose#	Source
Abutilon menziesii	F	C	Skolmen unpubl.; C. Corn pers. comm. 1993
Abutilon eremitopetalum	F	C	Skolmen unpubl.
Achyranthes mutica	?	C	Perlman pers. comm. 1993
Achyranthes splendens var. rotundata	R	M	L. Mehrhoff pers. obs. 1994
Argyroxiphium kauense	S	C	Hawaii Volcanoes National Park records 1993
Argyroxiphium sandwicense ssp. macrocephalum	S	C	Morris 1967
Argyroxiphium sandwicense ssp sandwicense	R	C	L. Mehrhoff pers. obs. 1993
Bonamia menziesii	S	C	C. Corn pers. comm. 1993
Caesalpinia kavaiensis	F,S	C,T	Morris 1967
Chamaesyce skottsbergii var. skottsbergii	F,S,R	M,C	L. Mehrhoff pers. obs, 1993 R. Fenstemaker pers. comm. 1993
Clermontia pyrularia	S	C	Jefferies pers. obs. 1994
Colubrina oppositifolia	F,S	T	Morris 1967
Cyanea stictophylla	S	C	J. Giffin pers. comm. 1995
Delissea undulata	S	C	J. Giffin pers. comm. 1995; L. Mehrhoff pers. obs. 1995
Gardenia brighamii	S	C	C. Corn pers. comm. 1993
Hibiscadelphus distans	S	C	C. Corn pers. comm. 1993
Hibiscadelphus giffardianus	S	C	Morris 1967
Hibiscadelphus hualalaiensis	S	C	Morris 1967
Hibiscus arnottianus ssp. immaculatus	S	C	C. Corn pers. comm. 1993
Hibiscus brackenridgei ssp. brackenridgei	S	C	L. Mehrhoff pers. obs. 1994

TABLE 5-3 (*Continued*)

Taxon	Outcome*	Purpose#	Source
Hibiscus brackenridgei ssp. mokuleianus	S	C	C. Corn pers. comm. 1994
Hibiscus clayi	S	C	L. Mehrhoff pers. obs. 1993
Hibiscus waimae ssp. hannerae	S	C	C. Corn pers. comm. 1993
Kokia cookei	F	C	C. Corn pers. comm. 1993
Kokia drynarioides	F,S	C	Morris 1967; C. Corn pers. comm. 1993; L. Mehrhoff pers. obs. 1995
Marsilia villosa	R	C	L. Mehrhoff pers. obs. 1995
Melicope zahlbruckneri	F	C	Hawaii Volcanoes National Park records 1993
Munroidendron racemosum	S	C	C. Corn pers. comm. 1993
Nothocestrum breviflorum	F	C	Morris 1967
Pleomele hawaiiensis	F	C	Morris 1967
Portulaca molokiniensis	S	C	L. Mehrhoff pers. comm. 1993
Pritchardia affinis	S	C	C. Corn pers. comm. 1993
Pritchardia aylmer-robinsonii	S	C	C. Corn pers. comm. 1993
Pritchardia napaliensis	S	C	C. Corn pers. comm. 1993
Scaevola coriacea	S	C	C. Corn pers. comm. 1993
Sesbania tomentosa	S	C	C. Corn pers. comm. 1993
Vicia menziesii	S	C	C. Corn pers. comm. 1993
Wilkesia hobdyi	S	C	C. Corn pers. comm. 1993

*Denotes the success of transplantations on taxon; S = survived, F = failed, R = regenerating.

#Denotes purpose of transplantation; C = conservation, M = mitigation, T = timber/watershed.

ARGYROXIPHIUM SANDWICENSE VAR. SANDWICENSE

The silverswords are among Hawaii's most well-known plants, and most are either endangered or threatened. The Mauna Kea silversword (*Argyroxiphium sandwicense* var. *sandwicense*) is a large, showy member of the Aster family that takes several years to flower (Figure 5-3). After flowering, the plant or branch that flowered dies. This taxon is restricted to the flanks of Mauna Kea on the island of Hawaii, where it was originally common. Overbrowsing by cattle, goats, and sheep has reduced this variety to a single population of around two dozen plants. In the 1980s, the state of Hawaii constructed a fence to protect this population and also initiated an augmentation program. Since then, hundreds of nursery-grown plants have been transplanted into the exclosure. Many of these transplants have survived, flowered, and produced juveniles. The population may eventually become self-sustaining, if alien herbivores continue to be kept out of the population.

KOKIA COOKEI

This medium-sized tree is one of Hawaii's most beautiful and most endangered plants. It is a member of the Hibiscus family and has very large, red, somewhat curved flowers. The *Kokia cookei* story (Woolliams and Gerum 1992) illustrates the effort involved in some Hawaiian conservation activities.

Kokia cookei was discovered on the island of Molokai in 1871, and it became extinct in the wild in 1918. The cause of extinction was probably a com-

FIGURE 5-3.   *Argyroxiphium sandwicense* ssp. *sandwicense*. The Mauna Kea silversword is an extremely endangered member of the Aster family. Fewer than two hundred plants remain, many of which, like this plant, were part of a reintroduction program.

bination of habitat loss, cattle and goat browsing, wildfire, rat predation of seeds, and possibly the effects of alien invertebrates. The original population consisted of three trees, and by 1910 only a single tree remained. In 1915, Joseph Rock collected four seeds from this last tree, which subsequently died three years later. The four seeds were distributed to three local biologists and the Bureau of Plant Industry gardens in Buena Vista, Florida. All four seeds germinated, but only one seedling, at a private residence on Molokai, survived for longer than a few years. This one seedling matured and produced many seeds, which were used to establish additional cultivated material and to provide material for territorial foresters to transplant into forest reserves. At least one hundred seedlings were introduced into Forest Reserves on the island of Oahu during the 1930s. None of these introductions survived.

By the 1950s there were cultivated plants in Hawaiian botanical gardens, Manuka State Park, and Hawaii Volcanoes National Park. Since the 1960s, the only successful germination of *Kokia cookei* seeds was in 1973 by the National Tropical Botanical Garden, which resulted in eight seedlings that were planted at the garden. These plants died in 1975. For some reason, seeds produced from the mid-1970s on were not viable. Even hand-pollination, tissue culture, cuttings, and air layering failed to produce viable material. In 1976, Waimea Arboretum was able to successfully graft a branch from the last known *Kokia cookei* onto a stock of *Kokia drynarioides*, another endangered endemic Hawaiian species. This plant flowered but also failed to produce viable seed.

In 1981, the State of Hawaii initiated a reintroduction program that established animal exclosures (to control the threat posed by alien deer and pigs) and preplanted the closely related *Kokia drynarioides* into the site to test site suitability. This was necessary because *Kokia cookei* transplants were grafted onto *Kokia drynarioides* stocks. In 1991, after it was determined that *Kokia drynarioides* could survive at the site, twenty-eight grafted *Kokia cookei* were transplanted back to Molokai. Most of these transplants are still alive but have not yet flowered.

ACHYRANTHES SPLENDENS VAR. ROTUNDATA

As part of mitigation for the development of an industrial park on the island of Oahu, an augmentation effort for this coastal shrub was funded. Several hundred nursery-grown plants were transplanted into a small, maintained preserve of approximately 0.4 ha (1 acre) (Figure 5-4). The augmentation increased the population from only a couple of plants to well over two hundred. This mitigation plan protected some of the natural populations within the development but allowed other populations to be destroyed. Six years after this transplantation, the effort appeared to have been successful. The planted *Achyranthes* have survived and the population is well into a second and third

FIGURE 5-4.   *Achyranthes splendens* var. *rotundata*. Most of the 0.4 ha (1 acre) augmentation site is shown in this photograph. The majority of plants visible are native species. Maintenance has removed most of the alien weed species.

generation of reproduction. Maintenance was (and still is) performed to remove alien weeds. It is likely that routine maintenance will be required well into the future.

This project may be a good example of a successful transplantation but an unsuccessful mitigation. The long-term stability of this site is probably poor, considering the small size of the preserve and proximity to major industrial activities.

CHAMAESYCE SKOTTSBERGII VAR. SKOTTSBERGII

A mitigation effort for this dryland shrub was part of a mitigation plan for a deep-draft harbor development located near the industrial development that spawned the *Achyranthes* transplantation effort. In 1979, 218 nursery-grown plants were outplanted into a relocation site situated ouside the industrial park and deep-draft harbor area. All transplants died soon after completion of the project. In 1980, 748 additional nursery-grown transplants of *Chamaesyce* were placed into an introduction site located several kilometers from the site of original population (which was destroyed by the development). These plants seem to have survived initially, but the transplantation failed after maintenance stopped. More specifically, while the watering schedule improved the survival of transplants, it also promoted the growth of competing weeds. The site was not adequately weeded and, at last count, fewer than ten

of the original transplants remained. In addition, no reproduction has been observed recently at the site (Mehrhoff, personal observation 1993). The lack of weeding has resulted in a dense growth of alien weeds that has overtopped or crowded out the *Chamaesyce* plantings.

Lessons Learned

What lessons have we learned from over eight years of transplanting endangered species into Hawaiian ecosystems?

TRANSPLANTATION FAILURES

The vast majority of Hawaiian transplantations have been failures, not just because the populations died or failed to reproduce but also because virtually none of the projects kept detailed records of the results. Thus, much of the knowledge that was or could have been gained from these transplants has been lost.

Pavlik (Chapter Six) defines a successful reintroduction as one that achieves goals relating to the abundance of the population, the area occupied by the population, the resilience of the population, and the population's persistence. Using these criteria, none of the Hawaiian reintroductions, introductions, or augmentations can at the present time be judged a success. However, if we relax our definition of success to simply be the establishment of a self-perpetuating population, then three endangered Hawaiian plants have had at least one successful reintroduction: *Achyranthes splendens* var. *rotundata*, *Argyroxiphium sandwicense* ssp. *sandwicense*, and *Marsilia villosa*. Most of the remaining thirty-four species that have been reintroduced have, at one time or another, at least survived to the point where they flowered. While virtually all of these appear to have produced viable seeds, none has successfully produced any offspring. This is presumably due to the presence of competing alien weeds and predation by alien insects, slugs, rats, goats, deer, pigs, and/or sheep.

TRANSPLANTATION NUMBERS

Committing large numbers of transplants to a project does not guarantee success. Some projects outplanted literally hundreds of *Chamaesyce skottsbergii* var. *skottsbergii*, *Caesalpinia kavaiensis*, and *Colubrina oppositifolia*; yet none of these reintroductions was successful. On the other hand, a large, expanding population of *Marsilia villosa* (M. Bruegmann personal communication 1994) and a small but regenerating population of *Chamaesyce skottsbergii* var. *skottsbergii* (R. Fenstemaker, personal communication Oct. 1993; K. Woolliams, personal communication 1994) were established using only a handful of plants. The *Marsilia* were transplanted by a private individual who simply wanted to reintroduce the species into an area where he thought

it would grow. We know very little about the specifics of this transplantation, but it has clearly succeeded to the point of expansion. The regenerating *Chamaesyce* transplants were placed by a local conservation organization into a semiwild garden that was cleared of alien weeds. The population quickly began to reproduce and establish itself. Regeneration continued until the site was developed for expansion by the University of Hawaii (R. Fenstemaker, personal communication Oct. 1993). Similarly, propagated *Chamaesyce* in a well-maintained botanical garden has also produced a viable reproducing population (K. Woolliams, personal communication 1994). The precise reason that some transplants failed and others survived is not known, but the establishment of regenerating populations appears to be more closely tied to the successful control of the factors preventing seedling establishment and survival (such as alien weeds) than the sheer number of transplants.

ELIMINATING THREATS

Elimination of threats is important to the survival of reintroduction efforts. It makes little sense to spend considerable time and resources to propagate transplants and then send these valuable transplants on a near-certain "death march" back into the wild if the threats endangering the species have not been eliminated or controlled (M. Buck, personal communication 1993). In the case of the successfully regenerating transplantations of *Achyranthes*, *Argyroxiphium*, and *Chamaesyce*, important threats were either intentionally or unintentionally controlled. For the *Achyranthes* and *Chamaesyce*, it was probably weeding of alien plants that allowed these populations to reproduce. The *Argyroxiphium* population was protected from cattle, goats, and pigs, and this protection was probably the pivotal factor in allowing regeneration. Why the transplanted *Marsilia* population expanded to such a degree is not known.

Controlling threats, however, is more easily said than done. This is especially true when the actual causes of endangerment are obscure or involve a large number of factors. As an example, the uhiuhi tree (*Caesalpinia kavaiensis*), is a federally listed endangered species found in dry and mesic forests throughout the Hawaiian archipelago. Probably fewer than fifty plants remain in the wild. The species is highly susceptible to wildfire and it is easily damaged or killed by goats and introduced insects. Rats are thought to be a major predator of seeds. Most populations have been heavily invaded by the introduced fountain grass, which is both a serious fire hazard and a potential competitor for seedlings. Even though the remaining plants produce many seeds, natural regeneration of this species in the wild is extremely rare. Controlling this diverse array of threats is not an easy task.

To put this difficulty into perspective, consider the costs of managing a large (for Hawaii), 40-hectare (100-acre) preserve. Fencing to exclude goats

and pigs would run approximately $30,000 for 2.5 kilometers (1.5 miles) of fence. This is a fairly large sum of money just to control one of several threats, but this cost pales compared to the cost of manually clearing fountain grass, estimated to be as much as $9,000 per acre (a total of $900,000). Even after spending this large amount of money, there would still be a need to continually control rat populations, remove new fountain grass invasions, and, somehow, prevent alien insect infestations.

HYBRIDIZATION

Hybridization between closely related species is more than just a theoretical problem. When two of the most endangered species in Hawaii (*Hibiscadelphus giffardianus* and *H. hualalaiensis*) were transplanted into the same area of Hawaii Volcanoes National Park, they produced a vigorous hybrid swarm (Baker and Allen 1976). This hybridization was considered to be a threat to the survival of *H. giffardianus*. The situation was finally rectified by eliminating all hybrid offspring and removing the endangered species that was foreign to the site (*H. hualalaiensis*). Since this hybridization event occurred, state and federal agencies have tended to minimize introductions of plants outside of their natural ranges and instead have focused more attention on ensuring that reintroductions use only stock material from adjacent or nearby populations. In at least a half-dozen instances, previous introductions have been removed in an effort to reduce the chances of inappropriate genetic mixing.

INTRODUCTIONS VERSUS
REINTRODUCTIONS VERSUS AUGMENTATIONS

The restricted ranges of Hawaiian plants have promoted a great deal of discussion about where transplantations should occur. Hawaii's experience with the hybridization of *Hibiscadelphus hualalaiensis* and *H. giffardianus* have clearly brought concerns about "genetic pollution" to the forefront of conservation discussions. While there is a general consensus as to the undesirability of introducing plant species outside of their known range, there is heated debate about transplanting populations from one island to another. Even in situations where populations of a species are thought to have been extirpated, there is concern that the introduction of plants derived from populations from another island may pollute "undiscovered" populations of the supposedly extirpated species. At present, there is so much concern about this "threat" that there is significant paralysis among conservation groups as to how to proceed. Concerns about genetic pollution of populations may be important, but they should not be the primary consideration in deciding when to re-establish populations of highly endangered plant species. The overriding concern should be the survival of the species, not the integrity of each island's gene pool.

There are two reasons why this is so. First, many populations have been reduced to such low numbers that it is unrealistic to re-establish those populations. For example, *Gardenia brighamii* is an endangered tree known from Hawaii, Maui, Molokai, Lanai, and Oahu. The Hawaii and Maui populations are extinct; Molokai has a single plant remaining; Oahu has three plants in two populations; and Lanai has twelve or thirteen plants in three populations. Theoretically, reintroduction efforts on Molokai could use only seeds derived by self-pollinating the lone remaining tree. However, it may be better to bring in other genotypes from Lanai in order to create the most diverse population possible. A better option may be to augment the original population with Molokai plants but establish new Molokai populations with more genetically diverse stock from other islands. The second reason why we need to be more amenable to interisland reintroductions, or even introductions, is that we have an incomplete record of "historical" distributions. Hawaii's vegetation has undergone massive changes since humans first arrived, and what we see today is not necessarily the way things were fifty, one hundred, two hundred, or two thousand years ago. Paleoecology studies on Oahu indicate that at least this island is very different today than it was nine hundred years ago (Athens, Ward, and Wickler 1992). Around A.D. 1000, Oahu's vast palm forests and an associated undescribed legume species declined to the point where they were lost from the pollen record. Recent exploration has found two living plants of an undescribed legume on the island of Kahoolawe whose pollen matches that of the undescribed Oahu legume species that was extirpated prior to European colonization. Should plants derived from this undescribed species be transplanted into suitable habitat on Oahu? I, for one, hope so.

Other concerns with interisland transplantation efforts deal with the unintentional introduction of diseases or alien pathogens. This is a more real threat and one that needs to be taken seriously, though I could not find any examples of when a reintroduction effort unintentionally introduced a pest species. Along these same lines are concerns that we should not be augmenting the "last" populations of species. Again, there is a legitimate concern that by transplanting nursery-grown plants next to the last known wild individuals we are placing those individuals at risk from introduced pathogens. A number of conservationists would prefer to restrict augmentations in favor of reintroductions into an area close (but not too close) to the original population. One potential problem with this approach is that if the original population dies out it is likely that any host-specific invertebrate species found on those plants will also die out (that is, become extinct). If we are interested in saving these host-specific species, it may be best to augment these populations in the hopes of maintaining host plants for Hawaii's many undescribed invertebrates.

MITIGATION

Mitigation efforts need to provide long-term maintenance and protection. Two endangered species have been transplanted as part of mitigation efforts (*Achyranthes splendens* var. *rotundata* and *Chamaesyce skottsbergii* var. *skottsbergii*). Both of these mitigation efforts placed several hundred individuals into apparently acceptable habitats, both involved short-term habitat maintenance to control threats, and both were monitored to at least some degree. However, the *Achyranthes* effort was largely successful, while the *Chamaesyce* project failed. The key difference was the level of maintenance aimed at controlling alien weed species. The project that continued maintenance to control threats succeeded, and the project that stopped maintenance failed. Since almost all of Hawaii's endangered plants are directly threatened by alien pest species (such as pigs, goats, weeds, and insects), mitigation efforts will almost always require long-term control of those alien pests. Exceptions to this generality will involve the few species that are more threatened by habitat loss from development than by alien pests (such as *Sesbania tomentosa* or *Scaevola coriacea*).

SURVIVAL RATES—SPECIES

Endangered and nonendangered species have similar survival rates. Almost 90 percent of the endangered species that have been reintroduced have shown some ability to survive and flower. Nonendangered species showed a similar survival rate (85 percent). The difference between these rates was not significant (chi-squared test, p=.635). The reason that transplanted endangered species generally fail to become self-sustaining populations is probably not due to differences in adult mortality but to differences in seed germination, seedling mortality, the presence of pollinators, or susceptibility to herbivory (most reintroductions are into fenced areas protected from cattle, goats, and pigs).

SURVIVAL RATES—HABITATS

Plants from different habitats have similar survival rates (Table 5-4). Species from different habitats had similar survival rates, though the small sample size precluded testing for statistical significance. In spite of the small sample size, one can generally make the same observations about habitat and transplantation survival as one can about the survival of transplanted endangered and nonendangered plants. If there is a difference, it does not appear to occur in juvenile and adult mortality of plants from different habitats.

TRANSPLANTATION CIRCUMSTANCES

Sometimes it is appropriate to transplant species even when threats cannot be controlled or when there is little chance of success. So little is known about

the biology of Hawaii's endangered plants that some risks and experimentation are warranted.

The summary of the *Kokia cookei* transplantation illustrates this point very
well. All known plants were in cultivation, none of the plants was producing
viable seeds, and the only method of propagation was by grafting onto another
species in the genus. The reintroduction effort involved transplanting over
two dozen plants into a small, fenced enclosure near the original, now extirpated population. Was this a successful project? We don't know yet. The
plants have survived, but they have not yet flowered. Was this a useful project?
Unquestionably, because there really were not many alternatives. *Kokia
cookei* was, and still is, extremely difficult to propagate in botanical gardens.
The odds are slim that we could keep this species in perpetuity as a series of
grafts onto another endangered species. This species has already been reduced to a single individual plant on three separate occasions. Can *Kokia
cookei* survive a fourth close call?

This project was also useful because there is the chance that by placing
Kokia cookei back into its native environment, something in that native
habitat may trigger successful reproduction. We really do not know why
Kokia cookei no longer produces viable seed. It may be due to long-term inbreeding; it may be a cyclic phenomenon; or it may have to do with the climate of the botanical garden it is growing in. The bottom line is that sometimes we are going to have to reintroduce endangered plants when the
situation seems hopeless or when we lack the information needed to provide
a reasonable chance of success.

In other instances we may need to conduct experimental reintroductions
into areas that have not been secured from all threats. Such reintroduction
may be necessary if common relatives are inappropriate as test subjects and if
there is a need to identify what the important threats are. These experimental
transplantations are acceptable if the transplants are "expendable" or if there
is no other recourse.

TABLE 5-4. Survival of transplanted endangered Hawaiian
plants from different habitats. Species were considered to have
survived if they flowered at the transplantation site.

Habitat type	Number of species		Percent survival
	Failed	Survived	
Coastal	1	7	86
Dry forest	6	36	85
Mesic forest	5	39	87
Wet forest	2	16	87
Subalpine areas	1	6	83
TOTAL	15	104	87

Future Needs

The Hawaiian extinction crisis is massive and all too real. The solutions to this problem will not be easy, cheap, or quick. We need to set in place an overall conservation strategy for the Hawaiian flora (Center for Plant Conservation 1994). The reintroduction of endangered plants needs to be an integral part of a strategy that attempts to preserve the remaining native habitat, provides for secure storage of genetically representative propagules, and tries to reduce the systemic factors (such as wildfire, alien weeds, and alien mammals) that threaten entire Hawaiian ecosystems. A statewide reintroduction policy could support this overall strategy by requiring that reintroductions are adequately planned in order to maximize genetic conservation, minimize hybridization or genetic pollution, utilize secured and appropriate habitat, and ensure adequate monitoring. Reintroduction, introduction, and augmentation projects should attempt to mimic pre-existing populations as much as possible and should strive to use stock material derived from the most closely related populations.

Even with these actions, many more of Hawaii's endemic species will at least temporarily become extinct in the wild. It will be necessary to develop and implement a long-term genetic conservation program to safeguard species that will eventually be used in transplantation efforts. But how long is long-term? Fifty years? One hundred years? Although there is no obvious answer to this question, it will be necessary to maintain these collections until we develop the social desire, political willpower, and technical know-how to control the many factors that have already caused the extinction of 101 Hawaiian plants and that threaten the survival of another 358 species.

ACKNOWLEDGMENTS

Many people contributed information for this chapter. I would like to particularly thank Carolyn Corn, Hawaii Division of Forestry and Wildlife; Lloyd Loope, National Biological Survey; Keith Woolliams and Shirley Gerum, Waimea Arboretum and Botanical Garden; Chris Zimmer, Hawaii Volcanoes National Park; and Winona Char, Char and Associates for the use of their data and files. Marie Bruegmann, Ron Fenstemacher, Robert Hobdy, Richard Nakagawa, John Obata, Linda Pratt, and Alvin Yoshinaga provided valuable comments on earlier drafts. Steve Perlman, of the National Tropical Botanical Garden, contributed frequent and numerous reminders about the real value of reintroducing endangered plants—saving species from extinction.

REFERENCES

Athens, J. S., J. V. Ward, and S. Wickler. 1992. "Late Holocene Lowland Vegetation, Oahu, Hawaii." *New Zealand Journal of Archaeology* 14: 9–34.

Baker, K., and S. Allen. 1976. "Studies on the Endemic Hawaiian Genus *Hibiscadelphus* (Haukauahiwi)." *Proceedings from the First Conference in Natural Sciences,* 19–22.

Carlquist, S. 1980. *Hawaii: A Natural History.* Lawai, Kauai, Hawaii: Pacific Tropical Botanical Garden.

Center for Plant Conservation. 1992. *Report on the Rare Plants of Hawaii.* St. Louis, Mo.

Center for Plant Conservation. 1994. *An Action Plan for Conserving Hawaiian Plant Diversity.* St. Louis, Mo.

Cuddihy, L. W., and C. P. Stone. 1990. *Alterations of Native Hawaiian Vegetation.* Honolulu: University of Hawaii Cooperative National Park Resources Unit.

Gagne, W. C. 1988. "Conservation Priorities in Hawaiian Natural Systems." *Bioscience* 38: 264–270.

James, H. F., and S. L. Olsen. 1991. "Descriptions of Thirty-Two New Species of Birds from the Hawaiian Islands." *Ornithological Monographs* 46: 1–88.

Kirch, P. V. 1985. *Feathered Gods and Fishhooks.* Honolulu: University of Hawaii Press.

Loope, L. L., and A. C. Medeiros. 1990. "Management and Research Efforts to Protect Vulnerable Endemic Plant Species of Kaleakala National Park, Maui, Hawaii." *Monogr. Syst. Bot.* 32: 251–264. Missouri Botanical Garden.

Loope, L. L., and A. C. Medeiros, and B. H. Gagne. 1991. "Aspects of the History and Biology of Montane Bogs." *Technical Report* 76. Cooperative National Resources Studies Unit, Honolulu.

Mehrhoff, L. A. 1993. "Rare Plants in Hawaii: A Status Report." *Plant Conservation* 7: 1–2.

Morris, D. K. 1967. *The History of Native Plant Propagation and Re-introduction in Hawaii Volcanoes National Park.* Unpublished paper.

Skolmen, R. G. 1979. *Plantings on the Forest Reserves of Hawaii, 1910–1960.* Honolulu: Institute of Pacific Island Forestry.

Smith, C. W. 1985. "Impact of Alien Plants on Hawaii's Native Biota." In *Hawaii's Terrestrial Ecosystems: Preservation and Management.* Edited by C. P. Stone and J. M. Scott. Honolulu: University of Hawaii Cooperative National Park Resources Unit.

Sohmer, S. H. 1990. "Elements of Pacific Phytodiversity." In *The Plant Diversity of Malesia.* Edited by P. Baas et al. The Netherlands: Kluwer Academic Publishers.

Solem, A. 1990. "How Many Land Snail Species Are Left and What Can We Do for Them?" *Bishop Museum Occasional Papers* 30: 27–40.

Stone, C. P. 1985. "Alien Animals in Hawaii's Native Ecosystems: Toward Controlling the Adverse Effects of Introduced Vertebrates." In *Hawaii's Terrestrial Ecosystems: Preservation and Management.* Edited by C. P. Stone and J. M. Scott. Honolulu: University of Hawaii Cooperative National Park Resources Unit.

Takeuchi, W. 1990. "Botanical Survey of Puuwaawaa: Draft Final Report." Unpublished report submitted to Department of Land and Natural Resources, State of Hawaii. Contract UC1988-1(ESP)-042.

Trask. H. K. 1992. "Another Delay Will Doom Center for Hawaiians." *The Honolulu Star-Bulletin.* 6 July, p. A3.

Vitousek, P. M., L. R. Walker, L. D. Whiteaker, D. Mueller-Dombois, and P. A. Matson. 1987. "Biological Invasion by *Myrica faya* Alters Ecosystem Development in Hawaii." *Science* 238: 802–804.

Wagner, W. L., D. R. Herbst, S. H. Sohmer. 1990. *Manual of the Flowering Plants of Hawaii.* Honolulu: Bishop Museum and University of Hawaii Press.

Woolliams, K. R., and S. B. Gerum. 1992. "*Kokia cookei:* A Chronology." *Notes from Waimea Arboretum & Botanical Garden* 19 (1): 7–12.

The Biology of
Rare Plant Reintroduction

Complexities of the political, legal, and social aspects of restoration may lead ecologists to lament that "the biology is the easy part." In fact, technical challenges to reintroduction remain immense. Given that the goal of reintroduction is to establish self-sustaining populations that are integrated functionally within natural communities, success hinges on seemingly endless biological requirements. Considerations include theoretical implications about gene-pool dynamics, reproductive behavior, niche requirements, demographic and successional relationships, population structure, and evolutionary resilience; they also include practical, horticultural aspects of collecting, planting, and culturing plants.

Rare plants present additional biological challenges to restorationists beyond those for widespread plants. Many rare plant species have extremely sensitive pollination, seed development, and germination and dispersal requirements; they are sensitive during establishment years, have reduced or altered gene pools due to fragmentation and/or small effective population sizes, may occupy only portions of potential or historical habitat, or may exist on marginal habitat due to anthropogenic exclusion. Small population sizes, low rates of population growth, and narrow environmental tolerances mean that there often is limited seed or propagule material for experimentation, let alone adequate material for reintroduction per se.

Given the amount of interest in rare plants and their restoration, significant gaps in knowledge remain that hinder successful rare plant reintroduction. Further, although considerable empirical information exists in agency files and other files, these data are uncompiled or analyzed, and access to them may be difficult. Many reintroductions have been done opportunistically or under tight time schedules and limited budgets imposed by legal imperatives. Many have not benefited by strategic planning, biological evaluation, or monitoring. Further, by their nature—and unlike commonly planted, widespread species such as forest trees or grasses—most reintroduction projects of rare plants will remain unique events or at least highly limited in geographic scope. Abundant (or even adequate) taxon-specific information will rarely accumulate about many rare plants we anticipate reintroducing. Thus, each reintroduction must be viewed as an important experiment, and each restorationist must depend on his or her own ability to understand and extrapolate general biological principles to specific projects and taxa.

Despite these difficulties, we gain valuable information for specific reintroductions in two ways. First, although an extensive database most likely will not develop for all rare plants we wish to reintroduce, we can learn much from case studies (see Part Four). Congeners, species with similar niche requirements, life histories, and genetic diversities can provide valuable information of direct use for reintroducing "virgin" taxa. Further, carefully de-

signed case studies inform us on the kinds of biological questions we should ask, the range of pitfalls we might encounter, and the nature of ecological solutions we might attempt.

Second, and perhaps more important, the general context for rare plant reintroduction provides many insights even when empirical studies are lacking. Our premise—that reintroduction is the establishment of self-sustaining populations functionally integrated in communities—provides a valuable framework for deriving specific recommendations. Successful reintroductions should be those that parallel natural ecological processes, and we encourage the use of a reintroduction process designed in this way. Collectively, the authors in this section develop processes and guidelines for reintroductions by considering topics that follow the ecological process from establishment to evolutionary adaptation. They highlight the thinking and the tools we must use if human-directed reintroductions are to work in concert with the natural genetic, ecological, and evolutionary processes.

Pavlik (Chapter Six) begins the section by defining the context for the reintroduction process and by developing ecological criteria for success. Many reintroductions fail do so because they lack such criteria, which would guide projects from planning through long-term monitoring. Pavlik defines four major topics to consider when constructing goals and objectives for reintroduction: abundance, extent, resilience, and persistence. By developing reintroduction objectives and goals that focus on these criteria—which address the entire spectrum of ecological processes relevant to plant establishment and long-term population adaptability—the effectiveness of reintroduction projects should improve.

Location of the reintroduction site is a critical factor for successful reintroduction. Fiedler and Laven (Chapter Seven) demonstrate how plants interact with their physical and biotic environment in ways that affect establishment, survival, reproduction, and integration into local ecosystem functions. The authors underscore the complexity and importance of site selection and develop practical principles from ecological theory to guide decisions. Fiedler and Laven group site selection criteria into four classes: physical, biological, logistical, and historical. Selection of sites may also be influenced by the type of rarity of the reintroduced taxon, such as natural endemics, sparsely distributed rarities, and anthropogenic rarities. Careful consideration of implications of each of these factors should favor successful reintroduction.

In nature, environments interact with plant propagules to determine successful establishment. In reintroduction, we improve our chances of success by basing our selection of founding populations (that is, the material for reintroduction) on sound ecological and genetic theory. Guerrant (Chapter Eight) discusses the essential genetic and demographic concepts that guide our design of the founding populations for reintroduction. He illustrates the

importance of these factors, both singly and in combination, through simulation models for plants with diverse life histories. Having developed guidelines based on demographic and genetic processes relevant to different types of species, Guerrant combines these concepts with practical horticultural options. Proper choice of these elements has far-reaching implications for long-term survival and reproduction of introduced plants.

The many ecological processes that influence dynamics of natural plant populations beyond establishment serve as models for later phases in the reintroduction process. Primack (Chapter Nine) develops lessons for rare plant reintroductions from ecological theory on seed dispersal, establishment, disturbance, and population structure. His chapter emphasizes the need for restoration biologists to consider temporal and spatial scales beyond those of the single reintroduction site or the lifespans of individual reintroduced plants. Considerations of taxon-specific requirements derived from life history and population-dynamic characteristics will favor survival and reproduction of reintroduced plants and should promote conditions suitable for post-reintroduced generations to colonize suitable habitat. Primack illustrates lessons based on ecological theory with an integrated case study from New England, where multi-species reintroductions were experimentally attempted.

Lack of monitoring and long-term evaluation of projects has been a major impediment to progress in the science and art of rare plant reintroduction; hence we conclude the conceptual chapters in this section with a detailed presentation on monitoring. Sutter (Chapter Ten) discusses the conceptual background for why monitoring is essential, what it can achieve, and which critical elements should be considered. Sutter presents an in-depth summary of the science of monitoring and develops the basic components of sampling design for rare plant monitoring. These include six major decisions: (1) the sampling universe, (2) placement of sample units in the sampling universe, (3) selection of sampling units, (4) selection of permanent or temporary sampling units, (5) sampling frequency, and (6) number of samples that need to be collected. Success in reintroduction can only be evaluated after appropriately long-term monitoring, evaluation, and feedback.

We illustrate the concepts in this section with a "focus" chapter by Ledig (Chapter Eleven) on the reintroduction of Torrey pine at Torrey Pines State Reserve in California. This population, one of only two naturally occurring for the species, has undergone many threats to its existence in the historic period. Most recently, beetle infections caused rapid mortality throughout the stand. Ledig examines the role of genetic variation, genetic conservation seed collections, and pheromone strategies in the reintroduction at Torrey Pines State Reserve.

The strategies developed in this section will be most usefully employed in

combination with each other and with empirical experience from case studies. The diverse life histories and ecological conditions of rare plants require that we lean heavily on biological theory for designing and conducting individual reintroductions. Long experience under controlled monitoring will eventually provide the real grist for planning reintroduction strategies.

CHAPTER 6

Defining and Measuring Success

Bruce M. Pavlik

⎯⎯⎯⎯➤⎯◉⎯◄⎯⎯⎯⎯

Success is relative:
It is what we make of the mess we have made of things.
                    —T. S. Eliot, *Family Reunion*

Rare plants are reintroduced to appropriate habitat and historic range for a variety of reasons. Reintroduction may play an important role in the recovery of species listed under the federal Endangered Species Act (1973, with amendments). In managing public and private natural areas, reintroduction is used to supplement the remnants of depleted or modified communities, to examine the effects of fire and other ecosystem manipulations, and to ensure sanctuary for rare biological resources. Finally, reintroduction (or its functional equivalents, translocation and augmentation) is frequently undertaken as mitigation for adverse impacts associated with land development. In all cases, reintroduction projects are intended to fulfill the same overall conservation purpose: to lessen the probability of extinction and encourage recovery of a particular rare plant through the creation of a new, self-sustaining population. Every project requires development of site- and taxon-specific tools. How are these tools of reintroduction evaluated, and how are populations determined to be self-sustaining? What are the success criteria for reintroducing a rare plant?

In a recent review of fifty-three reintroduction attempts in California, Fiedler (1991) assessed forty-six restoration, management, and mitigation-related projects. Project objectives, methods, and criteria for success were compiled and compared on a total of forty taxa of special concern. Fiedler found great variation in the design, execution, and follow-up of the projects but was concerned by the lack of performance measures for evaluation. "Only 15 of the 46 projects (33%) have explicitly defined criteria for success, and until quite recently, there was no consistency in these criteria. Without such

'industry' standards, [evaluations of the] success of translocation, relocation, and reintroduction projects cannot be made objectively" (p. 75). Fiedler goes on to suggest that insufficient or improper monitoring has been a primary contributor to misevaluation, a point also made by Hall (1987), Palmer (1987), and Pavlik (1994a).

How can "industry standards" for evaluating success in general and monitoring techniques in particular be established when there is such an enormous variety of plant life forms and environmental circumstances? Ideally, published literature on reintroduction projects would provide information on the relative merits of monitoring techniques or project designs. However, in reviewing the scant published literature and some of the abundant gray literature on reintroductions, it became clear to me that no consensus on standards is possible at this time (see also Maunder 1992). To begin generating that literature and those standards, however, a sound framework for defining and measuring success would encourage consistency in design and measurement so that projects for very different plants and habitats could be evaluated and compared. Without that framework, we have become mired in vague communication, confused purposes, and unconvincing data. We also risk losing many opportunities to deepen our understanding of the ecology of reintroduction and restoration.

The purpose of this chapter is to suggest a practical, biologically based framework for defining, measuring, and therefore determining success in reintroduction projects. To do this, I will offer an operational definition of success, suggest a flexible scheme of goals and objectives, and then show how the construction of taxon-specific objectives should define which measurements need to be made for purposes of evaluation. Although my selection of goals and objectives is derived from the best available models or paradigms, those models and paradigms are often untested, poorly supported by published data, or simply inadequate with respect to the life history and ecological diversity of rare plants. Reintroduction, therefore, provides us with an important opportunity for verifying those models and paradigms while we expand our practical knowledge of creating new populations and managing natural ecosystems.

Defining Success with Goals and Objectives

The concept of success, as applied to the restoration of species, presents several fundamental problems. Just as "progress" in evolution needed to be purged of any moral and teleonomic connotations (Dobzhansky et al. 1977; Mayr 1976), success in reintroduction must not imply finality, perfection, or victory. Respectively, *finality* implies that the future of a reintroduced popu-

lation can be reliably predicted; *perfection* implies that all variables are known and can be optimized; and *victory* implies that no further actions will be necessary once success has been achieved. Both reintroduced and natural populations are subject to many complex, interacting processes, some of which are random or chaotic, and make predictions of the future extremely problematic. The processes in turn are affected by many variables, only a fraction of which can be ascertained and manipulated (for optimization) for any one species and its ecosystem context. And, given the highly modified condition of many natural communities, it is likely that some planning and management will always be needed to ensure the availability of suitable habitat (Saunders et al. 1991; Pavlik 1994b). Consequently, reintroduction and recovery possess the characteristics of a "prickly problem": unpredictability, complexity, lack of optimization, and no clear endpoint.

In addition, defining success in reintroduction is problematic because success in the natural world is an abstract or relative concept. Living entities and natural forces have no plan for the future (the watchmaker is blind, according to Dawkins 1986) and do not evaluate themselves or their progress (that is, whether the "watches" are keeping the "proper" time). It is also true that over the eons, all biological systems "fail" (they go extinct or change—the watches run down or are salvaged for parts), so that success can only be defined with respect to a particular, arbitrary period of history. For these reasons, success is a concept that is avoided in most academic discourses on ecology, genetics, evolution, paleontology, and biogeography. Nevertheless, the literature of conservation biology, wildlife biology, and the biology of introduced plants and animals contains hypotheses or paradigms related to success. This is because these applied sciences involve human aspirations to restore endangered species; to provide large, harvestable populations of game animals; and to control infestations of pests. In other words, only when there are human-set, a priori goals can a particular system or endeavor be evaluated for success (Anderson and Hurley 1980; Kleiman 1989, 1990).

What then, are the goals for reintroduction? In the short term, a new, successful population should be able to carry on its basic life-history processes—establishment, reproduction, and dispersal—such that the probability of complete extinction by random or chaotic forces is low (Gilpin and Soule 1986; Menges 1992; Pavlik 1994a). Therefore, significant reductions in the probability of extinction are manifest in higher population growth rates and larger spatial distributions (see Pimm 1989 and Hengeveld 1989; see also Guerrant, Chapter Eight). In the long term, a reintroduced population should be as capable as its natural equivalent of integrating fully into ecosystem function and meeting the challenge of a changing environment through evolution or migration (Schonewald-Cox et al. 1983). Meeting the evolutionary challenge may require maximal genetic variation (Clegg and

Brown 1983; Barrett and Kohn 1991) and a landscape mosaic of both potential and actual habitat. These short- and long-term features of a viable population can serve as general goals for the design and evaluation of reintroductions. Abundance, extent, resilience, and persistence are general and interrelated goals distilled from paradigms regarding the importance of establishment, fecundity, population size, number of populations, genetic variation, self-sustainability (viability), and community invasiveness (Figure 6-1). Whereas abundance and extent can develop over short periods of time (one to ten years) and be directly influenced by the design factors of a project (such as the number of propagules used, selection of sites, spatial patterns of outplanting), resilience and persistence are only tested over long periods of time (one to several decades) by natural variation in the environment and in the new population itself.

The goals of abundance, extent, resilience, and persistence are generic: They can apply to most taxa under most circumstances. When a particular taxon is considered for reintroduction, the peculiarities of its life history, ecology, and current distribution must be used to specify exactly how the four goals are to be met. This is done by constructing one or more taxon-specific objectives for each goal. The objectives describe a particular condition of the new population(s) that should be achieved within a given time period.

Success in reintroduction, therefore, can be defined precisely as meeting taxon-specific objectives that fulfill the goals of abundance, extent, resilience, and persistence. In effect, success is an argument to be made with support from studies that make measurements related to each objective. The best argument is constructed when (1) objectives incorporate the crucial life history and environmental characteristics of a target taxon, (2) the reintroduction is designed as an experiment with enough statistical power to detect differences between important variables (such as management regimes and subpopulation performance), and (3) the monitoring program is designed to obtain data relevant to the experiment and the reintroduction objectives.

It is also necessary to distinguish between project success and biological success. *Biological success* includes only the performance of individuals, populations, and metapopulations of the target taxon (Figure 6-2). If biological success has been achieved, the new, vital population(s) can assume an appropriate function in the local ecosystem. *Project success* is much broader. With an experimental design and careful monitoring, a reintroduction project can be successful, even if its new populations fail, by contributing to our knowledge of rare and endangered plants or by developing new ecosystem–management techniques. (Keep in mind, however, that mitigation or restoration efforts are usually required to include some amount of biological success.) Reintroduction projects can also influence debate on conservation policy and educate the public on issues concerning biological

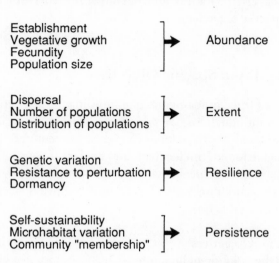

FIGURE 6-1. The goals of a rare plant reintroduction are derived from success paradigms in the literature of conservation biology, wildlife management, and biological invasions.

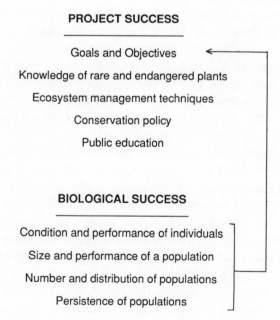

FIGURE 6-2. Project success includes much more than the biological success of a reintroduced population.

diversity and ecological restoration. The remainder of this chapter will focus on the biological criteria for defining and measuring success within the context of taxon-specific objectives.

Constructing Taxon-Specific Objectives

There are at least three fundamental questions that arise when constructing objectives for reintroductions: (1) Which levels in the biological hierarchy should be considered? (2) What kinds of analyses are required? (3) When will results be available for interim feedback and final evaluation? If we formalize and simplify these hierarchical, analytical, and temporal considerations, we can employ a standard framework of goals and objectives that can be applied to most plant taxa (Table 6-1).

A Framework for Objectives

In response to the question of hierarchy, objectives must be focused on a particular unit of biological organization (such as a population, taxon, or· ecosystem). Thus, population-level objectives (or simply *population objectives*) address single populations by prescribing a particular stage distribution, equilibrium finite growth rate (lambda), or level of allelic diversity. Objectives that apply to more than one population or to all populations are *taxon objectives*. These could prescribe metapopulation traits, the achievement of minimum viable population (MVP) sizes across historic range, or a value for the G_{st} statistic that represents an appropriate distribution of genetic variation between populations. Whereas most reintroduction efforts would include population objectives, the recovery of an endangered species would require a full palette of population and taxon objectives. Ultimately, we would include ecosystem objectives for all dominant species in a community undergoing extensive restoration (see Zedler, Chapter Fourteen).

Reintroduction objectives are also based on quantitative or qualitative analyses. Objectives that incorporate quanitative analyses of a target population, taxon, or ecosystem are referred to as *state objectives*. State objectives are expressed in terms of population size, seed/ovule ratio, polymorphic index, and microhabitat diversity. They require answers to questions such as how many? what frequency? or what is its value? Other objectives incorporate qualitative analyses and thus depend on a description or verification of a process within a target population, taxon, or ecosystem. These process objectives specify a desirable process or trend, a series of events, or a capability to be observed in a reintroduced population. They can include life-cycle completion, increases in population size, dispersal into new microsites, and the use of native pollinators. *Process objectives* require answers to questions such as

TABLE 6-1. Proximal (short-term) and distal (long-term) objectives for evaluating reintroductions of endangered plants. Most objectives address the performance of a single population, with some exceptions (as shown by notes below).

Goals	Objectives	
	Proximal	Distal
Abundance	1. Life cycle can be completed *in situ* (P)	1. Attain minimum viable population (MVP) size (s)
	2. Progressive increase in N_e relative to founders (P)	2. Maintain MVP size (P)
	3. Stage distribution matches reference population (s)	
	4. Seed production matches reference population (s)	
	5. Seed/ovule matches reference population or life history (s)	
Extent	6. Dispersal by native vectors beyond reintroduction site (P)	3. Population area matches historical record (s)
	7. Progressive increase in population area (P)	4. Metapopulations can self-establish (P)*
	8. Outcrossing satellite groups established (s)	5. Populations throughout range attain MVP (s)*
Resilience	9. Genetic variation is maximized within founders (P)	6. Taxon G_{st} appropriate for life history (s)*
	10. Lambda greater than 1.0 in at least one year (s)	7. MVP regained after perturbation (P)
	11. Seed bank density matches reference population (s)	
Persistence	12. Subpopulations are found greater than 1 microhabitat (s)	8. Max. microhabitat diversity among populations (P)*
	13. Native pollinators are utilized (P)	9. Regain structure/function role in community (P)
		10. Low coefficient of variation in N_e is maintained (s)

*Addresses the performance of taxon as a whole (including multiple populations)

sState objectives

PProcess objectives

of native pollinators. *Process objectives* require answers to questions such as has this phenomenon occurred? or what is the direction of change?

Feedback for modifying a project in progress (such as adaptive management) is crucial, because reintroduction is an iterative venture that can benefit from information obtained from novel experiments and observations. Although state and process objectives are usually established by the recovery plan, they may require modification during the design and execution of the reintroduction. This does not mean that objectives are modified to suit the results. Rather, objectives may require modification to reflect a better understanding of the species or to accommodate random or atypical events that occur during the project. For example, if the ratio of floral morphs becomes skewed in a new population of an outcrossing, heterostylous species, then a total population size (N) objective will have to be increased to ensure a larger effective population (N_e) *in situ*. Management techniques used during the reintroduction should also be modified depending on outcome and efficacy.

Consequently, any evaluation of the objectives, methods, and progress of the reintroduction project as a whole should be ongoing, so that short-term results are incorporated for the long-term benefit of the target taxon. In addition, there are some qualities of populations, taxa, and ecosystems that can be achieved only over long periods of time. A temporal dimension should thus be recognized by categorizing objectives as either proximal or distal (see Table 6-1). *Short-term* or *proximal objectives* are those that can be evaluated after a relatively short time period (one to five years for annuals and biennials; five to ten years for many perennials) to provide interim feedback and milestones for evaluation. *Long-term* or *distal objectives* are those requiring greater lengths of time (ten years to several decades, depending on the target taxon) and would only be achieved near the end of a reintroduction project. Proximal objectives are relative but readily evaluated with a small database. Distal objectives are absolute and draw their strength from robust models and large databases containing spatial and temporal variance in environmental and demographic parameters.

Proximal objectives prescribe a minimum level of performance of a new population in relation to the initial state (the founding population) or the best available reference population. In the case of an annual plant, a simple, proximal objective would be met if there were an upward trend in population size relative to the number of established founders. Other proximal objectives could be evaluated by comparing performance of the new population to one or more of the following reference populations: (1) large, apparently stable natural populations of the same taxon (if extant); (2) large populations of nonendangered congeners or conspecifics; (3) populations of nonendangered taxa that share similar life-history traits with the target; and (4) other reintroduced populations of the same taxon. Regarding the use of other rein-

troduced populations, nothing defines success better than a good solid failure. Failing populations are of great comparative value and need to be as closely monitored as those that appear to be successful (see also Simberloff 1989). Our current inability to construct a robust definition of success is due largely to our past unwillingness to document failure.

Distal objectives prescribe a minimum level of performance of a new population that is independent of its founding state or any reference population. These could be derived from reconstructions of a taxon based on historic records or remnant patterns of genetic variation. Distal objectives could also be based upon predictions from appropriate, well-tested theories and models regarding extinction, preservation of genetic diversity, and environmental variability. For example, models that predict extinction probabilities can be used to set a long-term abundance objective by determining the minimal viable population (MVP) size of a new population for its specific environmental context. One definition of MVP is the smallest number of individuals required for a 95 percent probability of survival over one hundred years (Mace and Lande 1991). But applying such model predictions to a practical conservation effort is often specious and always difficult (Botsford and Jain 1992). Assumptions and simplifications used in constructing the models currently exclude a broad array of plant life-forms (Pavlik 1994b) or do not take into account the genetic and demographic peculiarities of plant populations (Menges 1991). Worse yet, most models are untested—we do not know how well they integrate critical variables to simulate fluctuations in population size over long periods of time (Boyce 1992; Ralls et al. 1992). However, rareplant reintroductions, if well designed and demographically monitored, provide an empirical opportunity for testing existing models and building better ones in the future (see also Guerrant, Chapter Eight). So at this time it is possible to suggest distal objectives that appear to be rigorous and absolute but impossible to know if they truly are.

Biological Basis of Objectives
A reintroduction plan should begin with constructing a table of biologically based objectives using the population/taxon, state/process, and proximal/distal framework (see Table 6-1). Objectives within the framework must address the goals of abundance, extent, resilience, and persistence. Each objective should be a simple, focused statement that is congruent with the known life-history features of the target taxon. Among the most important life-history features to consider when constructing objectives are growth form (annual, perennial, herbaceous, woody), breeding system (selfing, outcrossing, mixed), and fecundity (seed production per ovary or individual, monocarpic versus polycarpic). Genetic and ecological attributes are also important to include (such as amount and distribution of extant genetic variation; direction

and intensity of gene flow; microhabitat preferences), but these may not be known at the onset of a reintroduction project and could be incorporated later. Well-constructed objectives will reflect the best available life-history information and be simple enough to suggest which attributes to measure for determining if the goals have been fulfilled.

The goal of *abundance* could be fulfilled in the short term by using a process objective that integrates several or many life-history events. *In situ* completion of the life cycle, for example, requires measurements of germination, survivorship, and reproductive output for comparison to available reference populations. Each of these parts of the process can be isolated and manipulated experimentally, allowing tests of population management techniques as well (Boucher 1981; Pavlik 1994a). However, for taxa that have an unusual life history or breeding system, several state objectives that each reflect a single demographic event will be more useful. For example, Mehrhoff's studies of the orchid *Isotria medeoloides* (Mehrhoff 1989) demonstrated that population success could be ascertained by using stage or size class distributions. With this species there was no clear progression between life-history stages within a cohort. Individuals could repeatedly flower, vegetatively proliferate, or appear in sterile or arrested states. He found that a series of proximal objectives separately examining stage distribution, survivorship, and fecundity were suitable for evaluating performance in the short term, but that stage class distribution alone was sufficient.

Attaining and maintaining MVP are distal objectives of abundance that are not likely to be achieved by a founding population. Several generations of seed production and establishment may have to transpire before a viable population size is attained. But how can MVP be adopted as an objective for taxa that do not fit the available models? There are no magic numbers for rare plant populations, but we do have a reasonable range in which to begin: fifty to twenty-five hundred individuals (CPC 1991; Mace and Lande 1991; Given 1994). We can select an MVP objective within this range based upon what is known about a taxon's life history (Figure 6-3). Generalizing from the work of Karron (1991) and others, obligate outcrossers require larger population sizes than selfing taxa to increase the probability of interplant pollen transfer. Smaller, shorter-lived taxa will also require MVP to be toward the high end of the range (Belovsky 1987; Soule 1987). Menges (1990, 1991) has demonstrated the importance of environmental variation in determining extinction probability of different-sized populations. Again, his models suggest a range of hundreds to thousands. Choosing MVP for an endangered species such as *Amsinckia grandiflora* would take into account a predominantly outcrossing breeding system; an herbaceous, short-lived, moderate fecundity life history; a tendency toward high survivorship and long seed duration; and a climax grassland habitat with modest variation in rainfall and temperature. Some of

50 ------- MVP ------> 2500

longevity:	perennial ---->	annual
breeding system:	selfing ------>	outcrossing
growth form:	woody ------>	herbaceous
fecundity:	high ------>	low
ramet production:	common ------>	rare or none
survivorship:	high ------>	low
seed duration:	long ------>	short
environmental variation:	low ------>	high
successional status:	climax ------>	seral or ruderal

FIGURE 6-3.   Selection of an objective for minimum viable population (MVP) size depends on the life-history characteristics of the target taxon. Long -lived, woody, self-fertile plants with high fecundity would have an MVP in the range of 50 to 250 individuals. MVP for short-lived, herbaceous outcrossers, however, would be in the range of 1,500 to 2,500+ individuals. See CPC (1991), Mace and Lande (1991), and Given (1994) for further discussion.

these life-history attributes suggest a lower MVP value, others a higher one; so fifteen hundred individuals were adopted in this case. Although theorists may be uncomfortable with such a qualitative selection, I argue again that this approach will be informative if reintroductions are conducted as experiments both for testing MVP concepts and for attempting to conserve rare taxa.

The goal of *extent* can largely be met in the short term by constructing objectives that focus on dispersal and subsequent establishment. Depending on the availability of appropriate habitat, the dispersal of propagules beyond a reintroduction site may be the proximal objective, with corresponding increases in population area (Hengeveld 1989). If animal vectors are involved in this process, observations of their abundance and activity patterns will be needed. When habitat is limited or patchy, or if the taxon is an ecological specialist, proximal objectives would emphasize a lack of dispersal and low rates of propagule mortality in areas of unsuitable habitat. In the case of outcrossers, effective dispersal (outlying individuals of opposite mating types) must be distinguished from ineffective dispersal (outlying individuals of the same mating type).

Distal objectives related to extent may simply be derived from historical

reconstructions. Herbarium records, photographs, and personal recollections can often portray the former area of an extinct natural population where a reintroduction is taking place. Re-establishing the new population throughout the area provides a logical endpoint for mitigation and maximizes occupation of any existing microhabitats. Other distal objectives may be related to achieving metapopulation characteristics (Bowles et al. 1993) or attaining MVP throughout the historical range of a species.

Resilience is the ability to recover from perturbation. The concept of resilience has been central to debates about community complexity and stability (Pimm 1986). In the case of a new rare plant population, recovery from low population numbers during unfavorable years or after catastrophic events may be the best indicator of performance at a given site. Since populations with high growth rates recover faster from low numbers than populations with low rates (Pimm 1989), values of lambda could be used as short-term indicators of resilience after perturbation (see also Akcakaya 1992). Decreases in extinction probability following natural disturbance or during community succession could also be taken as positive indicators of persistence in wild or properly managed reserves. For example, Menges (1990) addressed the relative persistence of *Pedicularis furbishiae* populations by linking low extinction probabilities to periodic flooding and removal of woody plant cover within the floodplain habitat of this species.

The potential for resilience might also be enhanced at the beginning of a reintroduction by maximizing genetic variation within the founding population (Clegg and Brown 1983). Although every effort should be made to maintain the integrity of genetic stocks of widespread species during seed collection and release back into the wild (Millar and Libby 1991), rare plants with restricted distributions and small populations seldom exhibit much ecogeographic variation (Brown and Briggs 1991; Hamrick et al. 1991). Therefore, the risk of breaking up coadapted gene complexes is probably low in rare plants, and mixing genetically different seed stocks could produce the heterogeneity required for maximizing reproductive success and stress tolerance (DeMauro 1989, 1993; Barrett and Kohn 1991). But this argument assumes that the consequences of inbreeding are worse than the consequences of outbreeding—an assumption that is not well supported by data from wild plant populations. In general, plants may be more tolerant than animals of inbreeding and low levels of intrapopulation genetic variation (Maunder 1992; Menges 1992), so that mixing genetic stocks could be irrelevant and possibly detrimental to a reintroduced population. Field experiments on the benefits or consequences of mixing seed sources from genetically depauperate populations of rare plants are urgently needed.

Long-term objectives related to resilience could also employ genetic criteria to ensure that evolutionary potential is maintained (Holsinger and Got-

tlieb 1991). Desirable patterns of genetic variation within and between populations (expressed as G_{st}) could be drawn from life-history characteristics of the target taxon (Clegg and Brown 1983). For example, Hamrick's comprehensive allozyme surveys (Hamrick 1983; Hamrick et al. 1991) have shown that selfing species usually have 20 to 40 percent of their allelic variation among populations (low gene movement, greater population differentiation), while outcrossers have less than 10 percent (greater mixing between populations). When composing founding populations from multiple propagule sources having different allozyme variants, such general patterns could be roughly mimicked. This might improve long-term resilience of inbreeding taxa by providing a potential for gene pool differentiation as populations are exposed to an array of selection forces. Maximizing heterozygosity within populations of outbreeding taxa may improve resilience by allowing a wider array of microhabitats to be exploited at any one location. Here again is an opportunity for using rare plant reintroduction to test hypotheses generated by genetic and ecological theory.

Ultimately, *persistence* is the result of resilience over long periods of time. The effects of deleterious environmental cycles or catastrophic events on a new population are ameliorated by the exploitation of microhabitat variation (Holsinger and Roughgarden 1985; Huenneke 1991; Menges 1990, 1991), incorporation into the function and structure of the natural community (Karron 1991; Reisberg 1991) and realization of a large, effective population size (Pimm 1989). Proximal and distal objectives regarding persistence would thus require the ecological differentiation of subpopulations or populations, pollen movement by native pollinators, assimilation into the vegetation canopy, or maintenance of a low coefficient of variation in N_e.

Can the goal of persistence be fulfilled with certainty? No, but there are documented and apparently successful rare plant translocations that date back more than eighty years in Britain (Birkinshaw 1991). A "quite natural" but new population of the calciphile *Arabis stricta* has persisted on a limestone outcrop since approximately 1913. More than twenty-five hundred individuals of *Trinia glauca* are now found at Goblin Coombe after six plants and forty seeds were introduced in 1955. After reviewing 144 such cases, Birkinshaw concluded that 29 percent did not persist, 15 percent were less than five years old and couldn't be evaluated for persistence, and 22 percent were more than five years old and could be called persistent. Unfortunately, there are little or no data to indicate why some of these populations persisted and why some didn't. Although it is possible to revisit the former and measure demographic and genetic characteristics, no measurements for comparative purposes can be made on the latter. Achieving a better understanding of the criteria for persistence, therefore, requires measurements from both ephemeral and persistent populations created during reintroduction.

Measurement and Evaluation

The measurements required to address taxon-specific objectives during rare plant reintroduction must have a demographic emphasis, with individuals as the primary unit of monitoring (see Pavlik 1994a; see also Sutter, Chapter Ten). Survivorship, fecundity, population size, genetic variation levels, and other population-level attributes are derived from following the fate and performance of marked or mapped plants. By using an experimental design, the effects of microsite, habitat manipulation, transplant treatment, seed source, or other relevant variables can be determined with individuals of the target species as sensing devises (phytometers, *sensu* Oosting 1956). This demographic, experimental approach to creating rare plant populations is being utilized on a wide variety of plant life forms, including annuals (Parenti and Guerrant 1990; Pavlik, Nickrent, and Howald 1993), herbaceous perennials (Meagher, Antonovics, and Primack 1978; Borchert 1989; Pavlik and Manning 1993; Bowles et al. 1993; DeMauro 1994), stem succulents (Olwell, Cully, and Knight 1990; Olwell et al. 1987) and woody shrubs (Boucher 1981; Wallace 1992).

As previously discussed, objectives take into account the life history of the target and thereby suggest the most relevant demographic attributes to measure for evaluation. Measurements on annual plants will occur within separate cohorts while those on perennial plants will, over time, integrate cohorts. Consequently, determining survivorship to reproduction during each year of a new population may be required to meet an establishment objective in an annual (Table 6-2), while stage-class distribution after several years would be appropriate for a perennial. Similarly, the small size of most annuals allows total seed production per plant to be measured for fecundity objectives. This would be impractical for large, well-branched perennials, so fecundity could be measured as seed production per unit of biomass (stem length, inflorescence length, and so on) or as a seed/ovule ratio within and between individuals (Weins 1984; Weins et al. 1987; Pavlik, Ferguson, and Nelson 1993). Other life-history characteristics, including breeding system, successional status, and pollination mechanism will also determine exactly which kinds of measurements should be chosen for each objective.

With proper controls, adequate replication, and appropriate statistical tests (see Hurlbert 1984; Travis and Sutter 1986; and Menges 1992; see also Sutter, Chapter Ten), objectives can be compared to results, models can be tested, tools can be evaluated, and conclusions can be reached. Even when the number of seeds, cuttings, or potted plants is extremely limited, a reintroduction should be designed to maximize the harvest of new data on the target, its habitat, and/or its methods of release (see Guerrant, Chapter Eight). Preserving the information content of the experiment ensures a modicum of project success, even when the probability of biological success is low.

TABLE 6-2. The measurements for determining if an objective has been met are chosen to reflect life-history characteristics of the target taxon.

Objective	Longevity	
	Annual	Perennial
Establishment	Survivorship to reproduction	Stage or size distribution
Vegetative growth	Biomass or size per individual	Biomass or size per year
Fecundity	Seed output per plant	Seed output per unit biomass, seed/ovule ratio

Objective	Breeding system	
	Selfing	Outcrossing
Genetic variation	Allelic diversity among populations (G_{st})	Allelic diversity within populations (H_s)
Population size	Total (N)	Effective (N_e)
Microhabitat variation	Diversity among populations	Diversity within populations
Community Membership	?	Visitation by pollinators/dispersers

Objective	Successional status	
	Early successional	Late successional
Dispersal	Dispersal among habitat patches	Dispersal within habitat patches
Population distribution	Patch distribution	Patch extent

Applying the Framework to an Endangered Annual

Amsinckia grandiflora A. Gray is an annual dicot of grasslands in and around the Diablo Mountains of Northern California. The species has declined across its original 650 km² range, largely due to the introduction of livestock grazing, fire suppression, and non-native plants. In 1987, only one population was known to exist, with fewer than two hundred individuals on a hillside that had supported thousands only twenty years earlier. The draft recovery plan for this federally listed species called for the creation of new, self-sustaining populations within its historical range. After two years of preliminary studies on propagation, germination, seed production, and allozyme variation, the design and installation of new populations began (see Pavlik, Nickrent, and Howald 1993 for details). Another large and apparently stable natural population (more than three thousand individuals) was subsequently discovered and used as a reference population.

Objectives for evaluating the performance of the A. *grandiflora* reintroductions are presented in Table 6-3. The proximal objectives reflect the taxon's annual growth form, sensitivity of size and seed production to competition from non-native grasses, and outcrossing breeding system. For this winter annual, abundance will immediately depend on life-cycle completion (objective 1) and fecundity (objective 4), while resilience will depend on the existence of an enduring seed bank (objective 7). Experiments with fire and grass-specific herbicides established effective management protocols, but these cannot be used intensively to "garden" the population. So, life-cycle completion must occur each year regardless of management (objective 1) by occupying low competition (that is, low grass cover) microsites (objective 9). A size distribution that includes the large, fecund plants observed in the stable natural population (objective 3) would also indicate successful performance. Outbreeding requires that N_e eventually be used to estimate abundance (objective 2), that native pollinators visit (objective 9), and that founder propagules contain the highest levels of available genetic variation (objective 6).

The distal objectives also integrate these life-history characteristics but rely upon the selection of a lower population size threshold of 1,500 plants from the MVP range (as mentioned previously). That objective needs to be met within each of five populations throughout historic range of A. *grandiflora* and must be regained after perturbation or detrimental variations in climate (especially certain rainfall patterns that accentuate competition with annual grasses). Over a ten- to fifteen-year period, progressive (but not necessarily consecutive) increases in N, population area, seed bank size, and microhabitat occupation should result in convergence of N_e and N, with low levels of year-to-year variation relative to environmental variation during the same period.

TABLE 6-3. Proximal (short-term) and distal (long-term) objectives for evaluating reintroductions of Amsinckia grandiflora. Most objectives address the performance of a single population, with some exceptions (as shown by notes below).

Goals	Objectives	
	Proximal	Distal
Abundance	1. Life cycle can be completed *in situ* with and without habitat management	1. Attain MVP = 1,500 plants/population
	2. Progressive increase in N, N_e relative to founders	
	3. Size distribution matches natural population	
	4. Seed output per plant matches natural population	
Extent	5. Progressive increase in population area	2. Five populations throughout historic range attain MVP = 1,500 plants*
Resilience	6. Genetic variation is maximized within founder populations	3. MVP regained after environmental and/or demographic perturbation (rainfall variations, herbivory, and so on)
	7. Seed-bank density matches natural population	
Persistence	8. Subpopulations are found greater than 1 grassland microhabitat	4. N_e remains a high proportion of N with low coefficient of variation
	9. Native pollinators are commonly utilized	

* Addresses the performance of taxon as a whole (including multiple populations).

Data from several reintroduced A. *grandiflora* populations are now available and can be used to begin short-term evaluation (Pavlik 1991, 1992, 1993; Pavlik, Nickrent, and Howald 1993). After electrophoretic screening, two seed sources (one wild, one cultivated) were mixed to maximize genetic variability at all reintroduction sites (fulfilling one proximal resilience objective). Detailed demographic analysis has not shown any significant differences in the survivorship or reproduction of plants derived from the two sources,. This suggests that the incorporation of subtle allozyme variants into the founding gene pool may not be an important component of population resilience in this species. *In situ* completion of the life cycle, including the production of viable seed, has been observed in all founding populations (fulfilling one abundance objective). The performance of these populations, however, varies greatly for different reasons at each site.

The oldest and largest reintroduced population at Lougher Ridge has experienced increases in population size, population area, and plant size (Figure 6-4). The 1,101 founders became established under an experimental management regime during the 1989–1990 growing season. The regime used controlled burning, hand clipping, and a grass-specific herbicide in twenty separate plots (including controls) to determine the effects of competition with exotic annual grasses on patterns of survivorship and reproductive output. A. *grandiflora* plants in plots that were burned prior to germination or sprayed with dilute herbicide after grass emergence were significantly more likely to survive and reproduce profusely than plants in control or clipped plots (Pavlik, Nickrent, and Howald 1993). These results provided the justification for subsequent burning and herbicide treatment in 1990–1991 and thus established an enhanced management regime. Consequently, 1,301 reproductive plants were found in a larger area of Lougher Ridge (Box B), with increases in mean and mean maximum sizes (Box C). The increases in plant size were correlated with increases in seed output throughout the population. No manipulation took place at Lougher Ridge in 1992 (a "natural" management regime), but more than 1,600 reproductive plants were produced over a larger area, with increases in mean and mean maximum plant size. The population dramatically decreased in size during the 1992–1993 season, largely because of a late fall–early winter drought that killed six hundred to one thousand A. *grandiflora* seedlings. The survivors, however, maintained a patchy, wide distribution and eventually attained the largest sizes yet recorded at any reintroduction site. This population must demonstrate an ability to recover from weather-related perturbation by resuming favorable patterns of growth, dispersion, and reproduction. Other measurements of genetic variation, seed bank size and variability, and microhabitat diversity are also being made and evaluated.

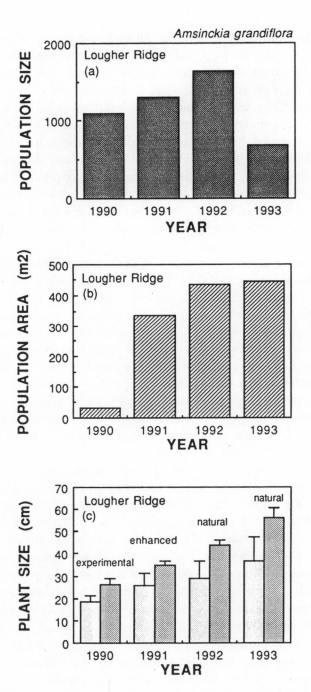

FIGURE 6-4. Population size (a), population area (b), and plant size (c) of *Amsinckia grandiflora* at the Lougher Ridge reintroduction site. Mean (light stipple; n = 100) and mean maximum (dark stipple; n = 10) plant sizes are shown with 1 standard deviation. Management regimes for each year are indicated as experimental, enhanced, and natural.

The other reintroduced *Amsinckia* populations, established across histor-ical ranges, are smaller, younger, and not uniformly successful. The popula-tion at Connolly Ranch, in the southern part of the range, is doing fairly well despite a large decrease in population size during 1992–1993 (Figure 6-5, box A). As at Lougher Ridge, seedling mortality was linked to late fall drought (even the nearby natural populations experienced significant decline). Population size and area increased, however, for 1991–1992, and mean and mean max-imum plant size remained constant and robust regardless of management regime (Figure 6-5, box C). If plant size and, therefore, seed production are sus-tained, this population should be able to resume a successful trajectory.

The populations at Los Vaqueros and Black Diamond II were also begun in 1990–1991, but they declined precipitously after the first year regime. Ex-tinction at Los Vaqueros (Figure 6-6, box A) was foreshadowed by signifi-cantly lower seed output per plant compared to the Lougher Ridge, Connolly Ranch, and Black Diamond II sites (Table 6-4). This was not due to differ-ences in the allometry of fitness components (such as inflorescence length per unit of stem, number of flowers per unit of inflorescence) but instead re-flected poor vegetative growth that produced diminutive, unbranched plants. Mean and mean maximum plant size were very small in response to unfavor-able site factors, especially cold winter temperatures. Evidence also suggested insufficient visitation by effective pollinators. Even with intensive second-year management, plant growth barely increased (Figure 6-6, box C). Based upon these growth and reproductive data, the Los Vaqueros population would be very unlikely to grow or persist, even if site factors or pollinator visi-tation improved in other years.

Black Diamond II is very close to extinction (Figure 6-7, box A), but not because of poor vegetative growth or reproduction per plant (Table 6-4). Some plants at this site became large, showy, and fecund during their first and second years (Figure 6-7, box C). Pocket gophers, however, severely disturbed the soil in ten of fourteen plots and killed many juvenile and flowering plants by destroying their root systems. There was also more intense competition with annual grasses at this site, even with second-year management. Plants did not establish in unmanaged areas or in control plots with high grass cover. It seems that low-competition microhabitats are rare at Black Diamond II, ex-cept when gophers are present.

Based upon these short-term evaluations, the populations at Los Vaqueros and Black Diamond II are failing because objectives concerning at least two goals (abundance and extent) are not being met. The populations at Lougher Ridge and Connolly Ranch, however, continue to perform in accordance with all proximal objectives. There is ongoing monitoring of all populations, failing and apparently successful, to further document the demographic and genetic characteristics associated with extinction and recovery.

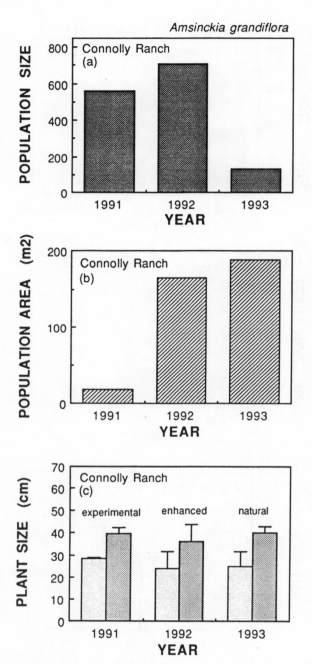

FIGURE 6-5. Population size (a), population area (b), and plant size (c) of *Amsinckia grandiflora* at the Connolly Ranch reintroduction site. Mean (light stipple; n = 50) and mean maximum (dark stipple; n = 10) plant sizes are shown with 1 standard deviation. Management regimes for each year are indicated as experimental, enhanced, and natural.

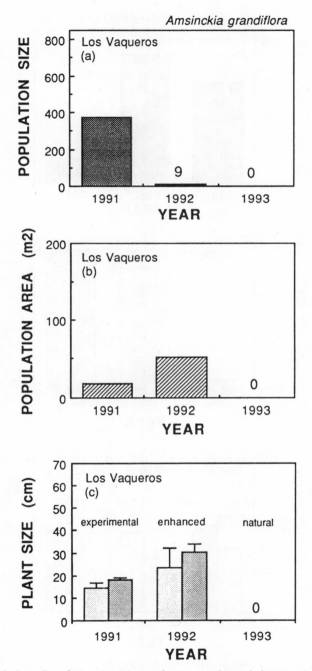

FIGURE 6-6. Population size (a), population area (b), and plant size (c) of *Amsinckia grandiflora* at the Los Vaqueros reintroduction site. Mean plant size (light stipple; n = 50 in 1991, n = 9 in 1992) and mean maximum plant size (dark stipple; n = 10 in 1991, n = 5 in 1992) are shown with 1 standard deviation. Management regimes for each year are indicated as experimental, enhanced, and natural.

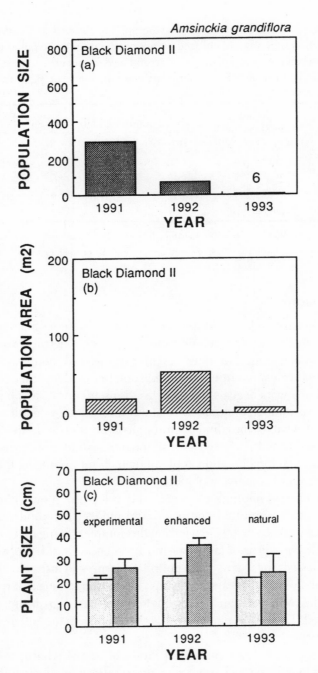

FIGURE 6-7.   Population size (a), population area (b), and plant size (c) of *Amsinckia grandiflora* at the Black Diamond II reintroduction site. Mean plant size (light stipple; n = 50 in 1991 and 1992, n = 6 in 1992) and mean maximum plant size (dark stipple; n = 10 in 1991 and 1992, n = 5 in 1993) are shown with 1 standard deviation. Management regimes for each year are indicated as experimental, enhanced, and natural.

TABLE 6-4. Comparison of seed output for *Amsinckia grandiflora* at four reintroduction sites, 1991. Mean seed output per plant (in treated plots only) and the linear correlation between plant size (cm) and seed output are shown. Values of seed output per plant (mean ± SD) followed by different letters are significantly different (P < 0.05, ANOVA).

Site	Mean seed output (number per plant)	Seed output vs. Plant size (cm)				n
		m	b	r	P	
LR	36.7 ± 8.4	4.60	−79.3	0.77	<0.01	18
CR	29.0 ± 7.5	3.42	−65.5	0.86	<0.01	10
LV	8.7 ± 1.4	0.92	−3.6	0.64	<0.05	10
BDII	44.5 ± 22.1	5.61	−93.1	0.85	<0.01	10

LR = Lougher Ridge.   CR = Connolly Ranch.   LV = Los Vaqueros.
BDII = Black Diamond.   m = slope, b = intercept, r = correlation coefficient.

Conclusions

The recovery of endangered species and the restoration of damaged ecosystems may be the greatest technical challenges in biological conservation. To meet these challenges, we must design, execute, and monitor plant reintroductions with clear intentions, thorough documentation, and rigorous evaluation. A framework of goals and objectives is useful during all phases of a project because it incorporates both what we think we know about conserving a species and what we must find out through observation and experimentation. Objectives that address the goals of abundance, extent, resilience, and persistence can, therefore, function as planning tools for improving the allocation and efficiency of conservation efforts.

The science of reintroduction cannot develop unless our collective experiences are communicated, compared, and applied. Time and money are too short to reinvent the same old wheel in the complex, living machinery we are trying to restore. Fear of failure or lust for success must be subverted (or at least harnessed) in favor of the penultimate desire — to learn about preserving biological diversity. No careful attempts at reintroduction are too shallow, no innovations too simple, and no lessons too apparent to go unsummarized and unreported. There is no other way to get this young science off the steep slope of the learning curve.

Success cannot come without risk. Some of the risk is borne by the taxa we are trying to reintroduce, as precious propagules are collected, stored, cultivated, and eventually placed back into the wild. Human institutions must bear the risk of funding potentially complicated, long-term projects with uncertain futures. Finally, researchers must bear a certain amount of professional risk when asked to report failure or hold themselves to a priori stan-

dards for judging success. But risk to all parties can be minimized by viewing reintroduction as an opportunity for conducting focused experiments that apply to ecosystem restoration while yielding the dividend of basic ecological knowledge.

REFERENCES

Akçakaya, H. R. 1992. "Population Viability Analysis and Risk Assessment." In *Wildlife 2001: Populations*. Edited by D. R. McCullough and R. H. Barrett. London: Elsevier Applied Science.

Anderson, K. H., and F. B. Hurley, Jr. 1980. "Wildlife Program Planning." In *Wildlife Management Techniques Manual*. 4th ed. Edited by S. D. Schemnitz. Washington, D.C.: Wildlife Society.

Barrett, S. C. H., and J. R. Kohn. 1991. "Genetic and Evolutionary Consequences of Small Population Size in Plants: Implications for Conservation." In *Genetics and Conservation of Rare Plants*. Edited by D. A. Falk and K. E. Holsinger. New York: Oxford University Press.

Belovsky, G. E. 1987. "Extinction Models and Mammalian Persistence." In *Viable Populations for Conservation*. Edited by M. E. Soulé. Cambridge: Cambridge University Press.

Birkinshaw, C. R. 1991. *Guidance Notes for Translocating Plants As Part of Recovery Plans*. CSD Report 1225. Peterborough, England: Nature Conservancy Council.

Borchert, M. 1989. "Postfire Demography of *Thermopsis macrophylla* H. A. var. *agina* J. T. Howell (Fabaceae), a Rare Perennial Herb in Chaparral." *American Midland Naturalist* 122: 120–132.

Botsford, L., and S. Jain. 1992. "Applying the Principles of Population Biology: Assessment and Recommendations." In *Applied Population Biology*. Edited by L. Botsford and S. Jain. The Netherlands: Kluwer Academic Publishers.

Boucher, C. 1981. "Autecological and Population Studies of *Orothamnus zeyheri* in the Cape of South Africa." In *The Biological Aspects of Rare Plant Conservation*. Edited by H. Synge London: J. Wiley and Sons.

Bowles, M., R. Flakne, K. McEachern, and N. Pavlovic. 1993. "Status and Restoration Planning for the Federally Threatened Pitcher's Thistle (*Circium pitcheri*) in Illinois." *Natural Areas Journal* 13: 164–176.

Boyce, M. S. 1992. "Population Viability Analysis." *Annual Review of Ecology and Systematics* 23: 481–506.

Brown, A. H. D., and J. D. Briggs. 1991. "Sampling Strategies for Genetic Variation in *Ex Situ* Collections of Endangered Plant Species." In *Genetics and Conservation of Rare Plants*. Edited by D. A. Falk and K. E. Holsinger. New York: Oxford University Press.

Center for Plant Conservation (CPC). 1991. "Genetic Sampling Guidelines for Conservation Collections of Endangered Plants." In *Genetics and Conservation of Rare Plants*. Edited by D. A. Falk and K. E. Holsinger. New York: Oxford University Press.

Clegg, M. T., and A. H. D. Brown. 1983. "The Founding of Plant Populations." In *Genetics and Conservation: A Reference for Managing Wild Animal and Plant Popula-*

tions. Edited by C. M. Schonewald-Cox, S. M. Chambers, B. MacBryde, and L. Thomas. London: Benjamin Cummings.

Dawkins, R. 1986. *The Blind Watchmaker.* New York: W. W. Norton.

DeMauro, M. M. 1989. *Aspects of the Reproductive Biology of the Endangered* Hymenoxys acaulis *var.* glabra: *Implications for Conservation.* Master's thesis, University of Illinois, Chicago.

DeMauro, M. M. 1993. "Relationship of Breeding System to Rarity in the Lakeside Daisy (*Hymenoxys acaulis* var. *glabra*)." *Conservation Biology* 7: 542–550.

DeMauro, M. M. 1994. "Development and Implementation of a Recovery Program for the Federally Threatened Lakeside Daisy (*Hymenoxys acaulis* var. *glabra*). In *Recovery and Restoration of Endangered Species.* Edited by M. Bowles and C. J. Whelan. Cambridge: Cambridge University Press.

Dobzhansky, T., F. J. Ayala, G. L. Stebbins, and J. W. Valentine. 1977. *Evolution.* San Francisco: W. H. Freeman.

Fiedler, P. L. 1991. *Mitigation-Related Transplantation, Relocation and Reintroduction Projects Involving Endangered, Threatened, and Rare Plant Species in California.* Sacramento: California Department of Fish and Game, Endangered Plant Program.

Gilpin, M. E., and M. E. Soule. 1986. "Minimum Viable Populations and Processes of Species Extinctions." In *Conservation Biology: The Science of Scarcity and Diversity.* Edited by M. E. Soule. Sunderland, Mass.: Sinauer Associates.

Given, D. R. 1994. *Principles and Practice of Plant Conservation.* Portland, Ore.: Timber Press.

Hall, L. A. 1987. "Transplantation of Sensitive Plants as Mitigation for Environmental Impacts." In *Conservation and Management of Rare and Endangered Plants.* Edited by T. Elias. Sacramento: California Native Plant Society.

Hamrick, J. L. 1983. "The Distribution of Genetic Variation Within and Among Natural Plant Populations." In *Genetics and Conservation: A Reference for Managing Wild Animal and Plant Populations.* Edited by C. M. Schonewald-Cox, S. M. Chambers, B. MacBryde, and L. Thomas. London: Benjamin Cummings.

Hamrick, J. L., M. J. W. Godt., D. A. Murawski, and M. D. Loveless. 1991. "Correlations Between Species Traits and Allozyme Diversity: Implications for Conservation Biology." In *Genetics and Conservation of Rare Plants.* Edited by D. A. Falk and K. E. Holsinger. New York: Oxford University Press.

Hengeveld, R. 1989. *Dynamics of Biological Invasions.* London: Chapman and Hall.

Holsinger, K. E., and L. D. Gottlieb. 1991. "Conservation of Rare and Endangered Plants: Principles and Prospects." In *Genetics and Conservation of Rare Plants.* Edited by D. A. Falk and K. E. Holsinger. New York: Oxford University Press.

Holsinger, K. E., and J. Roughgarden. 1985. "A Model for the Dynamics of an Annual Plant Population." *Theoretical Population Biology* 28: 288–313.

Huenneke, L. F. 1991. "Ecological Implications of Genetic Variation in Plant Populations." In *Genetics and Conservation of Rare Plants.* Edited by D. A. Falk and K. E. Holsinger. New York: Oxford University Press.

Hurlbert, S. H. 1984. "Pseudoreplication and the Design of Ecological Field Experiments." *Ecological Monographs* 54: 187–211.

Karron, J. D. 1991. "Patterns of Genetic Variation and Breeding Systems in Rare Plant

Species." In *Genetics and Conservation of Rare Plants*. Edited by D. A. Falk and K. E. Holsinger. New York: Oxford University Press.

Kleiman, D. G. 1989. "Reintroduction of Captive Mammals for Conservation." *Bioscience* 39: 152–161.

Kleiman, D. G. 1990. "Decision-Making About a Reintroduction: Do Appropriate Conditions Exist?" *Endangered Species Update* 8 (1): 18–19.

Mace, G. M., and R. Lande. 1991. "Assessing Extinction Threats: Toward a Reevaluation of IUCN Threatened Species Categories." *Conservation Biology* 5: 148–157.

Maunder, M. 1992. "Plant Reintroduction: An Overview." *Biodiversity and Conservation* 1: 21–62.

Mayr, E. 1976. *Evolution and the Diversity of Life: Selected Essays*. Cambridge, Mass.: Belknap Press, Harvard University.

Meagher, T. R., J. Antonovics, and R. Primack. 1978. "Experimental Ecological Genetics in *Plantago*. III. Genetic Variation and Demography in Relation to Survival of *Plantago cordata*, a Rare Species." *Biological Conservation* 14: 243–257.

Mehrhoff, L. A. 1989. "The Dynamics of Declining Populations of an Endangered Orchid, *Isotria medeoloides*." *Ecology* 70: 783–786.

Menges, E. S. 1990. "Population Viability Analysis for an Endangered Plant." *Conservation Biology* 4: 52–62.

Menges, E. S. 1991. "The Application of Minimum Viable Population Theory to Plants." In *Genetics and Conservation of Rare Plants*. Edited by D. A. Falk and K. E. Holsinger. New York: Oxford University Press.

Menges, E. S. 1992. "Stochastic Modeling of Extinction in Plant Populations." In *Conservation Biology: The Theory and Practice of Nature Conservation, Preservation, and Management*. Edited by P. L. Fiedler and S. Jain. New York: Chapman and Hall.

Millar, C. I., and W. J. Libby. 1991. "Strategies for Conserving Clinal, Ecotypic, and Disjunct Population Diversity in Widespread Species." In *Genetics and Conservation of Rare Plants*. Edited by D. A. Falk and K. E. Holsinger. New York: Oxford University Press.

Olwell, P., A. Cully, and P. Knight. 1990. "The Establishment of a New Population of *Pediocactus knowltonii*: Third Year Assessment." In *Ecosystem Management: Rare Species and Significant Habitats*. Edited by R. S. Mitchell, C. J. Sheviak, and D. J. Leopold. N.Y. State Museum Bulletin 471. New York: New York State Museum.

Olwell, P., A. Cully, P. Knight, and S. Brack. 1987. "*Pediocactus knowltonii* Recovery Efforts." In *Conservation and Management of Rare and Endangered Plants*. Edited by T. Elias. Sacramento: California Native Plant Society.

Oosting, H. J. 1956. *The Study of Plant Communities*. San Francisco: W. H. Freeman.

Palmer, M. E. 1987. A Critical Look at Rare Plant Monitoring in the United States. *Biological Conservation* 39: 113–127.

Parenti, R. L., and E. O. Guerrant Jr. 1990. "Down but Not Out: Reintroduction of the Extirpated Malheur Wirelettuce, *Stephanomeria malheurensis*." *Endangered Species Update* 8: 62–63.

Pavlik, B. M. 1991. *Reintroduction of* Amsinckia grandiflora *to Three Sites Across Historic Range*. Sacramento, Calif.: Department of Fish and Game, Endangered Plant Program.

Pavlik, B. M. 1992. *Inching Towards Recovery: Evaluating the Performance of* Amsinckia grandiflora *Populations Under Different Management Regimes.* Sacramento, Calif.: Department of Fish and Game, Endangered Plant Program.

Pavlik, B. M. 1993. *Current Status of Natural and Reintroduced Populations of* Amsinckia grandiflora *Under Different Management Regimes.* Sacramento, Calif.: Department of Fish and Game, Endangered Plant Program.

Pavlik, B. M. 1994a. "Demographic Monitoring and the Recovery of Endangered Plants." In *Recovery and Restoration of Endangered Species.* Edited by M. Bowles and C. J. Whelan. Cambridge: Cambridge University Press.

Pavlik, B. M. 1994b. "Conserving Plant Species Diversity: The Challenge of Recovery." In *Biodiversity in Managed Landscapes.* Edited by R. C. Szaro, New York: Oxford University Press.

Pavlik, B. M., and E. Manning. 1993. "Assessing Limitations on the Growth of Endangered Plant Populations. I. Experimental Demography of *Erysimum capitatum* ssp. *angustatum* and *Oenothera deltoides* ssp. *howellii.*" *Biological Conservation* 65: 257–265.

Pavlik, B. M., N. Ferguson, and M. Nelson. 1993. "Assessing Limitations on the Growth of Endangered Plant Populations. II. Seed Production and Seed Bank Dynamics of *Erysimum capitatum* ssp. *angustatum* and *Oenothera deltoides* ssp. *howellii.*" *Biological Conservation* 65: 267–278.

Pavlik, B. M., D. L. Nickrent, and A. M. Howald. 1993. "The Recovery of an Endangered Plant. I. Creating a New Population of *Amsinckia grandiflora.*" *Conservation Biology* 7: 510–526.

Pimm, S. L. 1986. "Community Stability and Structure." In *Conservation Biology: The Science of Scarcity and Diversity.* Edited by M. E. Soule. Sunderland, Mass.: Sinauer Associates.

Pimm, S. L. 1989. "Theories of Predicting Success and Impact of Introduced Species." In *Biological Invasions: A Global Perspective.* Edited by J. A. Drake et al. Chichester, England: J. Wiley and Sons.

Ralls, K., R. A. Garrott, D. B. Siniff, and A. M. Starfield. 1992. "Research on Threatened Populations." In *Wildlife 2001: Populations.* Edited by D. R. McCullough and R. H. Barrett. London: Elsevier Applied Science.

Rieseberg, L. H. 1991. "Hybridization in Rare Plants: Insights from Case Studies in *Cercocarpus* and *Helianthus.*" In *Genetics and Conservation of Rare Plants.* Edited by D. A. Falk and K. E. Holsinger. New York: Oxford University Press.

Saunders, D. A., R. J. Hobbs, and C. R. Margules. 1991. "Biological Consequences of Ecosystem Fragmentation: A Review." *Conservation Biology* 5: 18–24.

Schonewald-Cox, C. M., S. M. Chambers, B. MacBryde, and L. Thomas. 1983. *Genetics and Conservation: A Reference for Managing Wild Animal and Plant Populations.* London: Benjamin Cummings.

Simberloff, D. 1989. "Which Insect Introductions Succeed and Which Fail?" In *Biological Invasions: A Global Perspective.* Edited by J. A. Drake et al. Chichester, England: J. Wiley and Sons.

Soulé, M. E. 1987. "Introduction." In *Viable Populations for Conservation.* Edited by M. E. Soule. Cambridge: Cambridge University Press.

Travis, J., and R. Sutter. 1986. "Experimental Designs and Statistical Methods for Demographic Studies of Rare Plants." *Natural Areas Journal* 6 (3): 3–12.

Wallace, S. 1992. "Introduction of *Conradina glabra: A Pilot Project for the Conservation of an Endangered Florida Endemic.*" *Botanic Gardens Conservation News* 1 (10): 34–39.

Weins, D. 1984. "Ovule Survivorship, Brood Size, Life History, Breeding Systems, and Reproductive Success in Plants." *Oecologia* 64: 47–53.

Weins, D., C. L. Calvin, C. A. Wilson, C. I. Davern, D. Frank, and S. R. Seavey. 1987. "Reproductive Success, Spontaneous Embryo Abortion, and Genetic Load in Flowering Plants." *Oecologia* 71: 501–509.

Selecting Reintroduction Sites

Peggy L. Fiedler and Richard D. Laven

Plant species ought to be returned to the site from which they have been extirpated if the conservation management goal is to reintroduce a population of rare plants. If the term *reintroduction* is strictly interpreted, plants of special concern must be placed only in the locality of record. However, much of what rare plant conservationists do is not so strictly circumscribed. Manipulation of rare plant populations, particularly in the context of impact mitigation, involves not reintroduction per se but often the translocation (also known as transplantation) of a rare plant population. By collectively considering reintroduction, translocation, and introduction, a much broader, more inclusive universe of rare plant protection is encompassed. Here we are concerned with any placement of a plant population into suitable habitat, whether it is a former site from which the species currently is extirpated or potential habitat that might support a new population of rare plants in the future.

Given this broader, more inclusive perspective, we argue that the issue of "where" in the more inclusive context of introduction is not as self-evident as it might otherwise seem. Moreover, when determining what constitutes suitable habitat in the site-selection process, there are two broad issues to consider. The first is largely data rich and concerns a straightforward evaluation of the autecological and synecological characteristics of rare plant species. The second issue is driven less by readily obtainable data and more by the kind of rarity that is involved in the introduction attempt. Specifically, the age, habitat distribution, and population abundance of a rare plant species may be critical considerations (albeit in varying degrees) in site selection for rare plant introductions.

Four Classes of Selection Criteria

We have grouped the site-selection considerations for the introduction of rare plants into four categories: physical, biological, logistical, and historical. We

will consider each category separately and in some detail. Our purpose is not to provide a thorough review of transplantation efforts but to paint broadly a series of issues that should be addressed when a site is selected for the restoration of a rare plant population.

Physical Criteria

The physical criteria for site selection are often straightforward. The primary physical consideration is the landscape-specific selection of a similar geomorphic setting. For example, if the rare plant is typically found on narrow, densely shaded canyon slopes or on the floodplain of third-order streams, then a similar landscape position should be chosen.

Also at this scale is the issue of landscape matrix. For example, if the rare plant to be introduced is a species that typically is found in an (ecological) island habitat surrounded by a habitat that is inappropriate for the plant but essential to some life-history stage of its pollinator or disperser, then this landscape context will be critical to the long-term success of the introduced population.

At the site-specific level, soil type appears most crucial to successful location. This includes soil texture, pH, and other soil-chemistry factors that are at a much greater level of resolution than the typical country soil survey. In a review of mitigation-related introductions of rare plant species in California (Fiedler 1991), the majority of introduction failures was related to the inappropriateness of the soil characteristics at the receptor site. For example, a transplantation experiment performed by the U.S. Bureau of Land Management (BLM) for the stoloniferous pussytoes (*Antennaria flagellaris*) involved translocating a portion of a large population to assess whether transplantation would be an acceptable form of future mitigation for this rare species. The experiment was deemed a failure in that, of the more than four hundred plants transplanted to four different populations, only one population of seventeen individuals persisted after six years (Fiedler 1991). BLM personnel suggested that, among other issues, the transplantation failed because the receptor site supported a soil of inappropriate texture. In a mitigation-related transplantation project for the willowy monardella (*Monardella linoides* ssp. *viminea*) enacted in 1986 by the California Department of transportation, it was determined that salvaged *M. linoides* ssp. *viminea* plants required the parental soil to survive under nursery conditions until outplanting in the same substrate (Fiedler 1991).

Other site-specific characteristics that may be critical to introduction success include slope angle and position requirements, slope aspect, and the albedo effect of the substrate. These data are likely to be readily obtainable from known sites.

Biological Criteria

The biological criteria of concern are peculiar to both the autecology and synecology of a rare plant. These data are largely species-specific characteristics that allow an individual to grow and successfully reproduce on site. For example, it is important to consider the presence (or limitation thereof) of dispersal vectors, such as pollinators or fruit dispersers. Another consideration is the presence of mycorrhizal associates. Although the specific fungal mutualist(s) is largely unknown for most plant species, it may be of considerable importance in the successful establishment of the introduced plant population. This is particularly critical for rare plants in plant families, such as the Ericaceae and Orchidaceae, with known mycorrhizal specificities. Similar considerations regarding host plant species must be made for those families, such as the Scrophulariaceae, with known hemiparasitic rare taxa (such as *Castilleja* and *Cordylanthus*).

At the next level of organization, synecology of the taxon, biological site selection considerations should include an evaluation of the overall similarity of the proposed site to the salvage site. For example, is the proposed site (1) of a similar floristic composition and structure, (2) at a similar successional stage and appropriate trajectory, and (3) functioning in a mode similar to known habitat?

An additional consideration requires documentation—the presence (or absence) of potential competitors and herbivores. On-site presence of either significant plant competitors (such as choking, spreading weeds or dense, fast-growing exotics) or an overpopulation of generalist herbivores (such as deer, cottontails, and jackrabbits) can rapidly thwart the best of introduction efforts. A significant number of rare plant transplantation attempts in California as reviewed by Fiedler (1991) involved weeding and fencing the receptor site prior to transplantation. For example, site preparation for the mitigation-related transplantation of the rare thread-leaved brodiaea (*Brodiaea filifolia*) in northern San Diego County, California, involved the construction of rabbit enclosures to prevent grazing of the transplanted *B. filifolia* corms. In addition, the mitigation plan required the fencing of a nature preserve for the life of the project to exclude all local herbivores, as well as the creation of a stable, weed-free population of the thread-leaved brodiaea (WESTEC 1988). Fiedler (1991) determined that this rare plant transplantation effort was tentatively successful, although it was ongoing at the time of her review.

Another biological consideration of a less tangible nature is the on-site potential for the occurrence of intra- and interspecific hybridization, outbreeding depression, or other forms of genetic contamination. Such potential can be difficult to anticipate, but genetic considerations of closely related congeners ought to be included in receptor site selection. Genetic assimilation may be a legitimate concern for transplanted rare plant populations because small populations can lose genetic diversity by being assimilated by larger ones

(Cade 1983). For example, using isozyme evidence, Reiseberg et al. (1989) determined that the single population of *Cercocarpus traskiae*, a rare tree species confined to Wild Boar Gully on Santa Catalina Island, Los Angeles County, California, consists of five "pure" *C. traskiae* individuals, one "pure" individual of *Cercocarpus betuloides* var. *blancheae* (a more widespread species), and two hybrid *Cercocarpus* trees. Reiseberg (1991) argues persuasively that some floras, particularly those found on islands and those rapidly evolving (such as certain taxa in California), may be particularly susceptible to hybridization and introgression. Every effort should be taken to avoid the opportunity for such genetic events to occur when introducing a population of rare plants into a new location.

Logistical Criteria

No less important than physical or biological are the logistical criteria for site selection. These issues also fall into two broad groups: (1) those concerning the site protection and degree of human access, and (2) the responsibility of monitoring and remediation. With regard to the latter, land ownership is the first issue of concern because of the implications for preservation and monitoring. Also, the degree of protection, ranging from no access, such as in certain private nature preserves, to free public access, as in city and county parks, might influence receptor site selection. How long the site will be protected is of equal concern, particularly in mitigation-related translocation projects when the receptor site might be fenced and otherwise restricted during the construction phase but not protected after the project is completed. Such timing in site protection may be a critical consideration in the site-selection process.

A third issue in this broad category of site protection is the relative access for the monitoring of the introduction effort, for remediation if necessary, and for the potential for rare plant research. For example, if helicopter or boat access is required, then special logistical efforts must be included in the site protection planning effort. If these considerations are of concern to the responsible party (and at minimum, the issue of monitoring should be), then reasonable access for legitimate parties should be considered.

Of a less obvious nature is the issue of monitoring and management responsibility. If the person(s) responsible for monitoring or instituting corrective measures might be constrained by a site selection either too far away or too logistically complicated, or if they are unable to enter the site for legal reasons, then an alternative site may have to be chosen.

Historical Criteria

The final class of criteria for site selection is much less tractable and has controversial conservation implications. Two historical issues should be consid-

ered for site introduction. The first involves the selection of known site locations versus potential habitat. The second involves maintaining the potential for evolutionary change, which is elusive not only in time and space but also in terms of protection in perceived perpetuity.

To illustrate the potential dilemma of choosing known versus potential habitat, consider a rare plant species with a restricted distribution of four known populations, one large and three small. Nearby are four areas of never occupied but apparently suitable habitat. After the influence of environmental stochasticity resulting in the irreversible alteration of the habitats supporting known populations, there remains only one small population of the rare plant, one suitable historic location, and the four areas of unoccupied, suitable habitat. It is possible, and in some cases probable, that the greatest potential for persistence and evolutionary change is found in the never occupied but apparently suitable habitat.

What are the implications of this choice? First, site selection of unoccupied, suitable habitat—more specifically, habitat for which there are no records of occupation by the rare plant of concern—presents practical and strategic problems. For example, should a rare plant species characterized by a broad distribution and low population sizes (such as *Isotria medeoloides* [Pursh] Rafinesque) be introduced anywhere within the range of its potential habitat? Such consideration technically may imply that potential mitigation efforts will cross state (and federal) borders and thus involve a variety of jurisdictions. Second, there exists an enormous potential for misuse of this option—that is, introducing populations into potential versus known habitat potentially allows liable parties to transplant special-status plants far away from the project site, where monitoring, maintenance, or other activities might be difficult, forgotten, or, at best, construed to be someone else's concern. Third, the introduction of a rare plant population into a site never occupied might allow the taxon to become a weedy invader. This is rarely the case, as virtually no rare species have weedy or invasive attributes (Ehrlich 1989; Nobel 1989).

The second area of concern is conservation of a species' evolutionary potential. Consider a rare plant species found in a specific habitat along the banks of a particular river. Several viable populations exist within a protected reach, and several areas of suitable, unoccupied habitat also are found. Severe flooding occurs on an annual basis. During a one-hundred-year flood, only one of the populations remains; the historic locations no longer represent suitable but unoccupied habitat; and the unoccupied but apparently suitable habitat still exists. The choice here becomes not whether to introduce to unoccupied suitable habitat, but which patch of unoccupied suitable habitat should be designated as the receptor site.

The latter example is represented in theory by riverine rare species as

Furbish's lousewort (*Pedicularis furbishiae* S. Wats.; Scrophulariaceae). Furbish lousewort's is a federally listed rare plant species (U.S. Fish and Wildlife Service 1978), endemic to the St. John River valley of northern Maine and adjacent New Brunswick, Canada. Approximately twenty-eight colonies totaling about five thousand individuals have been documented along a 140-mile stretch of the river (Richards 1980; Stirrett 1980; Gawler 1983). The species' habitat requirements are defined by a midsuccessional shrub-dominated ecotone below undisturbed boreal forest and above a sparsely vegetated, frequently flooded riverine cobble zone (Menges 1990). Population viability of *P. furbishiae* is dictated primarily by annual spring ice scour and secondarily by bank slumping and other stochastic events along the riverbank. Thus populations of *P. furbishiae* do not necessarily persist very long, "winking" off and on as the St. John River destroys populations through flooding and ice scour, creating the conditions for suitable habitat to emerge by resetting the successional clock. Introducing a new population of *P. furbishiae* into a historical location—that is, one that was recently destroyed by ice scour or landslide—would not be successful in the long term. Therefore, in this case, and likely in a great many others, potential habitat should be considered, in addition to known or historical locations, in the site selection process. Menges (1990) suggests that such a landscape-level perspective is critical for the long-term survival of *P. furbishiae*.

In summary, in certain cases, potential habitat may be a poor choice and lead to the overall decline of the species. Primack and Miao (1992) suggest that the majority of apparently suitable but unoccupied sites are in reality not suitable or infrequently suitable for successful population establishment by seed. In other cases, selection of potential habitat may be the only way to ensure persistence and potential for evolutionary processes. This will become all too clear as global climate change becomes more of an anticipated, everyday management issue.

Types of Rarity and Selection of Receptor Sites

The type of species rarity also weighs into the site-selection equation. We have identified three broad types of rarity, with variations (Table 7-1). *Anthropogenic rarities* are species formerly common but through negative interactions with humans, are now considered rare. *Endemic rarities* are those restricted to a small geographic distribution; these have been discussed in great detail elsewhere (see Kruckeberg and Rabinowitz 1985). *Sparsely distributed rarities* are those species characterized by moderate to small population sizes, distributed over narrow or wide geographic ranges. Let us now examine each

general class of rarity and the ways in which the intrinsic form of rarity might influence receptor-site selection.

Anthropogenic Rarities

Two forms of anthropogenic rarities are distinguished based on whether the habitat is continuous or discontinuous. For example, Catalina Mariposa Lily (*Calochortus catalinae*; Liliaceae) is an anthropogenic rarity of Southern California, inhabiting open grassland and grassland/chaparral communities once continuously found along the Southern California coast. Much of this habitat has been converted to housing and retail businesses, and the remainder is in threat of future development. On the other hand, *Howellia aquatilis* (Campanulaceae), is a rare aquatic annual species found in fewer than ten locations through the Pacific Northwest (Lesica et al. 1988). The species has been found in new locations in the state of Montana, and the California collection is now in question (E. O. Guerrant, personal communication 1993). In fact, it is believed to be extinct in California and Oregon and thus extirpated from a large portion of historical distribution.

Anthropogenic rarity may be important in site selection because formerly widely distributed rarities may have fewer high-quality receptor sites available for consideration. This is particularly true for disjunct species, particularly wetland and aquatic taxa, where water quality, aquatic weed encroachment, and other factors degraded habitat quality.

Local Endemics

As for endemic rarities, three general forms are distinguished: paleoendemics, neoendemics, and those endemic species for which the age is unknown, uncertain, or not relevant to considerations of the species viability (please see the Introduction to this book). In examining paleoendemics, considering historical locations often does not make sense. Much of the historical range (such as Wyoming) of the Coast Redwood (*Sequoia sempervirens*), for example, no longer offers suitable habitat. For many other paleoendemics, such as the Shasta Snow Wreath (*Neviusia cliftonii*), viable historical habitat may not be possible for other reasons. Although this species is known from three locations, all limestone substrates in shaded canyons adjacent to small streams 60 to 80 km south of Mount Shasta, neither the historical location of British Columbia (Shevock, Ertter, and Taylor 1992) nor the immediate areas can be considered viable options. Interestingly, such site-selection considerations may be the likely fate for *Neviusia cliftonii*, as the area around Mount Shasta faces increasing pressures from development and resource extraction.

Neoendemics present similar site-selection difficulties. Often neoendemics

TABLE 7-1. Types of rarity, associated attributes, and relevant receptor-site considerations.

| Type of rarity | Associated attributes | Relevant receptor | | Example |
		Site considerations		
I Anthropogenic				
A. Continuous habitat	1. Formerly widespread 2. Habitat continuous 3. Habitat in decline and/or nearly gone	1. May require special site manipulation to be suitable habitat for introduction		*Calochortus catalinae* (Catalina Mariposa Lily)
B. Discontinuous habitat	1. Formerly widespread 2. Discontinuous, specialized habitat 3. Habitat declining and/or nearly gone	1. May require special site manipulation to be suitable habitat for introduction		*Howellia aquatilis*
II. Endemics				
A. Paleoendemic	1. Habitat requirements poorly known and possibly limited	1. Historic locations may not be feasible for political and logistical reasons		*Neviusia cliftonii* (Shasta Snow Wreath)
B. Neoendemic	1. Habitat specialized and restricted 2. Habitat generalized	1. Historic locations may not be possible for consideration 1. Known from two sites 2. Site typically altered by exotic invasions, change in fire regime, and so on		*Limnanthes floccosa* ssp. *californica* *Stephanomeria malheurensis* (Malheur Wire Lettuce)

C. Age unknown/uncertain	1. Habitat restricted 2. Habitat apparently unspecialized	1. May not understand the nature of endemism and therefore site selection includes a greater uncertainty	*Tetramolopium arenarium*
III. Sparsely Distributed			
A. Sparsely distributed/continuous habitat	1. Scattered populations in relatively restricted area 2. Moderately small population sizes 3. Habitat specialized and ephemeral	1. Potential habitat and minimum viable areas are probably concerns	*Pedicularis furbishiae* (Furbish's Lousewort)
B. Sparsely distributed/discontinuous habitat	1. Scattered populations in relatively restricted area 2. Population sizes unknown—clonal growth 3. Habitat restricted but requirements not overly specialized 4. Habitat ephemeral	1. Potential habitat and minimum viable areas are probably concerns, as are demographic considerations	*Lilaeopsis masonii* (Mason's Lilaeopsis)

are known from only one location, so historical or extirpated locations do not exist for consideration. The recent evolution of two of our more notorious rarities, *Clarkia fransciscana* and *Stephanomeria malheurensis* ssp. *coronaria*, are clear examples of the absence of opportunity in the site-selection process. However, vernal pool neoendemics might offer some insights regarding site selection because, despite being known from small geographic locations, several pools might be found within the area. *Limnanthes floccosa* Howell ssp. *californica*, restricted to vernal pools and streams in Butte County, California (Dole and Sun 1992) and *L. f.* ssp. *grandiflora* of the Table Rock region in central Oregon are good examples.

A bit problematic are endemics whose restricted distributions cannot be readily explained by age. These species, however, illustrate the general problem of endemic species; that is, species may be restricted for reasons we do not yet understand. For example, *Tetramolopium arenarium* is found on moderately old lava flows in the Pohakuloa Training Area on the island of Hawaii. It exists as a single population of fewer than 150 individuals, and the population currently is in decline (Laven et al. 1991). Potential habitat exists elsewhere on the Pohakuloa Training Area (Laven, personal observation); therefore it is likely that only strategy for the long-term preservation of *T. arenarium* is to introduce nursery-grown individuals or introduce seed to potential suitable, unoccupied habitat to establish new populations.

Sparsely Distributed Rarities

Consideration of potential habitat is perhaps most relevant with sparsely distributed rarities. As mentioned before, Furbish's lousewort is an example of rare plant species found in scattered populations in specialized and ephemeral habitats within a relatively restricted area. Menges (1990) has suggested that a landscape perspective is critical for the long-term survival of *P. furbishiae*, specifically, one that includes the protection and maintenance of known habitat, potential habitat, and areas that may become potential habitat because of the need to maintain the ecological processes within a minimum viable area.

Another example of equal relevance is *Lilaeopsis masonii*, a freshwater tidal rarity restricted to the littoral zone of the Sacramento–San Joaquin Delta (Delta) of California. This tiny perennial umbel forms dense clonal mats on a variety of substrates, including old pilings, rhizomes of *Scirpus californicus*, silt, sand and clay wave-cut shores, and even old riprap (Golden and Fiedler 1991). We distinguish it from Furbish's lousewort because population sizes are nearly unknowable, given the species' dense, clonal growth habit. Populations may exist of one to a few genets, as the founding of populations may occur by floating seed and population fragments within different por-

tions of the Delta. Here, too, a landscape perspective is essential to the maintenance of this species. Regular freshwater spring runoff from the Sacramento and San Joaquin Rivers is a critical dynamic feature of the Delta landscape. These river flows maintain a critical salinity balance and create rare plant habitat by a complex of geomorphic processes caused by tidal and riverine currents that change seasonally.

There are two distinguishing features for the two sparsely distributed rarities discussed here: (1) Furbish's lousewort survives in a specialized habitat along a 140-mile reach of a free-flowing river, while Mason's lilaeopsis grows in a specialized and rare ecosystem along the Pacific Coast of North America, and (2) Furbish's lousewort founds new populations primarily sexually, while Mason's lilaeopsis, although not well documented, may well establish new populations through the fragmentation and transport of clonal individuals by riverine currents. Both species, however, require not individual sites for restoration but portions of a larger landscape designated as a receptor "site" to allow for the natural extinction and founding of populations within the landscape.

Conclusions

Our overall conclusion is that selecting suitable habitat for rare plant introductions is far from self-evident. Site-selection considerations must be given to physical, biological, logistical and historical criteria. Additionally, the nature of rarity has to be incorporated into site-selection deliberations.

The examples discussed in this chapter (and in others in this book) also lead us to conclude that, in most cases, metapopulation dynamics must be assessed and utilized in selecting receptor-site locations. As suitable habitat and landscapes become increasingly fragmented, metapopulation analysis will play a much larger role in rare plant introductions.

Finally, Walters (1986) states that resource management is a "continual learning process that cannot conveniently be separated into functions like 're-search' and 'ongoing regulatory activities'" (p. 8). This is a critical point for mitigation-related activities involving the manipulation of rare plant species. Rarely will we know enough beforehand to guarantee the success of rare plant introduction attempts. However, if we approach the regulatory reality that "avoidance," although the preferred form of mitigation under virtually all circumstances, is not always possible, then we can ensure more rigorous, quantifiable, and repeatable manipulation procedures for the unavoidable project impacts. Receptor-site selection is but one of the critical decision processes in restoring diversity through rare plant introductions.

ACKNOWLEDGMENTS

We wish to thank E. O. Guerrant and M. Vasey for their thoughtful guidance, both verbal and written, during the reintroduction symposium and the evolution of this chapter. All mistakes, misinterpretations, or omissions are ours. We also wish to thank the organizers of the symposium, especially D. Falk and P. Olwell, for the opportunity to offer guidance in rare plant restoration efforts.

REFERENCES

Cade, T. J. 1983. Hybridization and gene exchange among birds in relation to conservation. In *Genetics and Conservation: A Reference for Managing Wild Animal and Plant Populations*, eds. C. M. Schonewald-Cox, S. M. Chambers, B. MacBryde, and W. L. Thomas. Menlo Park, Calif.: Benjamin-Cummings.

Dole, J. A., and M. Sun. 1992. Field and genetic survey of the endangered Butte County meadowfoam—*Limnanthes floccosa* subsp. *californica* (Limnanthaceae). *Cons. Bio.* 6: 549–559.

Erhlich, P. A. 1989. Attributes of invaders and the invading process: vertebrates. In *Biological Invasions: A Global Perspective*, eds. J. A. Drake and H. A. Mooney. New Delhi, India: Thomson Press.

Fiedler, P. L. 1991. *Mitigation-related transplantation, relocation, and reintroduction projects involving endangered, threatened, and rare plant species in California.* Unpublished final report submitted to the California Department of Fish and Game, Endangered Plant Program, Sacramento, Calif., June 14.

Gawler, S. C. 1983. *Furbish's lousewort* (Pedicularis furbishae S. Wats.) *in Maine and its relevance to the Critical Areas Program* (a revision of the 1976 report by Charles D. Richards). Planning Report no. 13. Augusta, Maine: State Planning Office, Maine Critical Areas Program.

Golden, M. L., and P. L. Fiedler. 1991. *Characterization of the habitat for Lilaeopsis masonii (Umbelliferae): A California state–listed rare plant species.* Unpublished final report submitted to the California Department of Fish and Game, Endangered Plant Program, Sacramento, Calif., June 3.

Kruckeberg, A. R., and D. Rabinowitz. 1985. Biological aspects of endemism in higher plants. *Ann. Rev. Ecol. Syst.* 16: 447–479.

Laven, R. D., R. B. Shaw, P. P. Douglas, and V. E. Diersing. 1991. Population structure of the recently rediscovered Hawaiian shrub *Tetramolopium arenarium* (Asteraceae). *Ann. Missouri Bot. Gard.* 78: 1073–1080.

Lesica, P., F. W. Allendorf, R. F. Leary, and D. E. Bilderback. 1988. Lack of genetic diversity within and among populations of an endangered plant *Howellia aquatilis*. *Cons. Bio.* 2: 275–282.

Menges. E. S. 1990. Population viability analysis for an endangered plant. *Cons. Bio.* 4: 52–62.

Nobel, P. 1989. Attributes of invaders and the invading process. In *Biological Invasions: A Global Perspective*, eds. J. A. Drake and H. A. Mooney. New Delhi, India: Thomson Press.

Primack, P. B., and S. L. Miao. 1992. Dispersal can limit local plant distribution. *Cons. Bio.* 6: 513–519.

Reiseberg, L. H. 1991. Hybridization in rare plants: insights from case studies in *Cercocarpus* and *Helianthus.* In *Genetics and Conservation of Rare Plants,* eds. D. A. Falk and K. E. Holsinger. New York: Oxford University Press.

Reiseberg, L. H., S. Zona, L. Aberbom, and T. D. Martin. 1989. Hybridization in the island endemic, Catalina Mahogany. *Conservation Biology* 3: 52–58.

Richards, C. D. 1980. Report on monitoring populations of Furbish's lousewort *Pedicularis furbishiae* along the St. John River in northern Maine and New Brunswick during the summer of 1980. Newton Corner, Mass.: USFWS, Region 5.

Shevock, J. R., B. Ertter, and D. W. Taylor. 1992. *Neviusia cliftonii* (Rosaceae: Kerrieae), an intriguing new relict species from California. *Novon* 2: 285–289.

Stirrett, G. M. 1980. The status of Furbish's lousewort, *Pedicularis furbishiae* S. Wats. in Canada and the United States. Ottawa, Ont.: Canadian Wildlife Service.

U.S. Fish and Wildlife Service. 1978. Determination that various plant taxa are endangered or threatened species. *Federal Register* 43 (81): 17910–17916.

Walters, C. 1986. *Adaptive management of renewable resources.* New York: Macmillan.

WESTEC Services. 1988. *Biological mitigation plan for the College Area Specific Plan, San Marcos, San Diego County, California.* Unpublished consultant's report, prepared April 29.

CHAPTER 8

Designing Populations: Demographic, Genetic, and Horticultural Dimensions

Edward O. Guerrant, Jr.

The goal of reintroduction is to establish resilient, self-sustaining populations that retain the genetic resources necessary to undergo adaptive evolutionary change. Choosing species most likely to benefit most from reintroduction is not an exclusively scientific enterprise. Rather, such choices are also based on a host of philosophical, political, strategic and practical reasons (see White, Chapter Three). Once the decision to reintroduce a species has been made, it is important to define the desired end product. Arguably, the best models to emulate are healthy, naturally occurring populations of the same species (see Pavlik, Chapter Six). However, to recreate the rich and subtle tapestry of genetic and demographic structure found in healthy populations, not to mention the ecological milieu within which each exists, would be a daunting challenge indeed. Fortunately, the task of designing populations for reintroduction may, at least in principle, be somewhat easier. I propose that a well-designed founding population has, by virtue of its initial composition, placement, and care, a high probability of surviving and growing rapidly. Thus, for the practice of reintroduction, there is an important distinction to be made between the desired ends and the means by which they are pursued.

To design a population for reintroduction is to evaluate the array of potential benefits and risks associated with different situations and courses of action (see Burgman, Ferson, and Akçakaya 1993). The strategy offered here aims to maximize the probability of long-term success by designing founding populations that, by their composition and care, will minimize extinction risk and maximize population growth rate in the short term. The ideal founding population may not be practical to deploy, so, in the real world, founding-population design will probably often involve compromises within limits set by practical considerations. Regardless of their founding size or composition, artificially initiated populations that grow should develop a complex demographic structure of their own over time. Hence, even though many aspects

of a population's structure are indicative of its health and can serve as useful measures of success (see Pavlik, Chapter Six), they may not have to be built in from inception. Rather than use the metaphor of a *blueprint* of a finished product as our guide—where we ourselves assemble all the parts in their desired configuration—it may be both more effective and easier to view reintroduction in terms of a recipe. We can thus view the task of population design as the strategic assembly and deployment of appropriate ingredients in a suitable environment (which includes active care giving), in order to achieve an acceptably high probability of fostering the dynamic process of population growth.

Attempts to understand the myriad implications of small population size have been central to the development of modern ecological and evolutionary thought. The effects of small population size on genetic diversity are especially well understood theoretically (see Wright 1969). It is hardly surprising, therefore, that addressing the corrosive effects of small population size on genetic diversity was one of the first major problems to be considered seriously by the discipline of conservation biology (Soulé and Wilcox 1980; Shaffer 1981; Schoenwald-Cox et al. 1983). The larger body of evolutionary and ecological theory, much of it supported by a solid base of empirical information, has been organized for the purposes of conservation biology into the concepts of the Minimum Viable Population (MVP) (Shaffer 1981, 1987; Nunney and Campbell 1993) and its derivative, Population Vulnerability/Viability Analysis (PVA) (Gilpin and Soulé 1986; Boyce 1992; Menges 1991, 1992), the process whereby the MVP can be estimated. Although these approaches are couched in seemingly precise, predictable, lawlike, genetic terms, they must be applied in the more mercurial arena of actual demographic performance, which is contingent on a host of unpredictable environmental inputs. Even though genetic factors are clearly important to the long-term success of populations, demographic factors are increasingly seen as being of more immediate concern (Lande 1988; Menges 1991, 1992; Caro and Laurenson 1994; Schemske et al. 1994). Without a demographically self-sustaining population, questions of genetic diversity are moot.

I begin this chapter with an examination, by means of computer simulation, of the effects on extinction risk and population growth rate of a single variable that is relatively easy to manipulate in the field—the size/stage distribution of individuals in the founding population. By *size/stage distribution*, I mean the mix of size and reproductive status of individuals composing the founding population. The genetic composition of a founder population and the maintenance of genetic diversity through time are important to a new population's success. There are many ways that genetic diversity in a newly founded population can be maintained, including equalizing founder representation, equalizing family size, and controlling immigration. Theory and

simulation models can inform us of the potential opportunitites and risks for founder populations having different size/stage compositions. But reintroduction/outplanting in practice will be influenced by the quantity, quality, and genetic characteristics of the available stock and the horticultural options available for each taxon. Other factors to consider in the design of founder populations are how large each founding population should be; whether single or mixed sources of founders should be used; and how many populations should be deployed. The selection of appropriate sites for outplanting is, of course, of paramount importance in influencing the probability of success. Site selection is a complex topic of its own, which is discussed in detail by Fiedler and Laven (Chapter Seven).

Demographic Models

To explore some of the potential effects that the size/stage distribution of individuals composing the founding population may have on extinction risk and population growth rate, I used empirically derived stage-based transition matrices from the published literature as a basis for stochastic modeling. Using this approach, inspired by Menges (1991, 1992), I looked at one example each across a range of life histories: an herbaceous iteroparous (polycarpic) perennial, *Calochortus pulchellus* (Fiedler 1987); a woody iteroparous perennial, *Astrocaryum mexicanum* (Piñero, Martinez-Ramos, and Sarukhán 1984); a semelparous (monocarpic) perennial, *Dipsacus sylvestris* (Werner and Caswell 1977; Caswell 1989), and two contrasting annuals (Table 8-1).

To model potential effects of founding population size/stage distribution on extinction risk and population growth rate, I used a flexible, commercially available program, RAMAS/stage (Ferson 1990). Briefly, I used empirically derived transition probabilities from a single matrix of each example (parts 1–3), as the mean values on which to base stochastic modeling. For the annual life history, (part 4), two contrasting examples were constructed for purposes of illustration. Each transition matrix summarizes the fates of individuals over a period of one year. For example, in *Calochortus pulchellus* (part 1), 85 percent of the individuals in the size class II in the first year were again size class II individuals the next year, while 11 percent had become large enough to enter the next larger size, class III, and 1 percent decreased in size to size class I. Together, these account for 97 percent of the individuals seen in the first year, the remaining 3 percent having presumably died. Projections of future population growth based on a single transition necessarily assume that each and every year will have the same transition probabilities. Such projections are referred to as *deterministic*. Deterministic projections yield a single, precise value for population sizes in each of the coming years. Clearly,

TABLE 8-1.   Matrices used in modeling and describing size or stage classes used.

PART 1: *Calochortus pulchellus*. Data from Fiedler (1987), Tables 2 and 5, 1982–1983 transition.

Size/Stage Classes: Size classes are defined by maximum width of basal leaf in cm (Fiedler Table 2). The only size class that did not have any reproductive individuals was the smallest.

I	< 0.10–0.20
II	0.25–1.30
III	1.35–2.0
IV	> 2.05

Base matrix: λ = 1.073

			1982		
		I	II	III	IV
1	I	0.50	0.01	1.99	7.03
9	II	0.07	0.85	0.02	0.12
8	III	0	0.11	0.89	0.04
3	IV	0	0	0.07	0.37

PART 2: *Astrocaryum mexicanum*. Data from Piñero, Martinez-Ramos, and Sarukhán (1984), Plot C, modified slightly from published form, values represent averages over a six-year period.

Size/Stage Classes:

F	Fruits
I	Infants
J	Juveniles
IMi	Immature individuals
Mi	Sexually mature individuals

Base Matrix: λ = 1.018

	F	I	J	IMa	IMb	IMc	Ma	Mb	Mc	Md	Me	Mf	Mg	Mh
F	0	0	0	0	0	0	7.48	16.00	7.64	7.77	16.00	15.00	20.00	27.81
I	0.023	0.888	0	0	0	0	0	0	0	0	0	0	0	0
J	0	0.034	0.929	0	0	0	0	0	0	0	0	0	0	0
IMa	0	0	0.051	0.975	0	0	0	0	0	0	0	0	0	0
IMb	0	0	0	0.025	0.978	0	0	0	0	0	0	0	0	0
IMc	0	0	0	0	0.022	0.799	0	0	0	0	0	0	0	0
Ma	0	0	0	0	0	0.201	0.852	0	0	0	0	0	0	0
Mb	0	0	0	0	0	0	0.148	0.881	0	0	0	0	0	0
Mc	0	0	0	0	0	0	0	0.119	0.857	0	0	0	0	0
Md	0	0	0	0	0	0	0	0	0.143	0.733	0	0	0	0
Me	0	0	0	0	0	0	0	0	0	0.267	0.697	0	0	0
Mf	0	0	0	0	0	0	0	0	0	0	0.303	0.889	0	0
Mg	0	0	0	0	0	0	0	0	0	0	0	0.111	0.615	0
Mh	0	0	0	0	0	0	0	0	0	0	0	0	0.385	0.999

TABLE 8-1. (*Continued*)

PART 3: *Dipsacus sylvestris*. Data from Werner and Caswell 1977, modified according to Caswell, 1989.

Size/Stage Classes:

S1	Dormant seeds, year 1
S2	Dormant seeds, year 2
SR	Small Rosettes
MR	Medium Rosettes
LR	Large Rosettes
FL	Flowering Plants

Base Matrix: $\lambda = 2.334$

	S1	S2	SR	MR	LR	FL
S1	0	0	0	0	0	322.28
S2	0.966	0	0	0	0	0
SR	0.013	0.010	0.125	0	0	3.448
MR	0.007	0	0.125	0.238	0	30.170
LR	0.008	0	0.038	0.245	0.167	0.862
FL	0	0	0	0.023	0.750	0

PART 4: Annuals

Stage Classes

P	Growing Plants
S1	Dormant Seed, one year old
S2	Dormant Seed, two years old
S3	Dormant Seed, three or more years old.

Base Matrices $\lambda = 1.060$

Low fecundity
High seedling survivorship

High fecundity
Low seedling survivorship

	P	S1	S2	S3		P	S1	S2	S3
P	0.96	0.64	0.64	0.64	P	0.96	0.064	0.064	0.064
S1	0.15	0	0	0	S1	1.5	0	0	0
S2	0	0.10	0	0	S2	0	0.10	0	0
S3	0	0	0.10	0.10	S3	0	0	0.10	0.10

Note: The P-to-P transitions refer to the plants that germinate from seed produced the previous year, as opposed to seed that has been dormant for one or more years.

the assumption that any two years will be the same as all others with respect to transition probabilities is absurd. Indeed, Menges (1991, 1992) and others have shown that environmentally induced year-to-year variation in transition probabilities, resulting from environmental stochasticity, is an extremely important factor influencing population fate. Arbitrary levels of environmental stochasticity were imposed on these simulation models by assigning each element in the matrix a given variance with a normal distribution around the published value, which was taken as the mean. For the sake of simplicity, the mean-to-variance ratio was held at a constant value of 10 within and among the matrices. For *Dipsacus*, a weedy colonizing species that had by far the highest population growth rate of all taxa studied, the value assigned was a less variable 30. The mean-to-variance ratios were chosen to generate enough but not too much variation in the results (that is, to avoid simulations resulting in either no extinction or universal extinction). As in Menges (1991, 1992), simulations were begun with two hundred individuals, and quasiextinction was considered to have occurred when the population dipped to ten individuals. The term *quasiextinction* refers to a population dropping below an arbitrary threshold (see Burgman, Ferson, and Akçakaya 1993). Ten replicate simulations were run of each size class of each example. Each simulation consisted of one hundred iterations over a period of one hundred years each (except for *Dipsacus*, which had one hundred iterations over a period of fifty years). Thus, each data point described here is based on a total of one thousand simulated populations, each modeled independently over a period of one hundred years (fifty for *Dipsacus*). Potential differences among means were examined by means of a Scheffe test using SYSTAT (Wilkinson 1990).

Two main aspects of the results are of interest here: the effects of founding population size/stage distribution on extinction risk and on population growth rate. I first describe simulation results of each life-history example and then throughout the rest of the chapter describe the implications for designing populations for reintroduction in the context of particular design questions.

Herbaceous, Iteroparous Perennial: *Calochortus pulchellus* (Fiedler 1987)

Populations founded with individuals of the smallest size class were at a significantly greater risk of extinction than were those of any of the other three size classes (which are statistically indistinguishable from one another: Figure 8-1, box A).

Significant population size differences among simulated populations founded with individuals of different size/stage classes had emerged within the first ten years of the hundred-year simulation (box B). Projected average population sizes at ten years paralleled the sizes of individuals used to found them: The largest founder individuals yielded the largest populations. Average popula-

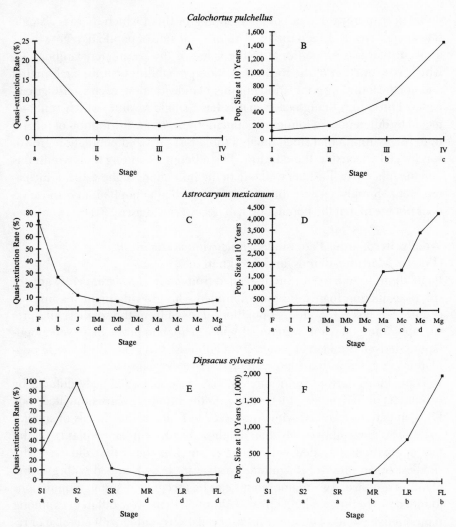

FIGURE 8-1. Quasi-extinction rates, expressed as an average percentage, and actual population sizes at ten years as a function of size/stage class for *Calochortus pulchellus* (boxes A and B), *Astrocaryum mexicanum* (boxes C and D), and *Dipsacus sylvestris* (boxes E and F). Letters below size/stage designations refer to results of Scheffe tests: Similar letters are not significantly different (even at $p < 0.05$); different letters differ at $p < 0.001$.

tion size decreased monotonically with decreasing size class of founder individuals, but the two smallest size classes resulted in populations that did not differ significantly from one another.

Here and elsewhere, the different population sizes at ten years into the simulation are used as an index of realized population growth rates that can be expected of the different situations modeled. These rates are distinct from

differences in intrinsic rates of population growth, of which there is a single rate characteristic of each matrix. The intrinsic rate of population growth for a given matrix is proportional to the value of the parameter, lambda (λ), which is a function of the mean transition probabilities within each matrix and not the founding population number of individuals in each size/stage category. Thus, even though each matrix has a single λ value, which is not affected by differences in founding population size/stage distribution or magnitude of environmental stochasticity (Table 8-1), realized population growth was indeed affected by these factors. The differences among trials within this and the other taxa illustrate well one of the limitations of the value of the parameter λ. Simply because the λ value for a particular population is above 1.0 does not mean that the population will necessarily increase in size over time.

Woody, Iteroparous Perennial: *Astrocaryum mexicanum* (Piñero, Martinez-Ramos, and Sarukhán 1984)

Populations founded with single-seeded fruits were at a significantly greater risk of extinction than were those founded with growing plants of any size (Figure 8-1, box C). So too, populations derived from the smallest and presumably youngest plants were much more extinction prone than were populations derived from larger plants. Extinction risk was generally lower in populations founded with the larger, reproductive size classes.

Populations derived from different size/stage classes already differed by two orders of magnitude at ten years into the hundred years simulated (box D). Populations founded with single-seeded fruits or individuals of smaller size classes grew more slowly than did those begun with larger plants. Of the few populations founded with seeds to survive, those that did were the smallest by an order of magnitude relative to those founded with growing plants of any size. Two major groups can be recognized among populations founded by seedlings or larger plants. Although there were differences among reproductive plants of different sizes, populations started with any size of reproductive plant grew larger than did those founded with any size of nonreproductive plant.

Semelparous Perennial: *Dipsacus sylvestris* (Werner and Caswell 1977; Caswell 1989)

Extinction risk at the end of the fifty-year simulation differed dramatically among populations founded with different propagule types (Figure 8-1, box E). Populations from founders that comprised either of the two seed stage classes were at a significantly greater risk of extinction than were those founded with any size/stage class of plant. Populations derived from two-year-old seed were less likely to survive than were those from year-old seed. Those from the smallest-size class of growing plant were at greater risk of extinction

than were populations derived from any of the other juvenile classes, which did not differ among themselves and were not significantly different than reproductive plants.

Again, at only ten years into the simulation the sizes of populations derived from founder plants of different size differed (box F), but in these, differed even more extremely than in *Astrocaryum* or *Calochortus*. Two groups are distinguished statistically, with the three largest-size classes of plants giving rise to larger populations than those of seeds or the smallest-plant-size class.

Annuals

Plants with an annual life history were modeled more abstractly, without reference to particular taxa, partly because the life history of annuals varies so tremendously. The parameters and values modeled were chosen to illustrate some of this variation and to fall within the naturally occurring range (see Symonides 1988). Both matrices used here had the same λ value (1.06, indicating a slightly growing population), but had contrasting patterns of fecundity and seedling survivorship.

HIGH FECUNDITY, LOW SEEDLING SURVIVORSHIP

Newly founded populations of annual plants that have high fecundity and low seedling survivorship were at much greater risk of extinction if started with (year-old dormant) seeds than if started with plants. In ten years, populations founded with plants grew to be almost two-and-one-half times their initial size of two hundred individuals, while those founded with seed were only about one-sixth of their initial size (Table 8-2, part 1).

LOW FECUNDITY, HIGH SEEDLING SURVIVORSHIP

Simulations of populations of annuals that have low fecundity but high seedling survivorship did not show any difference in extinction risks between those started with either seeds or plants (Table 8-2, part 2). The size of reintroduced/outplanted populations started with plants were, however, significantly larger than of those with seeds, with only those founded by plants exceeding at ten years after initiation the initial cohort size of two hundred individuals.

Demography and the Maintenance of Genetic Diversity

Reintroduction is not simply a question of relatively short-term demographic performance: Long-term success depends on perpetuating an appropriate amount and pattern of genetic diversity. In all but the most favorable cases, reintroduction efforts will probably be based on relatively small numbers of

TABLE 8-2   Average rate of quasi-extinction over the hundred-year duration simulated and average population size at ten years into the simulation for annuals.

PART 1: annual with high fecundity, low seedling survivorship.

Stage	Quasi-extinction rate (%)	Pop. size at 10 years
Plant	17.6 a*	496 a*
Seed (S1)	61.8 b	35 b

PART 2: annual with low fedundity, high seedling survivorship.

Stage	Quasi-extinction rate (%)	Pop. size at 10 years
Plant	20.1 a*	221 a*
Seed (S1)	22.4 a	153 b

*Scheffe test: Similar letters are not significantly different; different letters differ at p < 0.001

plants that retain only a subset of a taxon's genetic diversity. Preserving as much as possible of the remaining genetic diversity is, therefore, an important design and management task.

Reintroduction efforts will probably often be associated with bottleneck events in the history of these taxa, either because they are being reintroduced as a consequence of a population crash or because the effective population size (N_e) drops during the reintroduction itself. Nei, Maruyama, and Chakraborty (1975) showed that population growth rate following a founder event is an important determinant of how much genetic diversity is ultimately lost in the bottleneck. Populations that grow rapidly after a dip in population size retain a remarkably large portion of their genetic diversity relative to the severe losses that can occur if subsequent population growth rate is low. Templeton (1990, 1991) discussed this phenomenon in the context of a specific example with Speke's gazelle (*Gazella spekei*). A major charge (and challenge) for designing populations for reintroduction is therefore to create new populations that will increase in size as rapidly as possible. Rapid population growth will serve not only to buffer newly founded populations from extirpation due to demographic and environmental stochasticity but also tend to ameliorate the corrosive effects of bottlenecks on genetic diversity. However, as the earlier demographic simulations showed, it may be much easier to avoid extinction than to promote rapid population growth.

Implications of the Simulation Models
These simulations suggest founding-population stage distribution is an important determinant of extinction risk and realized population growth rate

following reintroduction. The challenge is using this information within the broader framework of theory and practical experience to better design populations for reintroduction.

The simulations generally showed that extinction risk is greatest when seeds or the smallest plants were used as founders, and that extinction risk dropped dramatically when even slightly larger size classes were used. This is a very fortuitous outcome from a practical point of view. Relative to the most extinction-prone stage (seeds or the smallest size class of plants), the biggest reduction in extinction risk was obtained in the simulations by planting just incrementally larger size classes of plants. In other words, if the goal is simply persistence, relatively small founder plants will do. Indeed, it appears that not much more, if any, extinction risk can be avoided by transplanting even the largest individuals.

Unfortunately, the demographic simulations suggest population growth rate will not behave in quite the same way. The marginal increases in realized population growth rate during the initial ten-year period, as a function of using larger founder individuals, were more gradual than were the precipitous decreases in extinction risk. This more sobering outcome of the simulations is well illustrated by *Calochortus pulchellus* in which, relative to the smallest size class, planting the next-to-smallest size class achieved an extinction risk indistinguishable from what would be obtained if the largest reproductive plants were used as founders. In contrast, the average population sizes founded with individuals of the two smallest size classes did not differ significantly. To achieve both low extinction risk and high population growth rate required the use of the larger size classes—a task that will in most cases presumably require more time and resources to achieve.

Maintenance of Genetic Diversity
Given limited biological (and financial) resources with which to work, I can envision many projects that may result in relatively small populations that persist but do not grow rapidly in size. In such cases, maintenance of genetic diversity will be a particularly serious concern. The consequences of low population growth rate underscore how reintroduction generally will not be a simple Johnny Appleseed affair—a matter of planting a founder population, walking away, and letting nature take its course. Reintroduction should be a long-term commitment, and long-term success may well require active management for years after outplanting, as indicated by a comprehensive monitoring program (which itself should be integral to any well-designed reintroduction effort [see Sutter, Chapter Ten]).

Fortunately, a variety of measures can be used at the time of outplanting and during subsequent generations to help ensure that the greatest proportion of

genetic diversity possible will persist. Many of these approaches are encapsu-
lated in a recent series of important papers involving *Drosophila
melanogaster*, by Frankham and colleagues, (Briscoe et al. 1992; Frankham
and Loebel 1992; Loebel et al. 1992; and Spielman and Frankham 1992 Borlase et al. 1993;). They empirically examined a number of theoretical pillars
of population genetics that bear on the maintenance of genetic diversity in
captive populations. Much of what they found is relevant to the design of
founder populations as well. Specifically, they looked at the effects of equalizing the contributions of founder individuals and equalizing family size derived from each wild caught individual. They also considered the effects of
immigration on the genetic diversity of captive populations. In a related
paper, Frankham and Loebel (1992) examined adaptation of populations to
captivity, which also has relevance to reintroduction. These measures, which
differ in their relevance and difficulty to use in reintroduction, will be described in turn. Finally, there appears to be a potential conflict between
maintaining the maximum genetic diversity and the evolutionary adaptation
of populations to local conditions.

Equalize Founder Representation

The beneficial effects of equalizing founder representation have been appreciated in theory for some time (see Haig, Ballou, and Derrickson 1990;
Loebel et al. 1992). Loebel and colleagues empirically examined the genetic
consequences of equalizing founder representation in captive *Drosophila*
populations over a span of nine generations. Relative to control lines, in
which founders were chosen randomly, they found that equalizing founder
representation resulted in lowered inbreeding and higher genetic variation,
as measured by degree of polymorphism, average heterozygosity, and number
of alleles. Haig, Ballou, and Derrickson (1990) reached similar conclusions
using computer simulation studies comparing several possible breeding
strategies that might be used in introducing Guam rails (*Rallus owstonii*) to
the island of Rota in the South Pacific. Templeton (1991) pointed out that to
obtain the benefits of equalizing founder representation, it must be pursued
right from the beginning, because the contributions of individual founders
can rapidly become irrecoverably lost, especially in small populations. Equalizing founder representation in plants at the time of reintroduction is relatively simple if the original wild collections of seed or other propagules from
each plant have been kept separate (Guerrant 1992); if not, then equalization
it is difficult to impossible.

Loebel and colleagues (1992) suggested even greater benefits than they actually achieved should be possible in a situation with overlapping generations, as opposed to the discrete generations they used experimentally. In
some plants, generation time may be artificially lengthened by horticultural

techniques, such as taking repeated cuttings over time to rejuvenate older, larger individuals, or by employing the use of an *ex situ* seed bank to store seeds or pollen for later use.

Loebel and colleagues (1992) speculated that even slightly greater benefits than they found empirically may be possible to achieve if matings in the captive/founding population were arranged strategically. Choosing the appropriate parents is neither a simple task, nor one without risk of reducing population growth rate or imposing artificial selection. Haig, Ballou, and Derrickson (1990) used computer simulation to compare six different breeding strategies that could be employed with the introduction of Guam rails. They compared the following six strategies, with respect to the maintenance of genetic diversity:

1. Random breeding among adults

2. Selectively breeding the most fecund adults

3. Selectively breeding to produce the most genetically diverse offspring, using allozyme data to choose breeding stock

4. Selectively breeding to equalize founder contribution

5. Selecting to maximize allelic diversity

6. Selectively breeding to maximize founder genome equivalents

They found that options 4, 5, and 6 would produce substantially more genetically diverse stock for reintroduction than would options 1, 2, and 3, with 6 the most diverse overall. Although these measures might be difficult and require additional resources to employ, even marginal benefits might be worthwhile in extreme situations where species have been reduced to a very few surviving individuals, as is sadly the case with many plants in Hawaii (see Mehrhoff, Chapter Five) and presumably elsewhere. Tonkyn (1993) recently outlined how mathematical optimization techniques can be used to address a variety of problems in genetic management of endangered species, including the choice of mates that will best equalize founder representation.

EQUALIZE FAMILY SIZE

As with founder representation, the beneficial effects of equalizing family size on maintaining genetic diversity in captive populations have been appreciated in theory for some time (see Franklin 1980). The concept of family size in this context includes the number and sex of offspring per parent that become parents themselves. Borlase et al. (1993) empirically examined the notion that once a captive population has been launched, equalizing family size of the founders should approximately double the effective population

size (N_e), reduce inbreeding, and retard the loss of genetic variation, relative to a population of breeding adults (N) in which mating is random. Although the benefits were striking—mean inbreeding coefficient (F) increased more slowly than in random mating; average heterozygosity and reproductive fitness were higher; and N_e was higher—equalizing family size is technically much more difficult to achieve than is equalizing founder representation at the time of outplanting. It may also have serious drawbacks for demographics and for the prospects of genetic adaptation to new circumstances. For example, although the number of seeds per plant can be counted or estimated, there is no easy way to estimate paternal reproductive contribution (Ellstrand 1984; Marshall and Ellstrand 1985; Meagher 1986; Schoen and Stewart 1986, 1987). Furthermore, removing seed production of the more fecund families to match the less fecund members in an attempt to equalize family representation would clearly retard population growth rate, which could be lethally counterproductive. One relatively easy way to approach the goal of equalizing family size might be to artificially pollinate members of larger, more fecund families with those of smaller, less fecund ones, or by employing a more sophisticated approach where breeding was controlled artificially (but see Haig, Ballou, and Derrickson 1990; Tonkyn 1993). A potential disadvantage of any approach using controlled breeding would be to risk imposing artificial selection for properties unknown and therefore erode the average absolute fitness of the population. Alternatively, such an active approach might merely relax selection and result in a greater number of alleles persisting for longer than would otherwise be the case. Thus, in the attempt to equalize family size, we begin to see possible tradeoffs between efforts to maintain the maximum amount of genetic diversity possible and allowing adaptation by means of natural selection to proceed unimpeded.

IMMIGRATION

In a commentary on the series of papers by Frankham and colleagues, Ralls and Meadows (1993, p. 690) note that "even 'properly managed' populations of captive *Drosophila* lost 74 per cent of their reproductive fitness after 11 generations and had lower genetic diversity than large wild populations." Spielman and Frankham (1992) showed that decreased reproductive fitness can be counteracted by immigration. They showed that immigration, even at a rate of a single immigrant into a small, isolated, partly inbred population (a characterization that may well describe many reintroduced populations) significantly alleviated inbreeding depression. They were able to eliminate F_1 hybrid vigor as an explanation, because experimental populations went through three generations of random mating before fitness tests were conducted.

To use immigration as a management tool assumes knowledge of what

constitutes a population in nature. The geographical extent and limits of populations in nature are, however, notoriously difficult to establish. We are often influenced most by what our senses can detect, and it is much easier to see established plants than dispersing seeds or pollen. Consequently, the definition of *population* offered by Ellstrand (1992a, p. 77) — "spatially discrete groups of conspecific individuals" — is probably close to how most of us think of populations, at least operationally. Implicit in this view is that gene flow is negligible. In a review of the implications for plant conservation of gene flow by pollen, Ellstrand characterized three views of gene flow in plants: (1) gene flow is highly restricted (see Levin 1981); this is a common view among plant evolutionists; (2) gene flow is extensive (see Govindaraju 1988); Ellstrand suggests that this is a common view among forest geneticists; (3) "gene flow in plants is idiosyncratic, ranging from very low to very high, and varying among species, populations, individual plants, and even over a season" (Ellstrand 1992a, p. 78). Ellstrand argues that existing estimates favor this third, relatively new view.

The potential impacts of gene flow, intentional or not, on newly established populations are as great as they are diverse. The magnitude of gene flow is determined by many factors, including the size, distance, and modes of dispersal of potential sources of seed and pollen. From a statistical point of view, smaller populations are relatively more affected by any given absolute amount of gene flow than are large ones (Ellstrand 1992a). Small, newly founded populations may thus be seriously at risk from genetic contamination, especially by pollen from intercompatible congeners (see Rieseberg et al. 1989). Conversely, controlled immigration, if done properly, could become a standard prescription to ameliorate the corrosive effects of random genetic drift and to supply additional genetic diversity to fledgling populations.

GENETIC DIVERSITY VERSUS NATURAL SELECTION

Efforts to maintain the maximum amount of genetic diversity in a reintroduced population may conflict with a population's ability to adapt evolutionarily to local conditions. Frankham and Loebel (1992) demonstrated empirically that captive populations can rapidly adapt genetically to captivity and offered a conceptual framework for predicting the rate of genetic adaptation to captivity that may find use in reintroduction. Adaptation to captivity is not in a taxon's best interest if it is to be released back into the wild. Nevertheless, assuming that a source of founders genetically representative of naturally occurring populations is available from, for example, an *ex situ* dormant seed collection (see Center for Plant Conservation 1991; Guerrant 1992), then rapid adaptation to new conditions may not be undesirable for the long-term prospects of a reintroduced population. However, because the initial year-to-year environmental variance experienced by the newly founded population

may not be representative of long-term conditions, any benefits of rapid adaptation in a small, newly founded population could well be short lived.

There appear to be several ways in which the short-term goal of maintaining maximum genetic diversity may conflict with the ability of outplanted populations to adapt evolutionarily to their new circumstances. Consistent with the findings of Borlase et al. (1993), Frankham and Loebel (1992) and Allendorf (1993) found that equalizing family size reduced the rate of adaptation to captivity. Introducing genes from wild populations, increasing generation interval (such as by seed or pollen storage), and reducing mortality rates (by cultivating plants) would all be expected to slow the pace of evolutionary adaptation. Thus active management to maintain maximum genetic diversity may ironically be at odds with rapid adaptation to local conditions, a situation that clearly deserves closer examination.

Properties of Founding Populations

I move now from theory and long-term considerations of the maintenance of genetic diversity to some initial concrete steps involved with designing the founding population.

Horticultural Options

There are, in theory, a wide variety of horticultural options from which to choose in a reintroduction program:

Seeds
Transplants
  Whole Plants
    Without associated soil
    With associated soil
  Plant Parts
    Cuttings
    Specialized stems and roots (bulbils, rhizomes, and so on)
    Grafting and budding
    In vitro micropropagation

In practice, however, only one or a few options might be realistic in any particular case. The basic choice is between sowing seeds or transplanting growing plants. Additional choices are embedded within each of the options presented in the list. For example, seeds can be either wild collected (and used fresh or stored in an *ex situ* seed bank) or propagated off-site for one or more generations. Transplants too can be either wild collected or propagated off-site. There are advantages and disadvantages associated with the use of

seeds or plants in a founding population and with the use of wild or propagated stock. The appropriate choices are strongly context dependent.

SEEDS AS FOUNDERS

Seeds have been used as founders in many projects with plants having a range of life histories, but they seem to be particularly favored for annuals. In their work with *Amsinckia grandiflora*, Pavlik, Nickrent, and Howald (1993) placed the single seeded nutlets (referred to hereafter simply as seeds) individually in a set pattern with the aid of a template, so that the fates of individual seeds could be monitored demographically. Pavlik and Espeland (1991) and Pavlik, Espeland, and Wittman (1992) also used wild collected seeds in a reintroduction of the serpentine endemic annual *Acanthomintha duttonii*. Wallace (1990) offers a cautionary tale of what we might expect with a preliminary reintroduction attempt of *Warea amplexifolia* in Florida that involved direct seeding into an apparently appropriate site at a favorable time shortly after a burn: two thousand seeds yielded six hundred seedlings, of which sixteen reproduced.

There are benefits and costs associated with the use either of wild or propagated stock. For example, in an especially well-designed and executed reintroduction project of the California grassland annual *Amsinckia grandiflora* at several sites, Pavlik (1990, 1991a, 1991b, 1992) and Pavlik, Nickrent, and Howald (1993) used seeds as founders. Founders included both seeds that had been collected directly from wild populations by R. Ornduff in the 1960s (Pavlik 1990) and seeds derived from the same original source but were a couple of generations removed from the wild. Although no genetic studies had been done on the wild populations, electrophoretic analyses of a small sample from each of the two 'sources' suggested that genetic variation had decreased after only two generations of propagation (Pavlik, Nickrent, and Howald 1993). A different measure of the potential biological costs of *ex situ* propagation is illustrated in another California annual, *Acanthomintha ilicifolia*, in which Mistretta and Burkhart (1990) showed a decrease in germination percentage from 95 percent to 45 percent in just one generation. They did, however, increase their seed supply from approximately 2 grams of wild collected seed to over 700 grams after one generation of *ex situ* propagation. But at what genetic cost?

Wild collected stock presumably has a better chance of being genetically representative of healthy, naturally occurring populations than does stock that has been propagated off-site for one or more generations. But there is a cost to the donor population to be considered each time seeds, plants, or plant parts are removed from wild populations. Insofar as the impact of seed collection is adequately modeled by an increase in environmental stochasticity in reproductive output, and the impact of collecting plants or plant parts is

modeled by increased environmental stochasticity in plant growth rate and mortality, then Menges's (1991, 1992) results indicate that collecting seeds will have much less impact on population growth rate than removing whole plants or plant parts. Large quantities of propagated stock can be produced off-site, but the selective pressures on plants grown off-site are presumably different than those experienced by wild populations. Other possible problems associated with growing plants off-site is the chance of introducing new pathogens to wild populations when they are outplanted (Maunder 1992; Falk and Olwell 1992) and of introducing foreign conspecific or congeneric genes as a result of unintended cross-pollination. Clearly, more work is needed before we can confidently evaluate the genetic costs and benefits of different strategies.

Seeds can be planted in a variety of ways, from simple hand broadcasting to manually or mechanically 'drilling' them into the ground (Bainbridge and Virginia 1990). Packard (1991) broadcast large numbers of seed from six artificially pollinated prairie fringed orchid plants (*Platanthera leucophaea*) into three disturbed prairie remnants in Illinois, with some success. Using this same broadcast technique with other small-seeded prairie plants, Packard (1991) had good results with four *Gentiana* species and two *Pedicularis* species but poor results with *Heuchera richardsonii* and *Potentilla arguta*.

Youtie (1992) used either seeds, transplants of rooted cuttings, or both of a large number of grasses, shrubs, and forbs in a restoration project along the Columbia River in Oregon. As part of a larger effort to restore disturbed high-elevation sites in Mt. Rainier National Park in Washington state, Rochefort and Gibbons (1992) used both seeds and transplants of many different species and provided much information of interest to this discussion of horticultural options.

TRANSPLANTS AS FOUNDERS

Transplantation is controversial, especially in a mitigation context. The notion of simply plucking up and moving offending plants to a more convenient location to make way for progress is as repugnant to some as it is appealing to others. Much of the controversy seems to center on a particular kind of transplantation: translocation of naturally occurring plants (Hall 1987; Fahselt 1988; Fiedler 1991; Gordon 1994). In this chapter, *transplantation* is used more broadly to refer to the founding of a new population with any propagule other than seeds. Transplantation can be done with or without associated soil; it can be done with whole plants, with material derived from parts of plants, such as shoot or root cuttings, or from roots or shoots that are morphologically specialized for asexual reproduction (such as bulbils, rhizomes, tubers, and so on, as well as other less specialized root-bearing shoots or shoot-bearing roots). As used here, *transplantation* can also include grafting plant parts onto

other plants. Perhaps the most esoteric origin of plant material to be transplanted comes from a variety of in vitro micropropagation procedures. Again, like seeds, material for transplant may be derived from plants that were either wild collected or have been propagated off-site for one or more generations.

Hall (1987) and Fiedler (1991) both noted the importance of site preparation and postestablishment care in their reviews of translocation projects. Watering (for years if necessary), mulching, shade cloth, grazing protection, insecticide, herbicide, and hand weeding have all been used. The object of reintroduction is to launch a self-sustaining entity, and, like child rearing, it may be a long, arduous process. Gordon (1994) has recently attempted to clarify the decision-making process surrounding the choice of where transplantation is and is not appropriate.

Whole Plants. Perhaps the most commonly practiced form of reintroduction involves germinating wild collected seed off-site and transplanting seedlings or older plants into the field. The seed can either be freshly collected for the purpose or taken from storage in an *ex situ* seed bank. Examples include experimental reintroductions of rare annual, biennial, and perennial plants by Sainz-Ollero and Hernandez-Bermejo (1979) in Spain, using seed that had been stored in their seed bank. Other examples include annuals (*Stephanomeria malheurensis*, Parenti and Guerrant 1990), semelparous (monocarpic) perennials (*Erysimum menziesii*, Ferreira and Smith 1987; *Circium pitcheri*, Bowles et al. 1993), and iteroparous (polycarpic) perennials (*Styrax texana*, Cox 1990; *Polemonium vanbruntiae*, Popp 1990, 1991; *Astragalus tennessensis*, Bowles et al. 1988, Bowles and DeMauro 1992; *Iris lacustris*, Simonich and Morgan 1990).

Translocation of living plants from nature is certainly one of the more contentious options available, in part because of its common use in a mitigation context (Hall 1987; Fahselt 1988; Fiedler 1991; Berg, Chapter Twelve; Howald, Chapter Thirteen). Translocation of living plants may offer one of those ironic situations in which the operation can be successful even if the patient dies. For example, Brumback and Fyler (1987) moved about 150 individuals of the terrestrial orchid *Isotria medeoloides* in New Hampshire out of harm's way to avoid certain destruction by development in 1986. Annual monitoring has shown consistently fewer individuals to have reappeared each year than were there the year before, and by 1992, fewer than 10 percent of them were still putting up shoots (Brumback, personal communication). This shows that transplanting living individuals of this species is not likely to be an effective conservation measure for this species, and that *in situ* efforts are imperative. This is useful (if not particularly encouraging) information. On a slightly more optimistic note, Ecker (1990) salvaged a number of plants of the rare cactus *Mammillaria thornberi* from a construction right-of-way in

Arizona before their habitat was developed. She used some of this material experimentally to test a number of hypotheses about how best to transplant it: planting cactus under nurse plants, especially creosotebush (*Larrea tridentata*) proved to be most successful. Her work exemplified how, with good experimental design, a series of explicit hypotheses can be evaluated, and useful information can come from the destruction of native habitat.

In some cases, transplantation has included not only the plants but also the surrounding soil in which they were growing. Examples include using a truck-mounted tree spade to move plugs of ground supporting dense stands of the annual tarweed *Holocarpha macradenia* away from land being developed in California (Havlik 1987; but see Howald, Chapter Thirteen for more recent data on this project). In another mitigation-related project, Stephenson (1992) used a front-end loader to move large chunks of sod containing the rhizomes of regionally rare horsetails (*Equisetum* spp.) in Ontario, Canada. Considered a success after two years (Stephenson 1992), the project was considered a failure after five, with the recommendation that more time be spent selecting or constructing a suitable site for transplantation (Hurkmans 1995). In an attempt to revegetate high-elevation gravel areas in Sequoia National Park, California, Ratliff and Westfall (1992) found direct planting (as opposed to planting in protective pots that would decompose *in situ*) of large diameter sod plugs (approximately 5 centimeters in diameter) containing unfertilized *Carex exserta* gave the best results among a series of conditions that were tested. In a comparison involving giant cane (*Arundinaria gigantea*) in Kentucky, Feeback (1992) reported very high survivorship when rhizome systems with surrounding soil were moved and complete mortality of those in which the soil and some roots were removed.

Plant Parts: Cuttings.   Horticulturists have long appreciated that parts of some plants can be removed and stimulated to produce roots and thus become capable of growing independently of the parent. Cuttings are a perhaps underutilized way in which new ramets can be generated for transplantation. However, taking cuttings may reduce the fitness of donor plants by reducing their fecundity or lowering their survivorship directly or indirectly (such as introducing pathogens). The only known population of *Pediocactus knowltonii* was used as a source of cuttings to establish a second population at a more well-protected site nearby (Olwell et al. 1987; Olwell, Cully, and Knight 1990). Other examples include the opportunistic experiment with *Penstemon barrettiae* near the Bonneville Dam, along the Columbia River in Oregon (Guerrant 1990). In this case, J. Kierstead (Nelson) took multiple cuttings from a large number of plants growing on a rock face that was slated for imminent destruction to make way for a new navigation lock. She took them

back to the Berry Botanic Garden, where they were rooted. Multiple ramets of each genet were transplanted back to another hillside near where they had been collected and have been monitored since. Backup cuttings have also been maintained at the garden. (Ironically, after six years, more clones have survived in the experimental outplanting than at the garden.) In another case involving *Conradina glabra* in Florida, Wallace (1992) was able to find material from which to take cuttings in the jumble following land-clearing activity. These have been propagated horticulturally at the Bok Tower Garden in Florida and reintroduced to suitable habitat. Plants from cuttings live only about three years, and the collection of growing plants at Bok Tower have been started over twice from cuttings of their plants. In Australia, Jusaitis (1991) reports on the transplantation of two species of the rutaceous shrub *Phebalium*, in which cuttings were used.

When cuttings are possible, multiple replicates of genetically identical populations can be placed into several environments simultaneously. In this way, the plants themselves can be used as "phytometers" to discriminate among possible reintroduction sites (Fowler and Antonovics 1981; Antonovics and Primack 1982). The experimental possibilities inherent in being able to put out multiple, genetically identical, replicate populations are enormous, as are the conservation rewards. But not all plants can always produce viable cuttings. As part of a mitigation project that involved translocating the Santa Susana tarplant (*Hemizonia minthornii*), Mangione and Vander Pluym (1993) translocated fifty-five plants, from which they obtained 1,430 cuttings: Only three took root.

Plants Parts: Specialized Stems and Roots.   Many plants have specialized stems and roots that would seem to facilitate their use as a "naturally" produced source of transplant material. Surprisingly, I found only one reference where the natural ability of some plants to reproduce asexually was consciously exploited to provide material for outplanting (Taylor 1991). Corms and seeds of thread-leaf brodiaea (*Brodiaea filifolia*) were gathered as part of a mitigation project associated with the destruction of a natural population to make way for a residential development in California. Attempts to establish plants from seed were entirely unsuccessful the first two years, and success the third year was described as "near zero." Establishment from hand-planted corms was more successful.

Plant Parts: Grafting.   Grafting, especially onto root stock of another species, would seem a desperate and poor way of conserving a rare species, but the situation is so bleak for at least one plant in Hawaii that this was the best available option. Wooliams and Gerum (1992) describe the saga of *Kokia cookei* in Hawaii, in which a variety of techniques have been used to perpetuate the last

remaining genetic individual: It now exists only as grafted scions on stock of
K. kauaiensis and *K. dryarioides*.

Micropropagation. In vitro propagation, or micropropagation, has been
used by many (Cervelli 1987; Ferguson and Pavlik 1990; Loope and Medeiros
forthcoming; Martínez-Vásquez and Rubulo 1989; McComb 1985; Rubulo,
Chávez, and Martinez 1989; Rubulo et al, 1993; Stewart 1993), and heralded
by some (Wochok 1981) as a potential boon to conservation. One perceived
advantage of these techniques is that they can generate large numbers of
plants from limited material. A disadvantage is that they are all derived mitot-
ically from a single individual, so they are nearly identical genetically. Nearly
identical, because the techniques apparently produce genetic damage called
somaclonal variation (Dodds 1991). The narrow genetic base and inherent
risk of inducing genetic damage are serious problems with the use of micropropagation in conservation. Nevertheless, in principle, these techniques
may have application in specific instances.

Caveats

Some of the potential demographic consequences of using seeds versus
growing plants have been explored with computer simulation. Despite the
apparently overwhelming advantage in both extinction risk and population
growth rate enjoyed, for example, by plants over seeds as founders in the simulation of annuals with high fecundity and low survivorship, it does not follow
that plants are necessarily better to use as founders than are seeds. Even in
such a case where, all else being equal, plants would be vastly superior to
seeds, it is necessary to consider other factors in coming to a decision, because
all else is rarely equal: Are a great many seeds available? What is the expected
reward for starting a population with more seeds than seedlings, relative to a
smaller number of seedlings? What would be the relative costs of each
choice, in terms of both biological and financial capital? Simply because deploying a given number of plants appears to be better than a similar number
of seeds does not mean that plants are necessarily a better option than seeds
under all circumstances.

Another important consideration is how well transplants fare relative to
naturally occurring individuals of similar sizes (T. Kaye and E. Alverson, personal communication). Implicit in the simulations is the notion that transplanted individuals will behave demographically like the naturally occurring
individuals from which the data were obtained. This assumption needs to be
addressed experimentally. In many cases, especially in the absence of
postoutplanting care, transplants will presumably not do as well as wild individuals. However, with appropriate care, some outplants could conceivably
have a higher survivorship, grow more quickly, and reach maturity earlier
than naturally occurring plants.

Using multiple ramets of one or (better) many genotypes is an option for reintroduction/outplanting with plants that can be implemented in many different ways. From a strictly demographic point of view, this would superficially seem to be a means of increasing founder population size and would have few drawbacks, especially where propagules available for reintroduction are very limited. Potential genetic consequences are not, however, all necessarily favorable. Perhaps the most obvious drawback is that opportunities for inbreeding (from a genetic point of view) might be increased. If multiple ramets are available, it may be possible to keep one or more sets in reserve, which would be available to replace individuals that died or otherwise failed to reproduce as vigorously as others. It may be also that the lifespan of a genet can be made longer in a garden setting than it would normally be in nature, for example, by taking repeated cuttings (such as in *Conradina glabra*, Wallace 1992). If so, "clonal" lineages of repeatedly propagated plants can be used either to plant out again many years after the original founding of the population or as pollen sources for artificial cross-pollinations. In these ways, the generation time of the founders could be extended, with its attendant effect of reducing the loss of genetic variation in a reintroduced population.

How Large Should Founding Populations Be?

The most serious short-term danger of low population size is probably extirpation due to chance environmental variation. Other long-term concerns of low population size are increased inbreeding and erosion of genetic variability due to random genetic drift (Gilpin and Soulé 1986; Barrett and Kohn 1991; Ellstrand and Elam 1993).

The rate at which populations lose genetic variability due to random genetic drift is correlated closely with population size; smaller populations lose genetic diversity at a faster rate than do larger ones. The critical measure of how fast a population is expected to lose variability due to random genetic drift is not the number of organisms or breeders that can be counted—the census size (N)—but rather its effective population size (N_e). The concept of the effective population size is used to estimate the rate of random genetic drift in terms of an ideal population that meets a variety of assumptions, such as random mating, one-to-one sex ratio, equal numbers of offspring per parent, and constant population size over the generations (see Wright 1969). In reality, N_e is generally less (often much less) than N—rarely is it larger (Lande and Barrowclough 1987; Bartley et al. 1992). Therefore, the population sizes projected in the simulations described previously probably overestimate the population size from a very important point of view: the rate at which random genetic drift will be expected to erode genetic diversity.

Unfortunately, N_e is not easy to estimate (Lande and Barrowclough 1987; Harris and Allendorf 1989; Bartlely et al. 1992; Grant and Grant 1992; Nunney and Elam 1994); for plants we have few empirical data that bear on it directly

(Govindaraju 1988; Barrett and Kohn 1991). The few available estimates of N_e in relation to N suggest that random genetic drift could be a significant problem even in relatively large populations. Mace and Lande (1991) state that the N_e to N ratio will often be in the range of 0.2 to 0.5. In a review of various genetic and ecological ways to estimate N_e, Nunney and Elam (1994) found that twice as many studies (43 percent) had overestimated the N_e to N ratio than had underestimated it. Even less encouraging are recent studies by Briscoe et al. (1992) in which they calculated the rate of genetic loss over time in a series of rather large captive *Drosophila* populations. They found the N_e to N ratio in two populations of N = 1,000 and N = 3,500 to be about 0.036 and 0.012, respectively. To what degree these extremely low estimates derived from captive fruit fly populations bear on plant populations is uncertain, but the message is clear: Random genetic drift can be significant even in what appear to be rather large populations, numbering in the thousands.

The implications for reintroductions/outplantings are clear and not particularly surprising. There are demographic and genetic reasons to support the conclusion that the founding population should be as large as possible, with the ceiling set primarily by practical and other strategic considerations (such as not using too much of the available stock in a single attempt).

Founding Populations from Single or Multiple Sources?

The choice of stock used to create the founding population is critical to the success of any reintroduction/outplanting project. The historic, ecological, and genetic relationships of the available stock to the potential reintroduction site(s) are important determinants of whether a single or mixed source founding population is more appropriate. The degree and spatial scale of local ecological adaptation and of the fitness effects of coadapted gene combinations are as important to making informed judgments as relevant information is difficult to obtain.

The relationship of the stock to the reintroduction site is of central importance (see Fiedler and Laven, Chapter Seven). Is the stock to be returned to the site from which it came? If so, how precisely do plants or propagules need be returned to the sites and microhabitats from which they or their ancestors were removed? How does the condition of the site compare with what it was when the samples were taken or, perhaps more important, with the site before the population began to decline? Alternatively, if nonindigenous stock is to be used, will it be placed into a site where the species is known to have lived historically, or will it be put into what is presumably a new site? The genetic constitution of the stock is a complex topic that can be viewed in both absolute and relative terms. In absolute terms, the genetic diversity of the available stock can be described quantitatively by a variety of means including but

not limited to electrophoretic analysis of isozymes. In relative terms, how well does the genetic complement of the stock compare with (possibly previously) healthy populations in the wild? There are some situations where either a single-or multiple-source founding population is clearly preferable, but for many situations there are no easy answers. This is an area where more theoretical and empirical work is needed before many useful generalizations can be offered.

At one extreme—where a single source is most appropriate—are cases where a large and genetically representative *ex situ* sample is available for reintroduction back into the site from which it was collected. Reintroducing *Stephanomeria malheurensis*, a genetically depauperate inbreeder, back into the site it occupied before it became extinct approaches this ideal (Parenti and Guerrant 1990). In this case, the question of single versus multiple sources was moot because the species is known from only a single site. Nevertheless, a population had become extinct in the wild, and there was a presumably representative *ex situ* seed sample that had come from the site, and was available for reintroduction. Even so, the sole native site was not in pristine condition. The aggressive exotic weedy annual cheatgrass (*Bromus tectorum*) had invaded the site after a fire, so, in some ways, the original habitat no longer existed. An example at the other extreme—where using multiple sources was necessary—can be found in the reintroduction of *Hymenoxys acaulis* var. *glabra* (DeMauro 1994). All remaining individuals in the last Illinois population of this taxon had failed to set any seed for nearly the entire decade in which they were monitored (1970–1979). Even artificial cross-pollinations failed to produce seed. The taxon was shown to have a sporophytic self-incompatibility system, and all individuals possessed the same S-allele (DeMauro 1993). To restore fertility and thus offer some hope of a genetic future for these plants, it was necessary to cross the Illinois plants with others from Ohio having alternative S-alleles.

Perhaps a more typical reintroduction situation is where only small scattered population fragments exist from which to take stock or where *ex situ* collections are not genetically representative (see Center for Plant Conservation 1991; Guerrant 1992) or where currently unoccupied but otherwise pristine original sites cannot be identified (see Fiedler and Laven, Chapter Seven). In such cases, the decision to use either single sources alien to the site or mixtures is contingent on many case-specific factors. There are both advantages and disadvantages associated with using either single or multiple sources as stock for reintroduction/outplanting (Huenneke 1991; Barrett and Kohn 1991; Holsinger and Gottlieb 1991).

Arguments in favor of using single-source reintroduction stocks include maintaining the integrity of lineages with coadapted gene complexes and of

stock that is specifically adapted to local biotic or abiotic conditions. In either case, crossing with "foreign" individuals may result in "outbreeding depression" (*sensu* Templeton 1986; see also Waser 1993). Using single sources for these reasons carries several assumptions. One is that plants can be placed precisely into spots where they are specifically adapted. Another is that the perceived benefits associated with a sole-source gene pool outweigh the expected costs of using either a seriously depleted or biased sample.

Arguments for using multiple stocks are varied, but many are associated with conditions favoring a single source not being met. Reintroduction in a conservation context often must deal with suboptimal conditions. For many taxa, only small, presumably biased, samples are available. Additionally, potential reintroduction sites are either unknown historical sites of the taxon or, if historical sites are available, indigenous stock is unavailable; it's also possible that the sites have been altered ecologically by, for example, the invasion of exotics. In such situations, where conditions for using sole-source stocks are not met, mixed samples may prove superior to any single source. Comprehensive genetic sampling guidelines for conservation collections of rare plants have only recently been proposed (Center for Plant Conservation 1991), and most conservation collections probably do not meet even these minimal conditions. It is likely that most *ex situ* samples of either seed or growing stock do not represent comprehensive or statistically unbiased samples of the populations from which they have been taken. Even where they do, many sampled populations are themselves only fragmented remnants of what they were historically. Assembling a founder population from two or more sources is one way that founding population genetic diversity can be increased initially. Arguments against using multiple-source stocks largely mirror those favoring the use of sole-source stocks: Mixing stocks can lead to outbreeding depression either for genetic reasons (disrupting coadapted gene complexes) or ecological reasons (crossing differentially adapted types). Philosophical arguments may also be marshaled against using mixed sources because they could consciously contaminate historically distinct lineages or somehow manipulate nature. Alternatively, pragmatic arguments may favor the use of mixed stocks. More empirical data and a more comprehensive theoretical base are needed to critically evaluate the question of when and where single or multiple stocks are more appropriate (see Leberg 1993).

Assessing a newly created population's potential to suffer outbreeding depression appears to be central to determining whether a sole-or mixed-source stock is preferable. Not only can there be large-scale differential adaptation among clearly discrete populations of a taxon (ecotypes), it appears that differential local adaptation can also evolve even within continuous populations. As evidence of local adaptation within continuous plant populations, Waser (1993) found fourteen transplant or similar experiments in which there

was an average selection coefficient of 0.5 against "foreign" individuals. In nine of these experiments, the distances involved were less than 10 meters. To be able to recreate accurately such populations from *ex situ* material would be a daunting challenge indeed. I suspect that few *ex situ* collections are sufficiently well documented to return progeny exactly to the same spot where they or the parent lived. Other experimental evidence has shown that individuals transplanted into the microhabitat from which they were collected have a higher fitness than those moved to new localities (Huenneke 1991; Schmidt and Levin 1985; Bradshaw 1984; Chapin and Chapin 1981; McGraw and Antonovics 1983; Silander 1985). It appears, therefore, that local adaptation may sometimes be important, even on a very fine scale. Insofar as different populations are locally adapted to their own sites, mixing sources may result in outbreeding depression (Templeton 1986; Barrett and Kohn 1991). Outbreeding depression can result not only from differential evolutionary adaptation to local ecological conditions, but also for genetic reasons having to do with disrupting coadapted gene complexes (Templeton 1986; Waser 1993). Waser (1993) assembled twenty-five studies of plants that looked at various components of fitness as a function of genetic similarity of mates (or at least used distance between mates as a correlate of genetic similarity). In fifteen of these studies, highest fitness accrued to crosses among individuals of intermediate distance to their mates, implying that there may be some broad optimum between the deleterious effects of close inbreeding and more distant outbreeding. It therefore appears that, for some species at least, there is significant potential for outbreeding depression among individuals derived from mixed sources. This is associated with their degree of genetic similarity and therefore due potentially to disruption of coadapted gene complexes. To what degree reduction in fitness would be a transient phenomenon, to be followed by newly adapted types, is unknown. How likely is it that such a forced march into an adaptive valley will, to use the imagery of Sewel Wright, set the stage for a selective ascent of a new adaptive peak?

To further complicate matters, the distinction between single and multiple sources may not always be as clear as it seems. As the studies described by Waser (1993) show, what appear to be single, continuous populations may be composed of a variety of differentially adapted genetic neighborhoods. Conversely, what appear spatially to be separate, discrete populations may in fact be genetically integrated by gene flow, especially by means of pollen. Plant populations spatially isolated by hundreds or even thousands of meters may frequently be in genetic contact at levels of gene flow by pollen sufficient to counteract random genetic drift and moderate levels of directional selection (Ellstrand 1992b). Because pollen can be carried long distances, relatively small, newly founded populations may receive "foreign" pollen from conspecifics or (more likely) interfertile congeners, which could disrupt the gene

pool of the newly founded population (Ellstrand and Marshall 1985; Ellstrand, Devlin, and Marshall 1989).

In closing this section, I repeat the call by Barrett and Kohn (1991) and others for the use of controlled experiments, in which single-and multiple-source reintroductions are compared. Not only will well-conceived experiments provide much-needed information about how to design populations; such projects, if done over multiple sites, would also have the virtue of reducing the overall risk of catastrophic loss of any single reintroduced population. That many plants can be vegetatively propagated means that different (genetically identical) ramets can be placed in several sites simultaneously, which would allow genetic and environmental effects on fitness differences to be distinguished experimentally. Adoption of an aggressive reintroduction program involving multiple sites where single and mixed sources are compared would serve both to help conserve rare plant species and to generate valuable information. Such experiments are necessary to develop better criteria for deciding how to choose between the use of single or mixed sources.

How Many Populations to Establish?
After all is said and done, reintroduction is at best a risky enterprise. Even in the most robust circumstances modeled, some populations succumbed. Indeed, even now after decades of intensive research and extensive practical experience, some large commercial plantings of even a major timber species such as Douglas Fir (*Pseudotsuga menziesii*) may fail. Currently, tree planting failure is presumably most often due to unpredictable environmental causes. However, in the first half of the twentieth century, genetic reasons (inappropriate sources) were a common cause of failure (C. I. Millar, personal communication).

Given this uncertainty, perhaps the single most effective design feature for reducing overall risk of failure is to reintroduce multiple populations as a standard procedure. These can then be managed as a metapopulation (which could also include extant naturally occurring populations). Similarly, the probability of failure can be reduced, both for single-and multiple-site projects, if repeated reintroduction/outplanting attempts are made in different years or even seasons (L. D. Gottlieb, personal communication).

Establishing multiple populations has various advantages over reintroducing/outplanting only one. The most obvious is that the taxon is better buffered against random loss of any one population due to catastrophic or other unpredictable environmental events (see Mangel and Tier 1994). For the sake of argument, consider each of four separately fenced experimental treatments used in the reintroduction of *Stephanomeria malheurensis* to be a population (Parenti and Guerrant 1990). Each of the treatments was designed to examine the effects on *S. malheurensis* of a different plant species abundant

in the area. All individuals in one of these treatments were quickly lost due to reasons completely unrelated to the experiment—small mammals presumably breached the fence and consumed the entire new population. An example of where different planting times prevented failure can be found with the reintroduction/outplanting of *Hymenoxys acaulis* var. *glabra* (DeMauro 1994). The first (Spring) planting was largely decimated (95 percent mortality) when it was followed by an unusually hot, dry summer; but a second (Fall) planting was much more successful. From a genetic perspective, establishing multiple populations has the advantage that, while each (sub)population would be expected to lose some alleles randomly, the chances are less that they would all lose the same allele at a given locus than they would if the population were maintained as one panmictic group. Templeton (1990, 1991) suggested that dividing a single captive population into semi-isolated subpopulations ensures that the most alleles are retained overall. By analogy, a single source could be used to establish multiple populations. By using these as sources of reciprocal migrants, the rate of genetic erosion (and local adaptation) would be slowed.

There are also, however, disadvantages associated with a multiple population strategy (Menges, personal communication). The most obvious is that multiple populations are more expensive to set up and to monitor than are a single population. Other necessary resources, such as appropriate protected sites and quantity of founders, may also be limiting.

Viewing reintroduction as a multipopulation effort further demonstrates that this is a long-term commitment that will involve not only an original deployment but subsequent monitoring and active management. Employing a metapopulation strategy also enlarges the scope of experimental work that can be done in any particular case. The use of multiple populations would allow more sophisticated hypotheses incorporating a larger number of variables to be subject to rigorous statistical analyses.

Conclusions

Designing populations for reintroduction is a complex task that begins well before and extends long past the act of outplanting itself. The ultimate goal is to establish self-sustaining populations that have adequate genetic diversity to allow them to adapt evolutionarily to changing circumstances. Evaluating the pool of potential founders and choosing among them is an important early step that will affect much of what follows. It is still too early to provide much insight into the consequences of establishing a new population from single or mixed sources. More experimental work on this and many other aspects of reintroduction/outplanting, in which explicit

hypotheses are compared empirically, will go a long way toward improving our success rate. Determining the founding-population stage distribution is one of many considerations that goes into the design of populations for reintroduction, but it appears it may greatly influence subsequent extinction risk and population growth rate. For good demographic and genetic reasons, we should do all we can to ensure the highest rate possible of population growth. Nevertheless, the simulation models and common sense suggest that we may end up establishing many populations that avoid extirpation for some time but do not grow rapidly. Concerns about the loss of genetic diversity will be particularly acute in such situations. Fortunately, there are many measures that can be used to reduce loss of genetic diversity, but all of them are time and resource intensive. Given the high rate of population failure due to unpredictable environmental causes, it is best to establish multiple populations from the beginning and to manage them as a metapopulation.

ACKNOWLEDGMENTS

I wish to thank Don Falk, Connie Millar, Peggy Olwell, and the Center for Plant Conservation for the opportunity both to participate in the symposium on reintroduction and to write this chapter. Their editorial assistance and patience are also much appreciated. I would also like to thank the many people who have helped to clarify my thoughts about this material and who have commented on earlier drafts: Ed Alverson, Paulette Bierzchudek, Peggy Fiedler, Margie and Chris Gardner, Tom Kaye, Dave Mayfield, Eric Menges, Bruce Rittenhouse, and Dan Salzer were especially helpful. The demographic simulations were made possible by the generous donation of the RAMAS/stage software by Applied Biomathematics. Let me also thank SYSTAT Inc. and Research Software Design for kindly donating statistical and bibliographic software, respectively. The Medford District of the U.S.D. A. Bureau of Land Management provided valuable cost-share funds used to develop the models of annual plant demography. To all of these people and organizations, I am most thankful.

REFERENCES

Allendorf, F. W. 1993. "Delay of Adaptation to Captive Breeding by Equalizing Family Size." *Conservation Biology* 7 (2): 416–419.

Antonovics, J., and R. B. Primack. 1982. "Experimental Ecological Genetics in *Plantago*. VI. The Demography of Seedling Transplants of *P. lanceolata*." *Journal of Ecology* 70: 55–75.

Bainbridge, D. A., and R. A. Virginia. 1990. "Restoration in the Sonoran Desert of California." *Restoration & Management Notes* 8 (1): 3–14.

Barrett, S. C. H., and J. R. Kohn. 1991. "Genetic and Evolutionary Consequences of Small Population Size in Plants: Implications for Conservation." In *Genetics and Conservation in Rare Plants.* Edited by D. A. Falk and K. E. Holsinger. New York: Oxford University Press.

Bartley, D., M. Bagley, G. Gall, and B. Bentley. 1992. "Use of Linkage Disequalibrium Data to Estimate Effective Size of Hatchery and Natural Fish Populations." *Conservation Biology* 6 (3): 365–375.

Borlase, S. C., D. A. Loebel, R. Frankham, R. K. Nurthen, D. A. Briscoe, and G. E. Daggard. 1993. "Modeling Problems in Conservation Genetics Using Captive *Drosophila* Populations: Consequences of Equalization of Family Sizes." *Conservation Biology* 7 (1): 122–131.

Bowles, M. L., K. R. Bachtell, M. M. DeMauro, L. G. Sykora, and C. R. Bautista. 1988. "Propagation Techniques Used in Establishing a Greenhouse Population of *Astragalus tennesseensis* Gray." *Natural Areas Journal* 8 (2): 121.

Bowles, M. L., and M. M. DeMauro. 1992. "Tennessee Milkvetch (*Astragalus tennesseensis*) Reintroduction." Unpublished poster displayed at the nineteenth annual Natural Areas Conference, Bloomington, Indiana, Oct. 27–30, 1992.

Bowles, M., R. Flakne, K. McEachern, and N. Pavlovic. 1993. "Recovery Planning an Reintroduction of the Federally Threatened Pitcher's Thistle (*Circium pitcheri*) in Illinois." *Natural Areas Journal* 13 (3): 164–176.

Boyce, M. S. 1992. "Population Viability Analysis." *Annual Review of Ecology and Systematics* 23: 481–506.

Bradshaw, A. D. 1984. "Ecological Significance of Genetic Variation Between Populations." In *Perspectives on Plant Population Ecology.* Edited by R. Dirzo and J. Sarukhán. Sunderland, Mass.: Sinauer Associates.

Briscoe, D. A., J. M. Malpica, A. Robertson, G. J. Smith, R. Frankham, R. G. Banks, and J. S. F. Barker. 1992. "Rapid Loss of Genetic Variation in Large Captive Populations of *Drosophila* Flies: Implications for the Genetic Management of Captive Populations." *Conservation Biology* 6 (3): 416–425.

Brumback, W. E., and C. W. Fyler. 1987. "Endangered Lesser Whorled Pogonia Transplanted Experimentally (New Hampshire)." *Restoration & Management Notes* 5 (2): 88.

Burgman, M. A., S. Ferson, and H. R. Akçakaya. 1993. *Risk Assessment in Conservation Biology.* London: Chapman and Hall.

Caro, T. M., and M. K. Laurenson. 1994. "Ecological and Genetic Factors in Conservation: A Cautionary Tale." *Science* 263: 485–486.

Caswell, H. 1989. *Matrix Population Models: Construction, Analysis and Interpretation.* Sunderland, Mass.: Sinauer Associates.

Center for Plant Conservation. 1991. "Genetic Sampling Guidelines for Conservation Collections of Endangered Plants." In *Genetics and Conservation of Rare Plants.* Edited by D. A. Falk and K. E. Holsinger. New York: Oxford University Press.

Cervelli, R. 1987. "In vitro Propagation of *Aconitum noveboracense* and *Aconitum napellus.*" *HortScience* 22 (2): 304–305.

Chapin, F. S. III, and M. C. Chapin. 1981. "Ecotypic Differentiation of Growth Processes in *Carex aquatilis* Along Latitudinal and Local Gradients." *Ecology* 62: 1000–1009.

Cox, P. 1990. "Reintroduction of the Texas Snobell (*Styrax texana*)." *Endangered Species UPDATE* 8 (1): 64–65.

DeMauro, M. M. 1993. "Relationship of Breeding System to Rarity in the Lakeside Daisy (*Hymenoxys acaulis* var. *glabra*)." *Conservation Biology* 7 (3): 542–550.

DeMauro, M. M. 1994. "Development and Implemenation of a Recovery Program for the Federal Threatened Lakeside Daisy (*Hymenoxys acaulis* var. *glabra*)." In *Recovery and Restoration of Endangered Species: Conceptual Issues, Planning, and Implementation*. Edited by M. Bowles and C. Whelan. New York: Cambridge University Press.

Dodds, J. H., ed. 1991. "*In vitro Methods for Conservation of Plant Genetic Resources*." New York: Chapman and Hall.

Ecker, L. S. 1990. "Population Enhancement of a Rare Arizona Cactus, *Mammillaria thornberi* Orcutt (Cactaceae)." Master's thesis, Arizona State University, Tempe, Ariz.

Ellstrand, N. C. 1984. "Multiple Paternity Within the Fruits of the Wild Radish, *Raphanus sativus*." *American Naturalist* 123 (6): 819–828.

Ellstrand, N. C. 1992a. "Gene Flow by Pollen: Implications for Plant Conservation Genetics." *Oikos* 63: 77–86.

Ellstrand, N. C. 1992b. "Gene Flow Among Seed Plant Populations." *New Forests* 6: 241–256.

Ellstrand, N. C., B. Devlin, and D. L. Marshall. 1989. "Gene Flow by Pollen into Small Populations: Data from Experimental and Natural Stands of Wild Radish." *Proceedings of the National Academy of Science* 86: 9044–9047.

Ellstrand, N. C., and D. R. Elam. 1993. "Population Genetic Consequences of Small Population Size: Implications for Plant Conservation." *Annual Review of Ecology and Systematics* 24: 217–242.

Ellstrand, N. C., and D. L. Marshall. 1985. "Interpopulation Gene Flow by Pollen in Wild Radish, *Raphanus sativus*." *American Naturalist* 126 (5): 606–616.

Fahselt, D. 1988. "The Dangers of Transplantation as a Conservation Technique." *Natural Areas Journal* 8 (4): 238–244.

Falk, D. A., and P. Olwell. 1992. "Scientific and Policy Considerations in Restoration and Reintroduction of Endangered Species." *Rhodora* 94 (879): 287–315.

Feeback, D. 1992. "Proper Transplanting Method Critical in Restoration of Canebrakes (Kentucky)." *Restoration & Management Notes* 10 (2): 195.

Ferguson, N., and B. Pavlik. 1990. "Endangered Contra Costa Wallflower Propagated by Tissue Culture (California)." *Restoration & Management Notes* 8 (1): 50–51.

Ferreira, J., and S. Smith. 1987. "Methods of Increasing Native Populations of *Erysimum menziesii*." In *Conservation and Management of Rare and Endangered Plants*. Edited by T. S. Elias. Sacramento: California Native Plant Society.

Ferson, S. 1990. *RAMAS/Stage: Generalized Stage-Based Modeling for Population Dynamics*. Setauket, New York: Applied Biomathematics.

Fiedler, P. L. 1987. "Life History and Population Dynamics of Rare and Common Mariposa Lilies (*Calochortus* Pursch: Liliaceae)." *Journal of Ecology* 75: 977–995.

Fiedler, P. L. 1991. *Mitigation-Related Transplantation, Relocation, and Reintroduction Projects Involving Endangered and Threatened, and Rare Plant Species in California*. Technical report to the California Department of Fish and Game Endangered Plant Program, Sacramento.

Fowler, N. L., and J. Antonovics. 1981. "Small-Scale Variability in the Demography of Transplants of Two Herbaceous Species." *Ecology* 62 (2): 1450–1457.

Frankham, R., and D. A. Loebel. 1992. "Modeling Problems in Conservation Genetics Using Captive *Drosophila* Populations: Rapid Genetic Adaptation to Captivity." *Zoo Biology* 11: 333–342.

Franklin, I. R. 1980. "Evolutionary Change in Small Populations." In *Conservation Biology: An Evolutionary-Ecological Perspective*. Edited by M. E. Soulé and B. A. Wilcox. Sunderland, Mass.: Sinauer Associates.

Gilpin, M. E., and M. E. Soulé. 1986. "Minimum Viable Populations: Processes of Species Extinction." In *Conservation Biology: The Science of Scarcity and Diversity*. Edited by M. E. Soulé. Sunderland, Mass.: Sinauer Associates.

Gordon, D. R. 1994. "Translocation of Species into Conservation Areas: A Key for Natural Resource Managers." *Natural Areas Journal* 14 (1): 31–37.

Govindaraju, D. R. 1988. "Life Histories, Neighborhood Sizes, and Variance Structure in Some North American Conifers." *Biological Journal of the Linnean Society* 35: 69–78.

Grant, P. R., and B. R. Grant. 1992. "Demography and the Genetically Effective Sizes of Two Populations of Darwin's Finches." *Ecology* 73 (3): 766–784.

Guerrant, E. O., Jr. 1990. "Transplantation of an Otherwise Doomed Population of Barrett's Penstemon, *Penstemon barrettiae*." *Endangered Species UPDATE* 8 (1): 66–67.

Guerrant, E. O., Jr. 1992. "Genetic and Demographic Considerations in the Sampling and Reintroduction of Rare Plants." In *Conservation Biology: The Theory and Practice of Nature Conservation, Preservation, and Management*. Edited by P. L. Fiedler and S. K. Jain. New York: Chapman and Hall.

Haig, S. M., J. D. Ballou, and S. R. Derrickson. 1990. "Management Options for Preserving Genetic Diversity: Reintroduction of Guam Rails to the Wild." *Conservation Biology* 4 (3): 290–300.

Hall, L. A. 1987. "Transplantation of Sensitive Plants as Mitigation for Environmental Impacts." In *Conservation and Management of Rare and Endangered Plants*. Edited by T. S. Elias. Sacrameto: California Native Plant Society.

Harris, R. B., and F. W. Allendorf. 1989. "Genetically Effective Population Size of Large Mammals: An Assessment of Estimators." *Conservation Biology* 3: 181–91.

Havlik, N. A. 1987. "The 1986 Santa Cruz Tarweed Relocation Project." In *Conservation and Management of Rare and Endangered Plants*. Edited by T. S. Elias. Sacramento: California Native Plant Society.

Holsinger, K. E., and L. D. Gottlieb. 1991. "Conservation of Rare and Endangered Plants: Principles and Prospects." In *Genetics and Conservation of Rare Plants*. Edited by D. A. Falk and K. E. Holsinger. New York: Oxford University Press.

Huenneke, L. F. 1991. "Ecological Implications of Genetic Variation in Plant Populations." In *Genetics and Conservation of Rare Plants*. Edited by D. A. Falk and K. E. Holsinger. New York: Oxford University Press.

Hurkmans, M. E. 1995. "Rare Horsetails in Trouble Five Years After Transplanting (Ontario)." *Restoration & Management Notes* 13 (1): 129.

Jusaitis, M. 1991. "Endangered *Phebalium* (Rutaceae) Species Returned to South Australia." *Re-introduction News* (3): 4.

Lande, R. 1988. "Genetics and Demography in Biological Conservation." *Science* 241: 1455–1460.

Lande, R., and G. F. Barrowclough. 1987. "Effective Population Size, Genetic Variation, and Their Use in Population Management." In *Viable Populations for Conservation*. Edited by M. E. Soulé. Cambridge: Cambridge University Press.

Leberg, P. L. 1993. "Strategies for Population Reintroduction: Effects of Genetic Variability on Population Growth and Size." *Conservation Biology* 7 (1): 194–199.

Levin, D. A. 1981. "Dispersal Versus Gene Flow in Plants." *Annals of the Missouri Botanical Garden* 68: 233–253.

Loebel, D. A., R. K. Nurthen, R. Frankham, D. A. Briscoe, and D. Craven. 1992. "Modeling Problems in Conservation Genetics Using Captive *Drosophila* Populations: Consequences of Equalizing Founder Representation." *Zoo Biology* 11 :319–332.

Loope, L. L., and A. C. Medeiros. 1994. "Impacts of Biological Invasions, Management Needs, and Recovery Efforts for Rare Species in Haleakala National Park, Maui, Hawaiian Islands." In *Recovery and Restoration of Enangered Species: Conceptual Issues, Planning, and Implementation*. Edited by M. Bowles and C. Whelan. New York: Cambridge University Press.

Mace, G. M., and R. Lande. 1991. "Assessing Extinction Threats: Toward a Reevaluation of IUCN Threatened Species Categories." *Conservation Biology* 5 (2): 148–157.

Mangel, M., and C. Tier. 1994. "Four Facts Every Conservation Biologist Should Know About Persistence." *Ecology* 75 (3): 607–614.

Mangione, L., and D. Vander Pluym. 1993. "Santa Susana Tarplant Revegetation on Public Utility Site (California)." *Restoration & Management Notes* 11 (2): 182–183.

Marshall, D. L., and N. C. Ellstrand. 1985. "Proximal Causes of Multiple Paternity in Wild Radish, *Raphanus sativus*." *American Naturalist* 126 (5): 596–605.

Martínez-Vásquez, O., and A. Rubluo. 1989. "In vitro Mass Propagation of the Near-Extinct *Mammillaria san-angelensis* Sanchez-Mejorada." *Journal of Horticultural Science* 64 (1): 99–105.

Maunder, M. 1992. "Plant Reintroduction: An Overview." *Biodiversity and Conservation* 1: 51–61.

McComb, J. A. 1985. "Micropropagation of the Rare Species *Sylidium coruniforme* and Other *Stylidium* Species." *Plant Cell Tissue and Organ Culture* 4: 151–158.

McGraw, J. B., and J. Antonovics. 1983. "Experimental Ecology of *Dryas octapetala* Ecotypes. I. Ecotypic Differentiation and Life-Cycle Stages of Selection." *Journal of Ecology* 71: 879–897.

Meagher, T. 1986. "Analysis of Paternity Within a Natural Population of *Chamaelirium luteum*. I. Identification of Most Likely Male Parents." *American Naturalist* 128: 199–215.

Menges, E. S. 1991. "The Application of Minimum Viable Population Theory to Plants." In *Genetics and Conservation of Rare Plants*. Edited by D. A. Falk and K. E. Holsinger. New York: Oxford University Press.

Menges, E. S. 1992. "Stochastic Modeling of Extinction in Plant Populations." In *Conservation Biology: The Theory and Practice of Nature Conservation, Preservation, and Management*. Edited by P. L. Fiedler and S. K. Jain. New York: Chapman and Hall.

Mistretta, O., and B. Burkhart. 1990. "San Diego Thornmint: Propagation, Cultivation Provides Clues to Ecology of Engangered Species (California)." *Restoration & Management Notes* 8 (1): 50.

Nei, M., T. Maruyama, and R. Chakraborty. 1975. "The Bottleneck Effect and Genetic Variability in Populations." *Evolution* 29: 1–10.

Nunney, L., and K. A. Campbell. 1993. "Assessing Minimum Viable Population Size: Demography Meets Population Genetics." *Trends in Ecology and Evolution* 8 (7): 234–239.

Nunney, L., and D. R. Elam. 1994. "Estimating the Effective Population Size of Conserved Populations." *Conservation Biology* 8 (1): 175–184.

Olwell, P., A. Cully, and P. Knight. 1990. "The Establishment of a New Population of *Pediocactus knowltonii*." *New York State Museum Bulletin* 471: 189–193.

Olwell, P. A., A. Cully, P. Knight, and S. Brack. 1987. "*Pediocactus knowltonii* Recovery Efforts." In *Conservation and Management of Rare and Endangered Plants*. Edited by T. S. Elias. Sacramento: California Native Plant Society.

Packard, S. 1991. "Broadcasting Seed Restores Prairie Fringed Orchid, Other Small-Seeded Forbs (Illinois)." *Restoration & Management Notes* 9 (2): 121–122.

Parenti, R. L., and E. O. Guerrant, Jr. 1990. "Down But Not Out: Reintroduction of the Extirpated Malheur Wirelettuce, *Stephanomeria malheurensis*." *Endangered Species UPDATE* 8 (1): 62–63.

Pavlik, B. M. 1990. *Reintroduction of* Amsinckia grandiflora *to Stewartville*. Unpublished technical report to the Endangered Plant Program, California Department of Fish and Game, Sacramento.

Pavlik, B. M. 1991a. *Reintroduction of* Amsinckia grandiflora *to Three Sites Across Its Historic Range*. Unpublished report to the Endangered Plant Program, California Department of Fish and Game, Sacramento.

Pavlik, B. M. 1991b. *Management of Reintroduced and Natural Populations of* Amsinckia grandiflora. Unpublished report to the Endangered Plant Program, California Department of Fish and Game, Sacramento.

Pavlik, B. M. 1992. *Inching Toward Recovery: Evaluating the Performance of* Amsinckia grandiflora *Populations Under Different Management Regimes*. Unpublished report to the Endangered Plant Program, California Department of Fish and Game. Sacramento.

Pavlik, B. M., and E. K. Espeland. 1991. *Creating New Populations of* Acanthomintha duttonii. I. Preliminary Laboratory and Field Studies. Unpublished report to the Endangered Plant Program, California Department of Fish and Game, Sacramento.

Pavlik, B. M., E. K. Espeland, and F. Wittman. 1992. *Creating New Populations of* Acanthomintha duttonii. II. Reintroduction at Pulgas Ridge. Unpublished report to the Endangered Plant Program, California Department of Fish and Game, Sacramento.

Pavlik, B. M., D. L. Nickrent, and A. M. Howald. 1993. "The Recovery of an Endangered Plant. I. Creating a New Population of *Amsinkia grandiflora*." *Conservation Biology* 7 (3): 510–526.

Piñero, D., M. Martinez-Ramos, and J. Sarukhán. 1984. "A Population Model of *Astrocaryum mexicanum* and a Sensitivity Analysis of Its Finite Rate of Increase." *Journal of Ecology* 72: 977–991.

Popp, R. 1990. "Reintroduction of Eastern Jacob's Ladder (*Polemonium vanbruntiae*)." Unpublished status report, Vermont Natural Heritage Program.

Popp, R. 1991. "Reintroduction of Eastern Jacob's Ladder (*Polemonium vanbruntiae*)." Unpublished status summary, Vermont Natural Heritage Program.

Ralls, K., and R. Meadows. 1993. "Breeding Like Flies." *Nature* 361: 689–690.

Ratliff, R. D., and S. E. Westfall. 1992. "Restoring Plant Cover on High-Elevation Gravel Areas, Sequoia National Park, California." *Biological Conservation* 60: 189–195.

Rieseberg, L. H., S. Zona, L. Aberbom, and T. D. Martin. 1989. "Hybridization in the Island Endemic, Catalina Mahogany." *Conservation Biology* 3 (1): 52–58.

Rochefort, R. M., and S. T. Gibbons. 1992. "Mending the Meadow: High Altitude Meadow Restoration in Mount Rainier National Park." *Restoration & Management Notes* 10 (2): 120–126.

Rubluo, A., V. Chávez, A. P. Martinez, and O. Martinez-Vásquez. 1993. "Strategies for the Recovery of Endangered Orchids and Cacti Through in vitro Culture." *Biological Conservation* 63: 163–169.

Rubluo, A., V. M. Chávez, and A. P. Martinez. 1989. "In vitro Seed Germination and Reintroduction of *Bletia urbana* (Orchidaceae) in its Natural Habitat." *Lindleyana* 4 (2): 68–73.

Sainz-Ollero, H., and J. E. Hernandez-Bermejo. 1979. "Experimental Reintroduction of Endangered Plant Species in Their Natural Habitats in Spain." *Biological Conservation* 16 (3): 195–206.

Schemske, D. W., B. C. Husband, M. H. Ruckelshaus, C. Goodwillie, I. M. Parker, and J. G. Bishop. 1994. "Evaluating Approaches to the Conservation of Rare and Endangered Plants. *Ecology* 75 (3): 584–606.

Schmidt, K. P., and D. A. Levin. 1985. "The Comparative Demography of Reciprocally Sown Populations of *Phlox drummondii* Hook. I. Survivorships, Fecundities, and Finite Rates of Increase." *Evolution* 39 (2): 396–404.

Schoen, D. J., and S. C. Stewart. 1986. "Variation in Male Reproductive Investment and Male Reproductive Success in White Spruce." *Evolution* 40: 1109–1120.

Schoen, D. J., and S. C. Stewart. 1987. "Variation in Male Fertilities and Pairwise Mating Probabilities in *Picea glauca*." *Genetics* 116: 141–152.

Schonewald-Cox, C. M., S. M. Chambers, F. MacBryde, and L. Thomas, eds. 1983. *Genetics and Conservation: A Reference for Managing Wild Animal and Plant Populations.* Menlo Park, Calif.: Benjamin-Cummings.

Shaffer, M. L. 1981. "Minimum Population Sizes for Species Conservation." *BioScience* 31: 131–134.

Shaffer, M. L. 1987. "Minimum Viable Populations: Coping with Uncertainty." In *Viable Populations for Conservation.* Edited by M. E. Soulé. Cambridge, England: Cambridge Univeristy Press.

Silander, J. A. 1985. "The Genetic Basis of the Ecological Amplitude of *Spartina patens.* II. Variance and Correlation Analysis." *Evolution* 39: 1034–1052.

Simonich, M. T., and M. D. Morgan. 1990. "Researchers Successful in Transplanting Dwarf Lake Iris Ramets (Wisconsin)." *Restoration & Management Notes* 8 (2): 131–132.

Soulé, M. E., and B. A. Wilcox, eds. 1980. *Conservation Biology: An Evolutionary-Ecological Perspective.* Sunderland, Mass.: Sinauer Associates.

Spielman, D., and R. Frankham. 1992. "Modeling Problems in Conservation Genetics Using Captive *Drosophila* Populations: Improvement of Reproductive Fitness due to Immigration of One Individual into Small Partially Inbred Populations." *Zoo Biology* 11: 343–351.

Stephenson, D. E. 1992. "Mats of Rare Horsetails Successfully Transplanted (Ontario)." *Restoration & Management Notes* 10 (2): 202–203.

Stewart, J. 1993. "The Sainsbury Orchid Conservation Project: The First Ten Years." *The Kew Magazine* 10 (1): 38–43.

Symonides, E. 1988. "Population Dynamics of Annual Plants." In *Plant Population Ecology*. Edited by A. J. Davy, M. J. Hutchings, and A. R. Watkinson. Oxford: Blackwell Scientific Publications.

Taylor, R. S. 1991. "Threadleaf Brodiaea Propagation, Restoration Technique Developed (California)." *Restoration & Management Notes* 9 (2): 135–136.

Templeton, A. R. 1986. "Coadaptation and Outbreeding Depression." In *Conservation Biology: The Science of Scarcity and Diversity*. Edited by M. E. Soulé. Sunderland, Mass.: Sinauer Associates.

Templeton, A. R. 1990. "The Role of Genetics in Captive Breeding and Reintroduction for Species Conservation." *Endangered Species UPDATE* 8 (1): 14–17.

Templeton, A. R. 1991. "Off-Site Breeding of Animals and Implications for Plant Conservation Strategies." In *Genetics and Conservation of Rare Plants*. Edited by D. A. Falk and K. E. Holsinger. New York: Oxford University Press.

Tonkyn, D. W. 1993. "Optimization Techniques for the Genetic Management of Endangered Species." *Endangered Species UPDATE* 10 (8): 1–9.

Wallace, S. R. 1990. "Central Florida Scrub: Trying to Save the Pieces." *Endangered Species UPDATE* 8 (1): 59–61.

Wallace, S. 1992. "Introduction of *Conradina glabra*, a Pilot Project for the Conservation of an Endangered Florida Endemic." *Botanic Gardens Conservation News* 1 (10): 34–39.

Waser, N. M. 1993. "Population Structure, Optimal Outbreeding, and Assortative Mating in Angiosperms." In *The Natural History of Inbreeding and Outbreeding: Theoretical and Empirical Perspectives*. Edited by N. W. Thornhill. Chicago: University of Chicago Press.

Werner, P. A., and H. Caswell. 1977. "Population Growth Rates and Age Versus Stage-Distribution Models for Teasel (*Dipsacus sylvestris* Huds.)." *Ecology* 58: 1103–1111.

Wilkinson, L. 1990. *systat: The System for Statistical Analysis*. Evanston, Ill.: SYSTAT.

Wochok, Z. S. 1981. "The Role of Tissue Culture in Preserving Threatened and Endangered Plants." *Biological Conservation* 20: 83–89.

Woolliams, K. R., and S. B. Gerum. 1992. "*Kokia cookei*: A Chronology." *Notes from Wiamea Arboretum & Botanica' Garden* 19 (1): 7–12.

Wright, S. 1969. *Evolution and the Genetics of Populations*. Vol. 2, *The Theory of Gene Frequencies*. Chicago: University of Chicago Press.

Youtie, B. A. 1992. "Biscuit Scabland Restoration Includes Propagation Studies (Oregon)." *Restoration & Management Notes* 10 (1): 79–80.

Lessons from Ecological Theory: Dispersal, Establishment, and Population Structure

Richard B. Primack

———————➤•◄———————

Reintroduction techniques have been increasingly advocated by conservation biologists as a way of ensuring the survival of endangered plant species. Many attempts at reintroduction, however, have not succeeded (Hall 1987). Part of the reason for these failures is that biologists have not appreciated that the process of successful reintroduction should mimic the natural process of plant dispersal and establishment. Conservation biologists would have a better chance of creating new populations if they used methods of reintroduction that were based on knowledge of the way in which plants naturally establish new populations. This knowledge would allow more effective site selection, site preparation prior to reintroduction, and choice of plant materials.

Relationship Between Dispersal and Distribution

Each plant species has its own niche within a biological community (Ricklefs 1993; Primack 1993). The niche consists of the characteristics of the environment needed for the existence of the species within the community. The characteristics may be physical, such as the local climate, soil conditions, and light levels; they may also be biological, including the abundance of competitors, predators, and mutualistic species. Many rare species appear to have specialized niche requirements and often are able to live only in a particular type of unusual habitat (Rabinowitz et al. 1986).

For many rare species, there may only be specific places in a habitat with conditions suitable for seed germination and seedling establishment, known both as *safe sites* and the *regeneration niche* (Grubb 1977; Harper 1977; Cook 1979; Urbanska and Schütz 1986). Conditions suitable for seed germination

and seedling establishment may occur only at particular times of the year or following some unusual event. For example, forest fires may release a nutrient pulse that stimulates seed germination; fire may also reduce plant competition, allowing seedlings to establish. For many desert perennial plants, seedling establishment may occur only after unusually rainy periods (Jordan and Nobel 1981). In some forested areas, hurricanes uproot trees and provide the exposed soil and high light conditions needed by some species for germination (Vasquez-Yanes and Orozco-Segovia 1993, 1994).

In the same way that there may be specific periods suitable for seed germination, successional changes in the biological community may gradually eliminate rare species from an area. Many rare plant species require a certain level of disturbance if they are to remain in a community (Denslow 1987). Without this disturbance, the species are outcompeted for light or nutrients by other plants. In many grasslands, for example, periodic burning removes the woody species that would otherwise grow up and outcompete the rare perennial herbaceous plants. In other cases, existing plants of common species act as "nurse plants," providing the conditions needed for the germination and establishment of rare species (Franco and Nobel 1988).

The natural processes of succession and disturbance create a pattern of change over the landscape. Plant species have adapted to this pattern in their dispersal characteristics (Howe 1984). Existing populations produce seeds that are carried by wind, water, and animals to new sites for colonization. In many species with dry fruits, the seeds simply fall to the ground and are dispersed more locally. This process of dispersal is crucial to the survival of the species because any existing population might be eliminated at a specific locality by succession or disturbance. Many species produce an abundance of seeds and have developed efficient dispersal mechanisms. Rare species may be particularly vulnerable during the dispersal and colonization process because they often produce relatively few seeds in comparison with common species; they may have less efficient means of dispersing to new sites; and the distance from existing populations to suitable new sites may be greater because of the specialized habitat required by the species. As a result, rare species may have difficulty colonizing new sites. Rare species may also have specialized relationships with other species in their community—such as orchids' needs for mycorrhizal fungi—which prevent independent colonization.

Colonization of newly available habitats may be more difficult for plants than for animals. Unlike animals, which can actively disperse over wide areas and seek out appropriate habitats, plants rely on wind, water, and animals to carry their seeds by chance to appropriate sites for new populations. In the past, such passive dispersal mechanisms were sufficient for rare plant species to continually replace declining local populations with populations on new sites. However, in recent decades the ability of plants to disperse and colonize

new sites has been greatly reduced due to several factors, all related to human activities. First, the populations of many vertebrate seed dispersers have been reduced or eliminated by hunting and habitat destruction, leading to a reduction in potential plant dispersal. For example, the decline and loss of many North American songbird populations has certainly led to reduced dispersal for many plant species (Terborgh 1989, 1992). Second, the modern landscape has many physical barriers such as fences, roads, farmlands, and human habitations that prevent or retard the dispersal of seeds (Peters and Darling 1985). Many bird and mammal seed dispersers do not attempt to cross open areas due to the increased likelihood of predation, so that disturbances that occupy even a small percentage of a landscape may seriously reduce dispersal rates (Lovejoy et al. 1986). Also, the seeds of many forest wildflowers are dispersed by ants, which are unable to cross areas disturbed by human activity. As the landscape is fragmented by human activities, and remaining parcels of intact habitat are even farther apart, the possibility of dispersal from one patch to another patch is further reduced. The inability of many rare plant species to disperse between landscape fragments is well illustrated by patches of forest in the British Isles. In one study, forests were classified according to whether they were ancient forests that had never been cleared or new forests that were up to several hundred years old (Peterkin and Game 1984). New forests that were adjacent to ancient forests had many more rare wildflower species than new forests that were isolated or were separated from the ancient forests. The study shows that many plant species are capable of dispersing slowly across an intact landscape but are incapable of crossing habitats dominated by human activities.

Many conservation areas, nature reserves, and other protected areas are becoming isolated habitat fragments as land around them becomes developed. In such fragments, species will gradually be eliminated over time due to successional processes, human disturbance, and chance events. These habitat fragments, isolated by human disturbance and a lack of dispersal agents, cannot be recolonized by native species. Rare plant species will be the first lost from isolated protected areas, and these species will be the least likely to recolonize the site from the nearest remaining populations elsewhere. As a result of poor dispersal, the number of species present, particularly rare species, will decline over time.

This scenario is illustrated by the Middlesex Fells, a 400-hectare conservation area in Medford, Massachusetts, northwest of Boston. Starting in the late nineteenth century, this woodland conservation area has become surrounded by suburban development, bisected by roads, altered by human-set ground fires and recreational activities, and increasingly occupied by exotic plant species. In 1894, this forest had 338 native species of flowering plants. By 1992, only 227 native species remained in the same area (Drayton and Primack

1995). Many of the wildflowers lost were formerly common and included many of the most well-known and attractive species of moist habitats, such as orchids and lobelias. The loss of native species from existing conservation areas may accelerate as a result of deterioration of the habitat by acid rain, increased anthropogenic nitrogen inputs, increased ultraviolet light, and altered weather patterns resulting from global climate change and deforestation (Peters and Darling 1985; Davis 1989).

In situations of precipitous decline in the species diversity of conservation areas, conservation biologists need to adopt an aggressive policy of eliminating the factors that lead to the decline of species in the first place and begin to manage the land in a way that provides suitable conditions for rare species to recolonize (Kleiman 1989). Conservation biologists can also facilitate the arrival of rare species at these sites by a program of reintroduction (Primack and Miao 1992).

Conservation biologists are recognizing that a strategy that protects isolated, individual populations of rare species is often too narrow an approach. The target of conservation efforts may also be the *metapopulation*, which consists of a network of populations in various stages of colonization and succession (Hanski 1982; Murphy, Freas, and Weiss 1990; Hanski and Gilpin 1991). Metapopulations also may consist of *core populations* that are fairly large and stable over time, and *satellite populations* that are smaller in size and are more temporary in nature. In favorable years, individuals from core populations may disperse to satellite areas and form new populations, while in unfavorable years, populations may only persist in the core areas. The metapopulation approach recognizes the temporary nature of many existing populations and attempts to protect future sites for colonization. In a large protected area, the metapopulation of a species will have one or more populations that are increasing or decreasing in size at any point in time. Existing populations will be producing propagules and offspring that colonize new sites and establish new populations. Over a wide area and over a period of years, the extinction of one local population may be balanced on average by the establishment of a new population. The metapopulation approach may be particularly appropriate for describing populations of some rare plant taxa, such as the Furbish's lousewort (*Pedicularis furbishiae*), that occupy temporary habitat patches along riverbanks (Menges 1990), but it is also appropriate for patches of wildflowers scattered across a landscape dissected by rivers and mountain ridges. Understanding the metapopulation dynamics of rare species, in particular the ability of propagules to disperse across inhospitable landscapes and colonize new sites, will assist in planning reintroduction efforts.

Many plant species are not only dispersing to new sites within their existing range but are continually dispersing to sites beyond their existing range. When these new populations become established, the range of the species ex-

pands. This process of range extension has often occurred in the past as world climates have changed, such as during the fluctuations caused by the Ice Ages (Webb 1992; Davis and Zabinski 1992). At present, species that can adapt to human disturbance and disperse in association with human activities are showing range expansions, while species that are unable to disperse and adapt to changes caused by human disturbances are showing range contractions. Many of the plant species that are rare today were formerly more widespread and common in the past but have gradually diminished in range.

Reintroduction: Mimicking the Natural Process of Dispersal and Colonization

A major goal of conservation biology is to prevent the untimely extinction of plant species by human activities. Reintroduction of rare species is one method currently being used to lower the risk of extinction in the wild. The assumption of reintroduction is that the probability of extinction is lower when there are more and larger populations of the rare species than occur at present (Terborgh 1974; Pimm, Jones, and Diamond 1988; see also Guerrant chapter Eight). In a certain sense, the goal of reintroduction efforts is to take a rare species in danger of extinction and transform it into a more common species. The reintroduction process permits biologists to take over the function of seed dispersal and establishment after the species appears to be unable to disperse on its own.

Successful reintroduction mimics the natural processes of dispersal and establishment. When a site appears suitable for a species, as described by Fiedler and Laven (see Chapter Seven), but the species is unable to disperse to the site on its own and establish a new population, there may be several reasons for the failure:

1. Perhaps the natural seed dispersal agent is no longer present. For example, birds might have been the principal dispersers in the past, but now their numbers are much lower and not as many seeds are dispersed. As a result, humans must take over the role of dispersing seeds to the suitable site.

2. Perhaps the habitat is fragmented to the extent that seed dispersal is ineffective. Seed dispersers are sometimes confined to one habitat fragment and are unable to cross human-dominated areas to reach other fragments where there is suitable unoccupied habitat for many plant species. In this situation, humans have to take over the role of seed dispersers and disperse seeds across barriers from one fragment to another.

3. Seed production of many rare species may be too low to colonize new sites at some distance away. In this case, humans act as efficient dispersers,

making sure that large numbers of high-quality seeds arrive at exactly the right location.

4. A site may appear to be suitable for the species, but it may lack suitable safe sites for seed germination and establishment. Sometimes "nurse plants" can be planted to facilitate the establishment of the targeted rare species (Urbanska, personal communication, April 1994).

Many reintroduction activities are based on the assumption that establishment of a new population should occur once the species has dispersed to a suitable site. In many cases, species are only able to colonize a site at a particular successional stage or following an environmental disturbance that occurs only every few years. Reintroduction efforts can mimic the dispersal and establishment processes by carrying out reintroduction efforts over several years, until a good year for colonization is encountered by chance. Another reintroduction stategy is to manipulate the site by burning, digging, or some other activity that mimics the natural processes of disturbance and succession, increasing the probability of seedling establishment (Pullin and Woodell 1987; Jacobson, Almquist-Jacobson, and Winne 1991). Such site manipulations and successive reintroduction efforts attempt to duplicate the natural processes that maintain the metapopulation of the rare species within its natural range. Reintroduction efforts involving site preparation and introduction of rare plant seeds and adults often attempt to establish new populations in only a few years, whereas the natural process of dispersal to a new site and establishment might take decades or centuries.

When conservation biologists introduce a species beyond its natural range they again take over the role of dispersal agents. Where rare species are introduced just beyond their range, but within the same biogeographical area, there is probably no danger that the species will become weedy or a pest. Rare species generally do not have the typical characteristics of weedy species (Ehrlich 1989; Nobel 1989). In any case, such gradual dispersal events have always been part of the natural history of species, but this pattern has been disrupted by human activity (Sauer 1988). The efforts of conservation biologists to establish new populations of a rare species just beyond the periphery of its range can be viewed as an attempt to recreate this former process.

From an ecological perspective, a reintroduction attempt can be judged an initial success when a second generation of plants establishes at a site following the initial reintroduction. Additional criteria involve the numbers of seeds produced, trends in the size of the population, and the area occupied (Primack and Miao 1992; Pavlik, Chapter Six). The final indicator of success occurs when seeds from this new population disperse beyond the reintroduction site and form satellite populations and even new, independent populations. The establishment of a dynamic metapopulation structure is an important goal of reintroduction efforts.

Attempts at reintroduction will have a greater chance of success when the ecological requirements of the species are considered. In a review of fifteen plant reintroduction projects, Hall (1987) identified five key elements of a successful project:

1. *The appropriateness of planting techniques and effectiveness of execution.* Each species has its own niche requirements, which must be met if a species is to survive at a new site. For example, seeds and transplants must be positioned at the correct depth in the soil, and water and shading must be given to minimize water stress. Timing is also crucial, with certain times of the year being more suitable than others for planting.

2. *Effective site selection.* Plants may survive and grow in places where the species is not normally found if competition with other plant species is reduced. However, for a new permanent population to develop, a locality must be selected that has *safe sites*—the set of specific environmental conditions in which the species can germinate, establish, and remain competitive with other plant species. For example, the Plymouth gentian, *Sabatia kennedyana*, is only found in coastal ponds with fluctuating water levels. Establishing this species in most other habitat types would probably be impossible. Key variables in site selection are soil type, soil moisture, temperature, and amount of shading. Indicator species with which a rare species is often found in nature can help to identify potentially suitable safe sites for reintroduction.

3. *Complete documentation.* The success of a reintroduction project cannot be determined if the location of the project is not known, the date of activities and the methods of planting are not recorded, and the number of individuals used not noted. It is crucial to keep good documentation of reintroduction projects in a secure, central location, and yet it is surprising how infrequently this is done. In commenting on the lack of documentation for 145 major animal reintroduction projects, Beck et al. (1994) state, "In two years of intense searching, we were able to acquire reasonably complete information on less than 50 percent of projects known to have released captive-born animals. Written information documenting reintroduction procedures and post-release outcomes for over 13,000,000 individuals fills less than one file drawer." The documentation for plant reintroductions is probably even more scanty, yet such information is critical if we are to determine which techniques are most effective.

4. *Appropriate maintenance.* A plant species may only become established at a certain successional stage and with certain environmental conditions. Successful plant reintroductions often require maintenance to keep the transplanted individuals alive and to alter the site to favor the

reintroduced species. In particular, plants may need to be watered until the root systems are established, straw and matting may need to be placed around the plants to retain soil moisture, and adjacent vegetation may have to be periodically cut back to reduce plant competition. Unfortunately, in the majority of cases listed by Hall (1987), there was little or no maintenance in the weeks, months, or years after transplanting, and the plants simply died. Whenever possible, maintenance should be done in a careful experimental manner with controls (such as water versus no water) to determine the significance of the treatment on establishment success.

5. *Long-term monitoring.* Reintroductions may be judged successful when a new expanding population is established. Careful censusing of the new population has to be done to determine if the number of individuals is stable, increasing, or decreasing over years and even longer periods of decades and centuries. Tagging and mapping individual plants for a demographic study can yield information on the growth rate of individuals, the frequency of reproduction, and the appearance of new individuals from seed or vegetative growth (Simberloff 1988; Primack 1993; Menges 1986; Sutter, Chapter Ten).

From a traditional horticultural or mitigation perspective, a reintroduction project involving perennial plants would be judged a success if a high proportion of transplanted plants survived for one or more years. This measure of success is very limited, however, because if no seedlings result from these transplanted adults, the population will gradually die out (Guerrant 1992; Primack and Miao 1992; Pavlik 1994 and forthcoming; see also Guerrant, Chapter Eight; Pavlik, Chapter Six). From an ecological perspective, success is considered to be the establishment of a new population. A completely successful reintroduction would result in a self-maintaining population in which successive generations of plants grow up on the site and persist amidst the range of environmental variation. In addition, seeds from this new population may eventually disperse naturally to establish new populations at nearby satellite sites. The complete measure of success may take many years to achieve, but there are population stages along the way that can be used to evaluate the potential success of the reintroduction projects. These stages are somewhat different for annual and perennial plants.

Monitoring Stages for Perennial Plants

1. *Seedling establishment.* If seeds are used to start the population, is there any evidence of seedlings? What percentage of seeds become seedlings? Do these seedlings persist and grow into adults?

2. *Survival.* Do transplanted individuals (or young plants, if using seedlings) survive at the site over the course of weeks, months, and years? What percentage of the plants survive? This measure of success is the one most often considered in mitigation efforts but is only of preliminary significance from an ecological perspective.

3. *Reproduction.* Do the transplanted individuals reproduce? How many years does it take plants growing from seeds to reach flowering age? When individuals flower, are they successfully pollinated by an appropriate agent? How many seeds are produced by the plants? Is fruit formed with viable seeds, and is the seed dispersed by an appropriate dispersal agent?

4. *Recruitment.* Are there any new seedlings in the population coming from seeds produced by transplanted individuals? Are the numbers of seedlings sufficient to replace the adult plants that have died in the population and to allow the population to expand in size? In other words, is there any evidence that the population is self-perpetuating through the development of a second generation?

5. *Population viability.* Does the population eventually take on the characteristics of a viable population? The clearest indication of population viability is a stable or growing population, in which the numbers of new individuals appearing in the population are equal to or greater than the numbers of individuals dying, and the population occupies a similar or larger area. Eventually, the population size stabilizes around a point where the birth rate equals the death rate.

The new field of population viability analysis and recent theories of minimum viable population size, based largely on theory and animal studies, have suggested that the long-term persistence of a species requires a population size of at least five hundred adult individuals (Franklin 1980; Soulé 1987; Menges 1991). For plants, minimum viable population size may depend on life form and habitat. If the population tends to have strongly fluctuating numbers of individuals, as might tend to occur for annual plants in unstable habitats such as river banks or talus slopes, a population size of ten thousand individuals might be considered necessary for long-term persistance (Lande 1988). Lower population sizes, closer to five hundred, might be sufficient for long-lived species of more stable habitats, such as trees of mature forests. Populations of this size range (five hundred to ten thousand) are thought to be sufficiently large to withstand the long-term problems of inbreeding depression and demographic stochasticity and the random effects of rare catastrophic events such as hurricanes, droughts, and disease epidemics.

Assignment of population viability is problematic in those situations in which recruitment is an infrequent event. A population may produce abundant seeds that accumulate in the soil seed bank, but no seedlings appear because the environmental conditions are unsuitable for seed germination. Such conditions might appear only once every few years, or once every few decades. A reintroduction experiment could also be judged successful if seeds produced at the site dispersed to nearby sites to form additional populations, even though no new plants formed at the original reintroduction site.

Monitoring Stage for Annual Plants

Evaluating the success of annual plant reintroduction projects is, in some cases, more straightforward because there are no overlapping generations of plants. On the other hand, annual species often have highly variable population sizes, particularly desert annuals that are dependent on unpredictable rainfall. If the number of plants in the population is stable or continues to increase over time, then the population presumably is viable. Another important indicator is the number of seeds produced each year by the plants. The number and density of viable seeds in soil seed banks can also give a measure of the species' ability to persist at the site during the years when seed production is poor.

Conservation biologists plant either seeds or transplants (juveniles or adults) to establish new populations of rare plant species. Seeds and transplants each have certain ecological and practical advantages and disadvantages in reintroduction efforts (see Guerrant, Chapter Eight). From an ecological perspective, however, reintroduction efforts involving seeds more closely mimic the natural dispersal process. In the great majority of species, new populations establish from seeds dispersing to a suitable but unoccupied site. Biologists sowing seeds on the site are mimicking the process of natural seed dispersal. Transplanting adult or juvenile plants from one place to another, or from the greenhouse to the field, resembles the less common process of dispersal and establishment by plant fragmentation that occurs in some species, particularly those in aquatic habitats. Seeds can even be reintroduced onto a site in a way that closely corresponds to the natural seed dispersal process, such as first passing the seeds of fleshy-fruited plants through the gut of a frugivorous bird to facilitate seed germination. The site may have to be disturbed in some way, such as by digging or burning the ground, to enhance seed germination and seedling establishment. The disturbance may temporarily reduce plant competition and mimic the effects of a disturbance that occurs naturally only every few years, decades, or centuries.

Using seeds in reintroduction activities also has certain genetic advantages. With adult transplants, only a few genotypes can be used because of the logistical difficulties of propagating, transporting, and transplanting large

numbers of individuals. It is possible that none of those transplants has a genotype that is suited to the new site, and all of the individuals die out. In contrast,
large numbers of seeds, if available, can be readily introduced onto a site with
minimal effort. Some of the seeds potentially will possess genotypes necessary
for surviving and even thriving at the new site. For genetic reasons, seeds are
preferable to transplants in reintroduction efforts. Use of seeds also is more effective at locating the exact microsites most suitable for the species at the site.
Seeds can be sown widely at a site; the places where seeds germinate and
where seedlings and new plants establish effectively demonstrate where the
species can grow. In contrast, transplanting plants into a site allows the species
to sample the site in a much more limited manner.

There are, nonetheless, ecological advantages to using transplanted plants
rather than seeds in reintroduction efforts. Plants, particularly adult plants,
have a higher likelihood of successful establishment than seeds if they are
planted into a suitable site and are well tended (Barkham 1992). These
plants have overcome the most vulnerable stages in the life cycle (seed germination and seedling establishment) so that their chances of surviving in
the new habitat are greatly increased. These individuals also have proven
genotypes that are free of lethal mutations and adapted to the general environmental conditions. When reintroduction efforts involve reproductively
mature adult plants, the new population has the potential to flower, produce
and disperse seeds, and create a second generation of plants within a year of
transplantation.

Ecological Lessons from Animal Reintroductions

Conservation biologists considering plant reintroductions can gain valuable
ecological insights by examining the literature on animal reintroductions.
The extensive data on animal reintroductions stand in marked contrast to the
meager literature on plant reintroductions. Large, well-funded projects involving such species as the California condor, the black-footed ferret, the
bison, the Arabian oryx, the golden lion tamarin, and the peregrine falcon
have no equivalents either in the botanical literature or in the priorities of
government agencies and grant institutions. It may be useful to review briefly
some of the ecological elements that are considered crucial to the success of
animal reintroduction efforts and consider whether they may be useful to
plant reintroductions.

A review of 198 bird and mammal reintroductions between 1973 and 1986
(Griffith et al. 1989) found that the success of projects was

- Greater for game species (86 percent) than for threatened, endangered,
  and sensitive species (44 percent)

- Greater for release in excellent-quality habitat (84 percent) than in poor-quality habitat (38 percent)

- Greater in the core of the historical range (78 percent) than at the periphery and outside of the historical range (48 percent)

- Greater with wild-caught (75 percent) than with captive-reared animals (38 percent)

- Greater for herbivores (77 percent) than for carnivores (48 percent)

- Greater when more animals are released than when fewer animals are released, with maximum benefit achieved with at least one hundred animals

A second survey of animal reintroduction projects (Beck et al., 1994) used a more restricted definition of reintroduction: the intentional release of captive-born animals into or near the historical range of the species. Also, the success of a project was more narrowly defined by Beck as a self-maintaining population having at least five hundred individuals. By this more restricted definition, only 16 out of 145 reintroduction projects were judged successful—a much lower proportion than in the earlier review. According to a statistical comparison of the successful projects with the less successful ones, the key element of a successful project is releasing large numbers of animals over many years (an average of 726 animals for successful programs versus 336 for unsuccessful programs; an average of 11.8 years of release for successful programs versus 4.7 years for unsuccessful programs). These results are also comparable to the results obtained from deliberate releases of animal species, primarily insect species (Crawley 1989), with the highest chances of success occurring when large numbers of animals are released at many sites over many years.

These and other extensive reviews of reintroduction methodology in animals have emphasized certain additional valuable techniques (Kleiman 1989; Kleiman, Price, and Beck 1994). First, animals should receive a careful veterinary screening before release to avoid transmitting diseases picked up in captivity into the wild population. Second, animals should be evaluated genetically to avoid creating a population that will suffer the deleterious effects of either inbreeding or outbreeding depression. The genetic makeup of the population should correspond as closely as possible to the characteristics of the habitat, most often achieved by using animals originating from near the release site. Third, animals should receive prerelease training to learn behaviors needed for recognizing and obtaining food, finding shelter, avoiding predators, interacting with members of the same species, and completing reproduction in the wild. Fourth, animals must be acclimatized to the site before release. Some degree of postrelease support should be given to the ani-

mals until they are able to survive on their own, a method known as a soft release.

Many of these lessons from animal reintroductions have direct and obvious ecological applications to plant reintroduction, as the following list illustrates:

Observations with animal reintroduction	Questions to be answered in plant reintroduction studies
1. Greater success with game species than threatened species	Are rare species harder to reintroduce than common species?
2. Greater success at core of the historical range than outside the range	Can rare plants be successfully introduced outside their historical range?
3. Greater success with wild-caught than captive-reared animals	Is there greater success using seeds collected in the wild or seeds from cultivated plants? If transplants are used, is success greater for plants collected in the wild or grown in cultivation?
4. Greater for herbivores than carnivores	How does life history and growth habit affect the probability of successful reintroductions? Are annual plants more difficult to reintroduce than perennial species?
5. Greater when more animals are released	Does the chance of success increase with increasing numbers of individuals transplanted and with greater numbers of seeds sown?

First, sites should be selected that are optimal for the species. Second, numerous disease-free individuals should be reintroduced over many years in order to maximize the chance of successful establishment. Third, the individuals should be selected to include the most suitable genotypes for the site, either by using numerous seeds from the nearest wild population or by transplanting wild-collected adults. Fourth, while plants, of course, cannot be trained before release, the comparable treatment should involve "hardening" cultivated plants by exposing them to outside conditions before transplanting, so that their leaves are tougher, better able to tolerate the natural environment, and less palatable to herbivores. Fifth, after transplantation, plants may require extensive postrelease support, comparable to a soft release in animals, in the form of fencing and screens to keep insects and mammalian herbivores

from eating the plants, weeding to reduce competition from other plant species, and shading to prevent desiccation while the root system establishes. Parallel studies need to be carried out on the reproductive biology of the species, particularly aspects of the mating system and seed-dispersal ecology. Hand-pollination of flowers might be necessary to ensure seed production if the species is self-incompatible and the flowers remain unvisited by pollinators. Artificial dispersal of seeds to the site might be necessary if natural seed dispersal is not taking place. Careful attention to these considerations before, during, and after the reintroduction attempt will increase the chance of a new population establishing.

Seed-Dispersal Experiments in Massachusetts

Most reintroduction projects involve transplanting wild-collected or captive-raised plants into apparently suitable sites. Even though seed dispersal is the typical way in which new populations are formed in nature, many botanists believe that creating new populations using seeds will not be effective. This perception is based on a twofold lack of (1) appearance of seedlings in many wild populations of rare plant species and (2) the specialized conditions required by many species for seed germination and initial seedling establishment, as observed by many wildflower gardeners and ecologists (Kenfield 1970; Grubb 1977; Barkham 1992). However, the probability of success when using seeds to establish new plant populations needs to be systematically investigated to determine the potential of this method for reintroduction and conservation efforts involving rare and endangered plant species. To provide such data, experiments were initiated in 1987 in Massachusetts and have continued to the present. The species used in these experiments are uncommon or absent at the field sites but are not rare on a larger geographical scale. Truly rare species would probably present even greater difficulties in reintoduction effects.

Most of the experiments were conducted in the Hammond Woods, a conservation area of approximately 80 hectares in Newton, Massachusetts, containing a mixture of deciduous woods, swamps, meadows, roads, and parking lots (Primack and Miao 1992). The present vegetation seems reasonably stable, but rock walls running through woods are lingering evidence of farming activities in the area over a century ago. A second field site is the Harvard Forest, a research station operated by Harvard University in central Massachusetts. At this site, investigations of the impact of hurricanes on ecosystem function have utilized a unique experimental approach in which trees are pulled down in a 1-hectare area to simulate the effects of a hurricane blowdown. The experimental blowdowns provided an opportunity to determine if extensive habitat disturbance is necessary for the establishment of new plant populations from seed.

All of the Massachusetts experiments used a similar design. Sites were located that appeared to be suitable for a particular target species but were lacking those species. Site selection was based on environmental characteristics (primarily soil characteristics), water availability, and the amount of vegetative cover. In addition, indicator plant species that are typically associated with the target species were used. At each experimental site selected for a species, a known number of seeds, typically thirty to one hundred, were sown onto a fixed, mapped quadrat or within a fixed distance of a stake, with an area varying from 0.25 to 3 m² depending on the species used and the number of seeds available. Quadrats were positioned within the site to be in the place that seemed most suitable for seed germination and seedling establishment. Often the quadrat was positioned to straddle a moisture or soil gradient to maximize the chance of including the right conditions at the microsite. Whenever possible, quadrats were also positioned to include an area of exposed soil to increase the possibility of seedling establishment. After seeds were sown, the area was raked lightly to bury the seeds. The seeds for about half of the species were obtained from plants growing in the wild (Table 9-1). Seeds for the other species came from plants grown by the New England Wildflower Society in Framingham, Massachusetts.

Experiments using seeds of six annual species were initiated in the late summer and fall of 1987 and 1988 (Table 9-1). Forty-eight unoccupied but apparently suitable sites were sown with seeds of particular species. Plants establishing at the sites in subsequent years were allowed to mature and disperse seeds. A census of the sites has been taken every year since the initiation of the experiments. The results represent a mix of outcomes. At twenty-seven sites, no seedlings appeared in any years following seed sowing. At twenty-one sites, some seedlings appeared, generally in the following spring, but the plants did not live to reproductive maturity. At three sites, some plants reached reproductive maturity and produced seeds in the first year, but no plants were seen in subsequent years. At five sites, where jewelweed and bracted plantain seeds were dispersed, two or more generations of plants occupied the site in successive years, but the populations did not persist until 1993. At one site alone, an expanding population of jewelweed became established and occupied an ever-increasing area as of 1993.

Several other observations related to the processes of dispersal and colonization were made in the course of these trials with annual species. First, seedlings almost exclusively appeared within the quadrats where seeds had been sown the previous year. That is, there is evidence that seeds are not moved around by animals, water, or wind following arrival at the soil surface. One extreme example of this was a site sown with one hundred pinweed seeds in the summer of 1987. No seedlings of this species were present in 1988 or 1989, but in 1990, thirty-eight plants grew exactly where the seeds had been sown more than two years previously. Second, sites in which several

generations of plants grew had highly variable rates of areal expansion: The successful jewelweed population spread outward from its initial site at a rate of about 2 meters per year; one plantain population expanded by 18 meters in two years before going extinct; and a second plantain population remained for five generations exactly within the quadrat in which it was sown. These results indicated that dispersal may be extremely limited for many species in particular situations.

Perennials at Hammond Woods

Seeds from a wide range of native perennial species were also planted at unoccupied, apparently suitable sites in the Hammond Woods (Table 9-1). A total of thirty-five species were used at 173 sites. These species had a variety of dispersal mechanisms: dry seeds in capsules; fleshy, bird-dispersed fruits; and hooked seeds. They also had different habitat requirements, including fields, forests, and bogs. The reason for choosing such disparate species was to try to develop generalizations about the dispersal and colonization process. If only one or a few species had been used, the results might not have had general applicability. Species used either already existed in small patches in the Hammond Woods, had previously occurred in the area but had gone extinct locally, or did not occur in the area but probably had occurred there in the past. One of the species used was an introduced European plant species, dame's rocket (*Hesperis matronalis*), locally present in thickets and field edges. For each site and each species, fifteen to six hundred seeds were sown, depending on the availability of seeds. While these numbers of seeds may seem small in comparison with the reproductive capacity of plants, they are probably comparable or even greater than the numbers of seeds that naturally disperse to an unoccupied site several kilometers away from an existing population. Seeds for most species were sown in the late summer of 1988, with some additional sites sown in 1989. Sites were monitored in the summers of 1989 through 1993.

From these extensive efforts to create new populations of perennial species using seeds, the overall outcome was a conspicuous lack of establishment. In 167 of 176 sites, no seedlings appeared in any year. No seedlings at all were seen in the wild for thirty-two of the thirty-six species sown at the Hammond Woods. However, there were a few events of note:

- Nine sites were planted with locally collected marsh marigold seeds (*Caltha palustris*) in June 1989. At one streamside site, two large healthy plants were noted in 1993. One of these plants finally reached flowering size in 1994.

- Eight sites were planted in 1989 with locally collected seeds of wild cranesbill(*Geranium maculatum*). Seedlings were present at four sites in 1990, but only one small, nonflowering plant persisted at one site by 1993.

- Six sites were planted in 1989 with New Hampshire–collected bluet seeds (*Houstonia caerulea*). A few flowering bluet plants were seen in 1990 at two sites, but no plants were seen in subsequent years.

Probably the most noticeable success was the introduced perennial, dame's rocket. Six sites were sown in 1989 with locally collected seeds. At each of two sites, many seedlings were present in 1990. Three of these plants reached flowering size in 1993, producing hundreds of new seeds each.

The reasons for such widespread failure of reintroductions in the Hammond Woods are hard to understand. For example, fireweed (*Epilobium angustifolium*) is a perennial of disturbed, moist roadsides, meadows, and recent burned areas, found scattered throughout Massachusetts. In the Hammond Woods, seven apparently suitable sites were sown with three hundred seeds each in 1989. No seedlings ever appeared at these sites during the next four years, even though the sites looked eminently suitable for the species. The seeds of the species were clearly viable, as seen by a 32 percent germination rate after one cold stratification, and the plants were suited to the area, as shown by the vigorous growth, flowering, and fruiting in a Newton garden of plants from the same seed sample.

Harvard Forest

The lack of success of the reintroduction experiments at the Hammond Woods could be due to an absence of disturbance needed to create conditions suitable for seed germination and seedling establishment. The Harvard Forest site provided a test of this hypothesis. In October 1989, seeds of fifteen perennial herbs were sown onto exposed ground of the experimental blowdown area at the Harvard Forest and onto adjacent areas of undisturbed forest. The hypothesis being tested was that reintroductions would be more successful if the site was disturbed, the soil was exposed, the light levels increased, and competition reduced. For each species, seeds were sown onto 0.25 m² quadrats, two quadrats in the blowdown area and two quadrats in the nearby forest. No seedlings of any species were seen in the summer of 1990 or 1991, but in 1992, one seedling of tall meadow rue, *Thalictrum polygamum*, was present on a quadrat in the blowdown site. This small, nonreproductive plant was evident again in 1993. These results again emphasize the difficulty of establishing new populations from seed, even when the site is experimentally disturbed to enhance the probability of colonization.

Using a Model Species

The results we have obtained so far using native and naturalized plants are so limited that Brian Drayton, and I have begun to initiate additional experiments using garlic mustard (*Alliaria petiolata*), an aggressive, introduced,

TABLE 9-1. Species used in reintroduction experiments at the Hammond Woods, Newton, Massachusetts. For each species, the number of sites used and the number of seeds sown per site are given. Also given is the number of sites with at least some seedlings and number of sites with at least a second generation of plants. The ecological requirements and growth habits of species are given by Fernald (1970).

Scientific name (Common name)	Number of sites	Number of seeds	Seed source*	Number of sites with seedlings	Number of sites with second generation
Annual species					
Abutilon theophrasti (Velvet leaf)	6	100	HW	1	0
Geranium carolinianum (Carolina cranesbill)	7	25	HW	1	0
Hypericum gentianoides (Pinweed)	7	100	HW	1	0
Impatiens capensis (Jewelweed)	14	100	HW	11	3
Plantago aristata (Bracted plantain)	7	100	HW	5	3
Sicyos angulatus (Bur-cucumber)	7	50	HW	2	0
Perennial species					
Actaea pachypoda (Doll's eyes)	7	50	NH	0	0
Actaea rubra (Red baneberry)	7	100	NH	0	0
Aquilegia canadensis (Columbine)	4	20	NEWS	0	0
Aralia hispida (Bristly aralia)	7	25	NH	0	0
Aralia racemosa (Spikenard)	7	100	NH	0	0
Asarum canadense (Wild ginger)	2	15	NEWS	0	0
Asclepias tuberosa (Butterflyweed)	4	25	NEWS	0	0
Calla palustris (Calla lily)	7	50	NH	0	0
Caltha palustris (Marsh marigold)	9	100	HW	1	0
Campanula rotundifolia (Harebell)	4	20	NEWS	0	0
Claytonia virginica (Spring beauty)	3	20	NEWS	0	0
Clintonia borealis (Blue-bead lily)	7	100	NH	0	0
Coptis groenlandica (Golden thread)	7	100	NH	0	0
Cornus canadensis (Dwarf dogwood)	8	25	NH	0	0

TABLE 9-1. *(Continues)*

Scientific name (Common name)	Number of sites	Number of seeds	Seed source*	Number of sites with seedlings	Number of sites with second generation
Epigaea repens (Trailing arbutus)	4	30	NEWS	0	0
Epilobium angustifolium (Fireweed)	7	300	NH	0	0
Gentiana clausa (Closed gentian)	4	50	NEWS	0	0
Geranium maculatum (Wild cranesbill)	4	75	HW	4	0
Heracleum lanatum (Cow parsnip)	2	25	NEWS	0	0
Hesperis matronalis (Rocket)	4	100	HW	2	0
Houstonia caerulea (Bluets)	4	100	NH	2	0
Lillium canadense (Canada lily)	4	10	NEWS	0	0
Lobelia cardinalis (Cardinal flower)	1	600	NEWS	0	0
Lobelia siphilitica (Great lobelia)	4	200	NEWS	0	0
Magnolia virginia (Sweet bay magnolia)	8	20	MA	0	0
Rhexia mariana (Meadow beauty)	2	15	NEWS	0	0
Sarracenia purpurea (Pitcher plant)	7	100	NH	0	0
Shortia galacifolia (Oconee bells)	4	30	NEWS	0	0
Streptopus roseus (Rose twisted stalk)	1	150	NH	0	0
Riarella cordifolia (Foam flower)	7	100	NH	0	0
Trillium erectum (Red trillium)	8	50	NH	0	0
Trillium undulatum (Painted trillium)	6	100	NH	0	0
Viola pedata (Bird's foot violet)	3	20	NEWS	0	0
Viola rotundifolia (Yellow violet)	2	10	NEWS	0	0
Zizia aptera (Golden alexander)	4	75	NEWS	0	0

*Seed Sources: HW = Wild plants at the Hammond Woods, Mass.; NH = Wild plants at Sunapee, N.H.; NEWS = Cultivated plants at New England Wildflower Society, Framingham, Mass.; MA = Wild plants at Magnolia, Mass.

invading species as a model species to test hypotheses on reintroduction. Even though this European species has no obvious dispersal mechanism, it has been gradually invading the U.S. woodlands, forming dense stands of plants. The species is valuable because it can provide an estimate of the greatest probability of success that we can expect in establishing new populations of species at unoccupied sites. There are many questions to be addressed with respect to garlic mustard: How many seeds must arrive at an unoccupied site for a new population to become established? What is the probability that a new population will form at an apparently suitable site once seeds arrive there? How important are genetic effects in determining the success of population establishment? Our preliminary results demonstrate that even in this aggressive weed species only a small percentage of seeds develop into seedlings, and new populations develop at only a minority of sites into which seeds are introduced. The final results with garlic mustard will provide a valuable baseline for dealing with more sensitive rare species.

Conclusions

Ecological theory and empirical studies highlight aspects of dispersal, establishment, and succession that are applicable to plant reintroduction efforts. If these components are better understood, the probability of success of reintroduction attempts may increase. At present, we can conclude that establishing new plant populations from seeds appears to be difficult, even when seeds are sown onto apparently suitable sites, and the sites are heavily disturbed by experimental treatments. Reintroductions using seeds have the potential to be successful, as shown by the few examples of success, but considerable difficulties have to be overcome. The lessons of the work discussed here and other published studies are as follows:

- Reintroduction projects should utilize many sites, because the majority of sites may not have the environmental and biological characteristics required by target species. Subtle and changing habitat characteristics that are difficult to predict in advance may determine the success or failure of establishment at any particular site. A site that is currently suitable for a species may soon not be suitable due to successional changes or further disturbance.

- Reintroduction efforts should use as many plants as possible, if using transplants, or as many seeds if possible, if using seeds. The success rate of seeds is so low that hundreds of seeds, if not more, should be sown at each potential site. Obtaining such large sample sizes of seeds may be problematic for many rare species. Experiments are particularly needed

in which the relative success at establishing new populations from seeds, transplanted seedlings, and adult transplants can be compared. While genetic factors should be considered in selecting material for reintroduction projects, ecological factors appear to play the dominant role in the success of establishing new populations.

- Predicting the probability of success of a reintroduction attempt for any species is difficult because the frequency of success may be very low. However, it appears that annual species have a greater likelihood of success than perennial species. Determining the importance of plant life form and habitat requirements to the probability of reintroduction success will be an important goal as more studies are published.

- Other potential factors contributing to success in germination and establishment need to be considered as well. The relative importance of these factors will probably vary depending on the species and particular habitat conditions. Feeding fleshy fruits to captive birds and other seed dispersers and planting the droppings may enhance seed germination. Planting seeds with soil from the site where the species is growing already may promote the formation of beneficial soil mycorrhizal associations with the seedlings. Applying mineral nutrients to the soil may promote germination and seedling survival in some species (Barkham 1992) but might eliminate other rare species that are dependent on low nutrient levels and low levels of competition. Fencing and screening the sites may reduce herbivory by insect or vertebrate grazers and enhance growth. Site manipulations, such as digging, burning, eliminating competing vegetation, and planting nurse plants of hardy species may be needed to create the safe site conditions needed for the germination and establishment of many rare plant species (Pullin and Woodell 1987; Franco and Nobel 1988).

- Finally, reintroduction should be conducted during several successive years, because conditions suitable for germination and establishment may only occur occasionally. Experiments are urgently needed to test the value of these treatments in enhancing the success of plant reintroductions. These experiments should be conducted with appropriate controls and experimental designs.

As a final note, the need for careful monitoring and documentation should be emphasized. While plant reintroductions are becoming increasingly commonplace among conservation biologists, the published literature on the subject is surprisingly limited. The literature on plant reintroductions is particularly poor in comparison with the literature on animals. More experiments will have to be undertaken, analyzed, and published before botanists can

confidently determine the best methods for plant reintroductions and make reasonable predictions of the chance for success. However, efforts that mimic the natural processes of dispersal and establishemnt will almost certainly be the most effective approach to establish new populations of rare species.

ACKNOWLEDGMENTS

Funding for this project came from the National Science Foundation, Conservation Biology Program (to RBP) and the Long Term Ecological Research Program (to the Harvard Forest). Permission to use the field sites came from the Newton Conservation Commission and the Harvard Forest. Miao Shili and Brian Drayton assisted in the field work. This manuscript benefited from the comments of S. Dayanandan, L. DeLissio, B. Drayton, S. L. Miao, C. Millar, B. Pavlik, E. Platt, M. Pomeroy, D. Ramseier, and K. Urbanska.

REFERENCES

Barkham, J. P. 1992. "Population Dynamics of the Wild Daffodil (*Narcissus pseudonarcissus*). IV. Clumps and Gaps." *Journal of Ecology* 80: 797–808.

Beck, B. B., L. G. Rapport, M. Stanley Price, and A. Wilson, 1994. "Reintroduction of Captive-Born Animals." In *Creative Conservation: Interactive Management of Wild and Captive Animals*. Edited by G. Mace, P. Olney, and A. Feistner. London: Chapman and Hall.

Cook, R. E. 1979. "Patterns of Juvenile Mortality and Recruitment in Plants." In *Topics in Plant Population Biology*. Edited by O. T. Solbrig, S. Jain, G. B. Johnson, and P. Raven. New York: Columbia University Press.

Crawley, M. J. 1989. "Chance and Timing in Biological Invasions." In *Biological Invasions: A Global Perspective*. Edited by J. A. Drake and H. A. Mooney. New Delhi, India: Thomson Press.

Davis, M. B. 1989. "Lags in Vegetation Response to Greenhouse Warming." *Climatic Change* 15: 75–82.

Davis, M. B., and C. Zabinski. 1992. "Changes in Geographical Range Resulting from Greenhouse Warming: Effects on Biodiversity in Forests." In *Global Warming and Biological Diversity*. Edited by R. Peters and T. Lovejoy. New Haven, Conn.: Yale University Press.

Denslow, J. S. 1987. "Tropical Forest Gaps and Tree Species Diversity." *Annual Review of Ecology and Systematics* 18: 431–451.

Drayton, B., and R. Primack. 1995. "Plant Species Loss from 1894 to 1993 in an Isolated Conservation Area in Metropolitan Boston." *Conservation Biology*, in press.

Ehrlich, P. 1989. "Attributes of Invaders and the Invading Process: Vertebrates." In *Biological Invasions: A Global Perspective*. Edited by J. A. Drake and H. A. Mooney. New Delhi, India: Thomson Press.

Fernald, M. L. 1970. *Gray's Manual of Botany*. New York: Van Nostrand Reinhold.

Franco, A. C., and P. S. Nobel. 1988. "Interactions Between Seedlings of *Agave deserti* and the Nurse Plant *Hilaria rigida*." *Ecology* 69: 1731–1740.

Franklin, I. R. 1980. "Evolutionary Change in Small Populations." In *Conservation Biology: An Evolutionary-Ecological Perspective*. Edited by M. E. Soulé and B. A. Wilcox. Sunderland, Mass.: Sinauer Associates.

Griffith, B., J. M. Scott, J. W. Carpenter, and C. Reed. 1989. "Translocation as a Species Conservation Tool: Status and Strategy." *Science* 245: 477–480.

Grubb, P. J. 1977. "The Maintenance of Species Richness in Plant Communities: The Importance of the Regeneration Niche." *Biological Review* 52: 107–145.

Guerrant, E. O. 1992. "Genetic and Demographic Considerations in the Sampling and Reintroduction of Rare Plants." In *Conservation Biology: The Theory and Practice of Nature Conservation, Preservation and Management*. Edited by P. L. Fiedler and S. K. Jain. New York: Chapman and Hall.

Hall, L. A. 1987. "Transplantation of Sensitive Plants As Mitigation for Environmental Impacts." In *Conservation and Management of Rare and Endangered Plants*. Edited by T. S. Elias. Sacramento: California Native Plant Society.

Hanski, I. 1982. "Dynamics of Regional Distribution: The Core and Satellite Species Hypothesis." *Oikos* 38: 210–221.

Hanski, I., and M. Gilpin. 1991. "Metapopulation Dynamics: Brief History and Conceptual Domain." *Biological Journal of the Linnean Society* 42: 3–16.

Harper, J. L. 1977. *Population Biology of Plants*. London: Academic Press.

Howe, H. F. 1984. "Implications of Seed Dispersal by Animals for Tropical Reserve Management." *Biological Conservation* 30: 261–281.

Jacobson, G. L. Jr., H. Almquist-Jacobson, and J. C. Winne. 1991. "Conservation of Rare Plant Habitat: Insights from the Recent History of Vegetation and Fire at Crystal Fen, Northern Maine, USA." *Biological Conservation* 57: 287–314.

Jordan, P. W., and P. S. Nobel. 1981. "Seedling Establishment of *Ferrocactus acanthodes* in Relation to Drought." *Ecology* 62: 901–906.

Kenfield, W. 1970. *The Wild Gardener in the Wild Landscape*. New York: Hafner.

Kleiman, D. G. 1989. "Reintroduction of Captive Mammals for Conservation." *BioScience* 39: 152–161.

Kleiman, D. G., M. Stanley Price, and B. B. Beck, 1994. "Criteria for Reintroductions." In *Creative Conservation: Interactive Management of Wild and Captive Animals*. Edited by P. Olney, G. Mace, and A. Feistner. London: Chapman and Hall.

Lande, R. 1988. "Genetics and Demography in Biological Conservation." *Science* 241: 1455–1460.

Lovejoy, T. E. et al. 1986. "Edge and Other Effects of Isolation on Amazon Forest Fragments." In *Conservation Biology: The Science of Scarcity and Diversity*. Edited by M. E. Soulé. Sunderland, Mass.: Sinauer Associates.

Menges, E. S. 1986. "Predicting the Future of Rare Plant Populations: Demographic Monitoring and Modeling." *Natural Areas Journal* 6: 13–25.

Menges. E. S. 1990. "Population Viability Analysis for an Endangered Plant." *Conservation Biology* 4: 52–62.

Menges, E. S. 1991. "The Application of Minimum Viable Population Theory to Plants." In *Genetics and Conservation of Rare Plants*. Edited by D. A. Falk and K. E. Holsinger. New York: Oxford University Press.

Murphy, D. D., K. E. Freas, and S. B. Weiss. 1990. "An Environment-Metapopulation Approach to Population Viability Analysis for a Threatened Invertebrate." *Conservation Biology* 4: 41–51.

Nobel, I. 1989. "Attributes of Invaders and the Invading Process." In *Biological Invasion: A Global Process*. Edited by J. A. Drake and H. A. Mooney. New Delhi, India: Thomson Press.

Pavlik, B. M., 1994. "Demographic Monitoring and the Recovery of Endangered Plants." In *Recovery and Restoration of Endangered Species: Conceptual Issues, Planning, and Implementation*. Edited by M. Bowler and C. Whelin. Cambridge, England: Cambridge University Press.

Pavlik, B. M., forthcoming. "Conserving Plant Species Diversity: The Challenge of Recovery." In *Biodiversity in Managed Landscapes*. Edited by R. C. Szaro. Oxford University Press.

Peters, R., and D. S. Darling. 1985. "The Greenhouse Effect and Nature Reserves." *BioScience* 35: 707–717.

Peterkin, G. F., and M. Game. 1984. "Historical Factors Affecting the Number and Distribution of Vascular Plant Species in the Woodlands of Central Lincolnshire." *Journal of Ecology* 72: 155–184.

Pimm, S. L., H. L. Jones, and J. Diamond. 1988. "On the Risk of Extinction." *American Naturalist* 132: 757–785.

Primack, R. 1993. *Essentials of Conservation Biology*. Sunderland, Mass.: Sinauer Associates.

Primack, R., and S. L. Miao. 1992. "Dispersal Can Limit Local Plant Distribution." *Conservation Biology* 6: 513–519.

Pullin, A. S., and S. R. J. Woodell. 1987. "Response of the Fen Violet, *Viola persicifolia* Schreber, to Different Management Regimes at Woodwalton Fen National Nature Reserve, Cambridgeshire, England." *Biological Conservation* 41: 203–217.

Rabinowitz, D., S. Cairns, and T. Dillon. 1986. "Seven Forms of Rarity and Their Frequency in the Flora of the British Isles." In *Conservation Biology: The Science of Scarcity and Diversity*. Edited by M. Soulé. Sunderland, Mass.: Sinauer Associates.

Ricklefs, R. E. 1993. *The Economy of Nature*. New York: W. H. Freeman.

Sauer, J. 1988. *Plant Migration*. Berkeley: University of California Press.

Simberloff, D. 1988. "The Contribution of Population and Community Biology to Conservation Science." *Annual Review of Ecology and Systematics* 19: 473–511.

Soulé, M. E., ed. 1987. *Viable Populations for Conservation*. Cambridge, England: Cambridge University Press.

Terborgh, J. 1974. "Preservation of Natural Diversity: The Problem of Extinction-Prone Species." *BioScience* 24: 715–722.

Terborgh, J. 1989. *Where Have All the Birds Gone? Essays on the Biology and Conservation of Birds That Migrate to the American Tropics*. Princeton, N.J.: Princeton University Press.

Terborgh, J. 1992. "Why American Songbirds Are Vanishing." *Scientific American* 264: 98–104.

Urbanska, K. M., and M. Schütz. 1986. "Reproduction by Seed in Alpine Plants and Revegetation Research Above Timberline." *Botanica Helvetica* 96: 43–60.

Vasquez-Yanes, C., and A. Orozco-Segovia. 1993. "Patterns of Seed Longevity and Germination in the Tropical Rainforest." *Annual Review of Ecology and Systematics* 24: 69–87.

Vasquez-Yanes, C., and A. Orozco-Segovia. 1994. "Signals for Seeds to Sense and Respond to Gaps." In *Exploitation of Environmental Heterogeneity by Plants*. Edited by M. Caldwell and R. Pearcy. San Diego, Calif.: Academic Press.

Webb, T. III. 1992. "Past Changes in Vegetation and Climate: Lessons for the Future." In *Global Warming and Biodiversity*. Edited by R. Peters and T. Lovejoy. New Haven, Conn.: Yale University Press.

CHAPTER 10

Monitoring

Robert D. Sutter

The successful reintroduction of a rare species into a conservation area is a complex and protracted process. Viewed in its entirety, the process consists of many overlapping components. Some of these are concluded early in a project, while others may take years or even decades to complete. Much of the recent literature and many of the other chapters in this book focus on the decision-making steps that are necessary before a reintroduction project is initiated, including the strategic context for reintroduction and restoration, the selection of sites for reintroduction, the source and collection of propagules, and the technical feasibility of propagating and transplanting a species. However, if reintroduction is to be meaningful in the long term, its proponents and practitioners must acknowledge that design and planting are but the first steps in a commitment that extends for many years.

Monitoring is a process that provides information for management decisions. It involves detecting changes in managed and unmanaged populations, communities, and ecological systems over time. The repetition of measurements over time for the purpose of quantifying change distinguishes monitoring from inventory (MacDonald 1991). Monitoring is a component of all scientific research and is an essential component of conservation, as information on the status of populations, communities, and ecosystems assumes increasing importance in management and decision making (White and Bratton 1980; Sutter 1986; Travis and Sutter 1986; Pavlik 1987; Stewart-Oaten 1993). Monitoring can be used to obtain basic biological information on populations, species, and communities. This includes life-history processes such as seed production, pollination, herbivory, predation, dispersal, and seed and plant dormancy. Monitoring can also be used in an experimental context where different management techniques can be tested (see Pavlik, Chapter Six). The inclusion of monitoring in an experimental approach can result in a project that brings twofold success: success in biological terms by reintroducing a rare species to a site, and success in general project terms by obtaining information about species biology, community

processes, management techniques, and conservation policy (see Pavlik, Chapter Six).

Monitoring is essential to any reintroduction project to determine the abundance, spatial extent, resilience, and persistence of the reintroduced population (Table 10-1). In the short term, monitoring is needed to document the establishment and survivorship of reintroduced plants and the basic life-history processes of growth and reproduction. In the long term, monitoring is essential to determine the population's response to management, the occurrence of recruitment, and the natural variability in population size and processes. In addition, information about future population states and the self-sustainability of the population is provided through consistent, sustained monitoring. Monitoring of a reference population is also necessary to obtain information on life-history and demographic characteristics of a naturally functioning population. Without monitoring, there can be no assessment of a reintroduction's success or failure and little insight into the management needed to enhance and ensure the survival of populations. Monitoring is the foundation of success in a good reintroduction project; it is not a luxury.

This chapter addresses this latter stage of the reintroduction process and examines the monitoring of reintroduced populations. It is the stage that will eventually require the greatest amount of time in any reintroduction project. Monitoring will be necessary for years, perhaps decades, until there is confidence that the reintroduced population is self-sustaining within its maintained or restored ecosystem. Monitoring will also be needed even further into the future, to document the continued status of the population and ecosystem. Thus the monitoring component of reintroduction requires a substantial long-term commitment, which appears to be absent from many restoration projects (Hall 1987; Howald, Chapter Thirteen; Palmer 1987). In addition, poorly designed monitoring may be resulting in the erroneous evaluation in a substantial number of other projects (Palmer 1987; Fiedler 1991; Pavlik 1993).

The Science of Monitoring

In order to reliably and precisely detect change in populations and communities, monitoring techniques and processes must meet four criteria:

1. Monitoring data must have a known and acceptable level of precision. Monitoring should detect real changes in a population with a level of accuracy that investigators find acceptable. Factors that guide this type of decision making include the extent of change expected in the population, the variability inherent in the population, and the time and funding available.

TABLE 10-1.   Monitoring objectives and data needs for a reintroduced population, grouped by goals of a rare plant reintroduction (Pavlik, Chapter Six).

Characteristics of abundance and extent

1. Reintroduced plants and established seedlings
    Monitoring objective:
        survival and performance of reintroduced and established seedlings
    Data collected:
        demographic data, performance data, life-history data
2. Recruitment of new individuals into the population
    Monitoring objective:
        establishing seedlings and safe sites for seedling establishment
    Data collected:
        presence of seedlings, occurrence of safe sites

Characteristics of resilience and persistence

3. Community condition and functioning
    Monitoring objective:
        processes and conditions that influence the reintroduced population
    Data collected:
        species composition (floristic lists, richness, diversity, evenness)
        species abundance (cover, density, frequency)
        community structure (vertical structure, horizontal structure)
        community function (productivity, effect of disturbance, nutrient cycling)
4. Genetic variability
    Monitoring objective:
        maintenance of the natural range of genetic diversity
    Data collected:
        allelic and genotypic data or their surrogates

2. Data collection techniques need to be repeatable over years and across personnel. By definition, monitoring consists of multiple samples collected over time and analyzed to detect change. To enable accurate, repeated measurements in a population over time requires that sampling sites can be found repeatedly; that sampling units (individuals, transects, or plots) can also be relocated; and that data from the sampling units can be repeatedly and consistently collected (that it is clear which data were originally collected and how they were collected). These procedures will enable different personnel to be trained to find sampling sites and units and to collect data. New personnel need to be able to replicate monitoring because land managers and researchers are often transient, while preserves and populations are permanent. It is also important that all necessary directions, maps, and data are archived and accessible.

3. Data must be collected over a long enough period of time to capture important natural processes and responses to management (Likens 1983,

1989; Magnuson 1990). This could be five years or fifty, but it is clear that in most cases two or three years of data are rarely enough and may at times be misleading. Specifically, long-term monitoring is needed to measure processes that

- Have long return intervals or are rare events (such as drought, fire, floods, ice or wind storms, and hurricanes)

- Vary from year to year (such as hydrology, rainfall, and average temperatures)

- Respond slowly (such as succession, disturbance, and effects of fragmentation)

- Have subtle changes that are impossible to measure in the short term

- Take a long time to equilibrate after disturbance

There are also cumulative responses of populations and communities that can only be understood through extended observation. The important dynamics in populations, such as recruitment, survivorship/mortality, seed set, predation, and herbivory, are usually dependent on natural processes that vary across time. The ultimate performance criterion for reintroduced populations is the continued recruitment and survival of individuals through a range of environmental conditions; the execution of this criterion depends upon long-term monitoring.

4. Efficiency must be considered an integral component of monitoring. No matter how well-designed a monitoring study is, if it is too expensive or too complex to continue over time, its utility is diminished. Efficiency means that the program can answer questions of pertinent conservation value while incurring reasonable cost. A monitoring program should be feasible, realistic, and inexpensive enough to be maintained over the long term.

Given these criteria, a critical aspect of designing a project is to specify the monitoring objectives (Falk and Olwell 1992; Travis and Sutter 1986). The clarity of the objectives will determine the focus and quality of the subsequent monitoring design and analysis. Monitoring objectives should possess various characteristics: They should (1) be specific and quantifiable; (2) provide the framework for defining specific tasks; (3) specify the variables to be measured and the frequency of the measurements; (4) specify how success or failure will be assessed in various phases of the project (see Pavlik, Chapter Six); (5) help communicate and justify the project and provide a historical record of the project.

Monitoring objectives, when populations are sampled, also should in-

clude specified levels of the false-change error rate (Type I error), the missed-change error rate (Type II error), and the magnitude of change one is interested in detecting (Salzer 1993). A *false-change error* occurs when the monitoring detects a change, but no change has taken place. This error is addressed when one sets an alpha or *p* value for a statistical test. A *missed-change error* occurs when the monitoring detects no change, but in reality there has been a change. While missed-change errors are an important consideration in all research, missing change in a population of an endangered species could result in extinction. (For discussion of the importance of Type II errors in conservation science see Eberhardt and Thomas 1991, Fairweather 1991, and Peterman 1990.) *Power* is the complement of the missed-change error (Sokal and Rohlf 1969). The false-change error rate is inversely related to the missed-change error rate: As one increases the other decreases (Salzer 1993). Thus, setting both of these error rates requires a compromise between these two types of error.

Monitoring objectives will also define which design will be used to detect change in reintroduced populations. There is a continuum of monitoring designs, ranging from observational monitoring to experimental monitoring, the choice of which is determined by the monitoring objectives and resources available.

Observational monitoring is conducted to collect quantitative data but is not intended to test specific hypotheses about the nature of the relationship between variables. This is equivalent to the measurative experiment of Hurlbert (1984) and the analytical sampling of Eberhardt and Thomas (1991). Included in this group are studies that investigate changes in variables (population size, individual growth, or reproduction) that occur as a function of time.

In contrast, *experimental monitoring* is conducted to investigate the specific effects of one or more independent variables on one or more dependent variables. An experimental design requires that the investigator has control over the definitions and levels of the independent variables (treatments) and that individuals are randomly placed in treatment groups. For example, an experimental design might be used to test the effect of different frequencies of fire on the growth and reproduction of a species. The goal of an experimental design, therefore, is to determine cause and effect.

Experimental designs are rigorous by definition and may be too difficult and expensive to use in the field-oriented work of conservation and environmental sciences for many reasons (Eberhardt and Thomas 1991). Populations of rare species are often small (Barrett and Kohn 1991), and provide limited offerings for large sample sizes or replication of experimental treatments. In many cases, monitoring is also initiated *ex post facto* to detect the effects of trampling, grazing, fire, or changes in hydrology. The treatments in these studies often may be uncontrolled and unreplicable.

In conservation, detecting trends or correlations may be the more appropriate approach. In these cases, a *quasiexperimental design* may be used. In this design, the independent variable is typically less rigorously controlled, and/or the investigator cannot randomly assign subjects to the treatment groups. An example of a quasiexperimental design is a study designed to determine the effect of trampling on a low-health plant community in the Southern Appalachians (Sutter et al. 1993). The objective of the study was to determine the impact of trampling on specific species within a defined area after a boardwalk was constructed. Quasiexperimental studies should be recognized in conservation and environmental studies as providing important insights into the biology of species and different management regimes (Stewart-Oaten 1993).

The Basic Components of Sampling Design

Sampling design is the selection and spatial arrangement of sampling units used to measure specific variables in a population or community. An appropriate sampling design is essential for the efficient collection of accurate monitoring data when sampling a population or community. The sampling design used in a monitoring study maximizes the ability to distinguish real changes, trends, or differences from random variation (Greig-Smith 1983; Green 1979; Kenkel, Juhasz-Nagy, and Podani 1989). It appears, however, that in current practice most sampling design decisions are made arbitrarily, uncritically, or by following general sampling procedures recommended in the literature that are specific to certain research conditions (Salzer, pers. comm.; Kenkel et al. 1989; Hurlbert 1984).

When developing a monitoring study, there are six major sampling design decisions. The decisions involve determining (1) the sampling universe, (2) the placement of sample units in the sampling universe, (3) the selection of sampling units, (4) the selection of permanent or temporary sampling units, (5) the sampling frequency, and (6) the number of samples that need to be collected. Each of these decisions will be discussed in more detail.

1. The Sampling Universe
The *sampling universe* refers to any collection of individuals or circumscribed area from which biological data are to be collected and about which inferences are to be made. Examples include all the individuals of a single species in a specific area, all species within a specific plant community, or a portion of a plant community occurring in an area that is being managed. Delineating the sampling universe defines the population of biological importance or interest.

For a reintroduced population, there are several levels of sampling universes. These are the specific universe of plants that have been reintroduced into a site, the broader universe of habitat appropriate for seedlings within the site, and the inclusive universe of the community or ecosystem that influences the study population. In a reintroduction project the sampling universe will also include new individuals that are recruited into the population or changes in the occurrences of a community type over time. Each of the sampling universes represents a population of interest that will be examined.

2. Placement of Sample Units in the Sampling Universe

The intent of sampling is to collect data that are representative of the population from which they were taken. In other words, the plants or the area included in the sample should not differ significantly in the variables of interest from the total sampling universe of individuals or areas. There are three primary ways of selecting samples from a population. *Preferential selection* involves subjectively selecting individuals or sites because they are typical, homogeneous, representative, undisturbed, or otherwise of interest because of phenomena they exhibit. *Systematic sampling* is the selection of samples from a regular pattern. *Random selection* means that each individual or sample position within the population has an equal probability of being selected for inclusion in the sample.

Random sampling is highly recommended for most long-term monitoring projects (Goldsmith 1991; Greig-Smith 1983; Stuart 1976) because it eliminates possible selection bias on the part of the investigator and technically allows the use of statistics to make inferences about the larger population. There are several types of random sampling (Figure 10-1): simple random sampling, systematic random sampling (Greig-Smith 1983), stratified random sampling (Usher 1991), two-stage sampling (Cochran 1977; Greig-Smith 1983), and adaptive cluster sampling (Thompson 1991). The most commonly used and one of the most efficient methods is stratified random sampling. With this approach, the population is divided into units, or strata, prior to sampling. Each stratum has characteristics relevant to the monitoring question. Strata can be based on natural subdivisions of the population or community, such as slope, aspect, moisture gradient, canopy, or demographic characteristics; it can also follow other subdivisions such as management units, timber units, or areas of previous disturbance. Once the strata are defined and designated, individuals or areas are randomly chosen from each stratum, and samples are allocated in proportion to the size of the stratum, number of individuals in each stratum, or the variability of selected characteristics within each stratum (Greig-Smith 1983). Stratified random sampling is used to ensure that each stratum is represented in the sample and to increase the precision of overall population estimates.

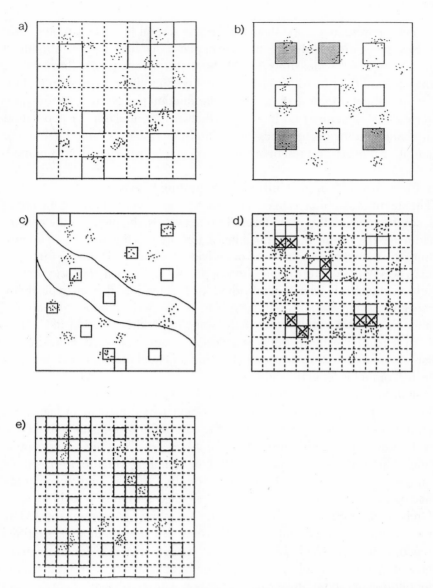

Figure 10-1. Examples of five of the six types of random sampling: (a) simple random sampling (squares outlined in bold are sampled plots); (b) systematic random sampling—sampling units are first chosen systematically (represented by the squares), then a random subset of these are chosen (represented by shaded squares); (c) stratified random sampling (sampled units are randomly chosen within each of the three strata); (d) two-stage sampling—sampling units containing the species of interest (marked with an X) are chosen within a larger, randomly chosen sample unit (the four-square clusters); (e) adaptive cluster sampling—if the species of interest is found in the randomly chosen sampling unit then all adjacent sampling units are sampled, continuing until the outer sampling units do not contain the species of interest.

3. Selection of Sampling Units

Sampling units are the areas from which samples are taken. There are many types, including individuals, points, lines, and plots. Each of these sampling units is appropriate for the collection of certain types of data in specific vegetation types.

The most commonly used sampling unit is the plot or quadrat. These can be of various sizes, with the most frequently used size and shape being a 1-by-1 meter square. Plots can be used to measure different measures of abundance (cover, frequency, and density), measures of performance (vegetative and/or reproductive vigor measurements and biomass), and demographic measures (survivorship and fecundity). Plot size and shape have been found to significantly influence the accuracy and precision of the data that is collected (Kenkel, Juhasz-Nagy, and Podani 1989). This finding is surprising and noteworthy when one considers that most biological studies use 1-by-1 meter plots (Salzer 1993).

To illustrate how important plot size and shape are for estimating population parameters, data are presented from an exercise done in monitoring workshops conducted by the Nature Conservancy. The exercise, which has been repeated numerous times with different species and populations with different spatial distributions, involves the use of plots to estimate the total number of plants of one species within a defined 50-by-50 meter area. Prior to the exercise, all the plants of a particular species within the designated area are counted. Workshop participants are divided into teams, and each team is given a different size or shape plot; plot sizes that have been used in the exercise include (in meters) 1 by 1, 3 by 3, 5 by 5, .5 by 2, 1 by 10, and 1 by 50. Teams are given an hour to sample the population using a plot of specific dimensions. Since the total population size is known, the accuracy of each method can be compared.

The results show that plots with different sizes and shapes give very different estimates of the total population (Figure 10-2). In the eight populations sampled in the exercise, large square plots or long rectangular plots (such as 5 by 5 or 1 by 50 meters) consistently obtained the most accurate estimates of population size. Computer-generated sampling done by Salzer (unpublished data) shows a similar pattern.

The spatial distribution of a population is the primary factor influencing the estimates obtained from the various plot sizes and shapes. If a population is randomly distributed, then any plot size and shape will give an equally precise estimate of a population parameter with an adequate sample size. If a population is clumped, plot size and shape will strongly influence sampling variance and the precision of population estimates. Thus, plot sizes that are slightly larger than the estimated mean clump size of the target species are recommended (Kenkel, Juhasz-Nagy, and Podani 1989).

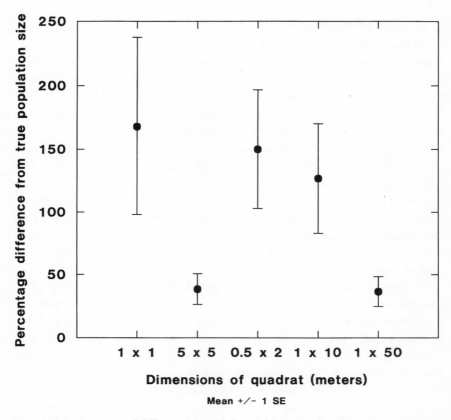

Figure 10-2. Accuracy of different size and shape plots using data from Nature Conservancy monitoring workshops. Data points show percentage difference from true population size according to plot size.

There are, of course, other factors to consider when choosing the size and shape of plots: (1) size of the plant being monitored, (2) density of the population, (3) spatial distribution of the population, (4) edge effects in sampling, and (5) investigator impact. Capturing the variability of a population within rather than among plots results in improved precision in estimating population parameters and detecting change over time. This is true because reducing the variability among plots reduces the variance of the data. Capturing the variability of a population within plots also reduces the sampling size needed to reach predetermined levels of precision.

One must also consider efficiency in establishing plots. Larger (and therefore fewer) plots are often more efficient because less preparation and transport of materials is necessary (consider the weight difference, for example, between twenty and two hundred pieces of rebar), and plots are easier to establish, locate, and relocate (the latter often the most time-consuming component of sampling). Thus, the most appropriate plot size and shape may be a compromise between precision and sampling efficiency.

Plots are not the only sample units. Individual plants are the most appropriate sampling unit when measuring specific characteristics such as plant height, number of fruits, survivorship, and fecundity. Sample lines can be effectively used to measure cover (line intercept) in populations. Points can be used to measure cover or frequency and can be used along a line transect or within a plot. In addition, there are many plotless methods that can be used to measure density and cover (Bonham 1989).

4. Selection of Permanent or Temporary Sampling Units

Permanent sampling units, within which variables are measured over time, are most commonly employed when monitoring populations and communities. Measuring the same individuals or area over time provides the most precise estimate of temporal change and eliminates errors that might be attributed to the selection of different sampling units during different sampling periods (Greig-Smith 1983). Variables measuring temporal change include vegetative growth, reproduction of individual plants, demographic parameters of mortality and survivorship, changes in species composition, and the effects of management.

While permanent sampling units are appropriate for detecting temporal changes, they do not work as well in detecting spatial changes in populations and communities. Populations of annuals, biennials, and short-lived perennials can change their locations, spatial dimensions, and densities over a period of years. Populations of rapidly expanding native species or invasive exotics can also rapidly change their size and density. If permanent sampling units are randomly established in a spatially changing population, these sampling units will yield a declining ability to make accurate inferences about the whole population in subsequent years. The establishment of new sampling units each year may be necessary to solve such problems.

Additionally, there are two statistical considerations associated with the sampling of permanent plots. The first is the lack of independence of the data collected from the same plot over multiple years. This violates one of the basic assumptions of standard parametric statistical tests. Another difficulty arises when there are more than two years of data, in that changes between two subsequent years are not completely independent of changes in previous years. Two-group comparisons or repeated measures analysis techniques are needed to remove the autocorrelation component from the comparisons (Salzer 1993). Autocorrelation can also be avoided through the partial replacement of existing sampling units during each sampling period (Usher 1991).

Permanent sampling units have several potential disadvantages: They are susceptible to loss or damage (the removal of stakes and tags), repeated investigator impacts (trampling), and the permanent marker impacts (attracting

other species). Relocating permanent sampling units can be one of the most time-consuming efforts in monitoring.

5. Sampling Frequency

The frequency with which data is collected depends on the specific objectives of the monitoring project and the rate with which change is expected to take place. More frequent sampling reduces the effect of temporal variation. For example, monitoring the presence and growth of tree species would require a lower frequency of monitoring than monitoring the presence and growth of herbaceous perennials or annuals. A higher frequency of monitoring allows a potentially greater correlation with environmental events such as the effects of drought, wind storms, predation, and herbivory. Some suggested frequencies for different plant population types and population characteristics are given by Spellerberg (1991).

6. Number of Samples to Be Collected

The precision of population estimates increases with the number of sample units that are used. The number of samples needed to obtain a specified level of precision is determined by the amount of variation among sampling units, the size of the effect that one wants to detect, the designated levels of false-change (Type I) errors and missed-change (Type II) errors, and the resources of the investigator (Salzer 1993). While precision increases with larger sample sizes, the increase is not linear. The statistical benefits of increasing sample size are reduced as more samples are taken (Figure 10-3) (Salzer 1993).

There are several sample size formulas that can be used to determine an adequate sample size (Greig-Smith 1983; Eckblad 1991; Kupper and Hafner 1989; Sokal and Rohlf 1969; Salzer 1993), as well as a number of computer software packages (Dallal 1988). All of these formulas require an estimate of variability; thus, data are needed from a pilot study or the first year of the project before an appropriate long-term sample size can be determined. Most of the formulas will give misleading information if the data are extremely skewed and the sample size is small. In these cases transformation of the data may be required to achieve normality. Several formulas also include a calculation of statistical power into the sample size estimate (Kupper and Hafner 1989; Dallal 1988).

Approaches to Monitoring Reintroduced Populations

The scope and implied significance of a reintroduction project require that monitoring measure more than just changes in species abundance over time. To evaluate the success of establishing a self-sustaining population, four char-

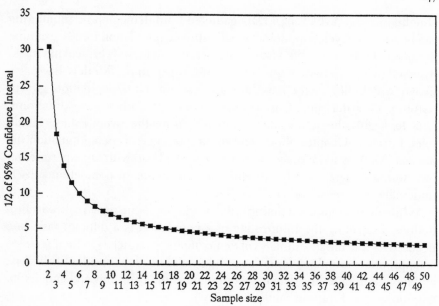

Figure 10-3. Influence of sample size on level of precision with one-half of the 95 percent confidence interval used to measure levels of precision (Salzer 1993).

acteristics of a reintroduced population need to be monitored: (1) plants reintroduced to the site (2) recruitment of new individuals into the population (3) condition and processes of the community and ecosystem, and (4) genetic variability of the population. I describe these characteristics separately, because each involves different monitoring objectives, variables, sampling designs, sampling frequencies, and measures of success.

1. Monitoring Plants Reintroduced to the Site

Demographic monitoring is the most sensitive and effective method of detecting change in populations. This approach follows marked or mapped individuals over time to estimate the survivorship and fecundity of different age or stage classes. These data provide a detailed assessment of the population, including mortality and survivorship, plant condition, seed production, and recruitment; they also provide an overview of population structure. When expanded to include the collection of additional data on environmental variables and attributes of the community, demographic monitoring can provide insights into the factors that control the abundance and distribution of a rare species.

Demographic data, when integrated into demographic models, can provide insights into the long-term sustainability of populations (Crouse, Crowder, and Caswell 1987; Lande 1988; Menges 1986, 1990, 1991; Waite and

Hutchings 1991; Pavlik 1993). Data gathered from demographic monitoring can be used to develop models that will estimate population trends, examine the effect of variance in life-history parameters, and identify critical life stages (Caswell 1989; Menges 1990; Stacey and Taper 1992). Models have also shown which life stages have the greatest effect on population growth (Crouse, Crowder, and Caswell 1987; Noon and Biles 1990; Guerrant, Chapter Eight), the significance of dispersal during the juvenile stage (Lande 1988; Primack, Chapter Nine), and the importance of metapopulation dynamics (Menges 1990; Stacey and Taper 1992). Used with an experimental monitoring design, models can also compare trends in population subsets under different treatments (Pavlik, Chapter Six).

While demographic modeling is a powerful tool, one must realize its limitations. Projecting the future state of a population is a difficult task; one needs to have accurate estimates of demographic parameters (which requires an appropriate sampling design), an understanding of the species life history, and long-term sampling data to record significant environmental and demographic events (King and Sutter, in review).

Demographic data can also be used to gain an understanding of a population's minimum viable size (MVP), the population size that is large enough to survive the systematic and stochastic events that threaten populations with extinction (Soulé 1987; Menges 1991; Guerrant, Chapter Eight; Pavlik, Chapter Six). Numerical estimates of MVPs require substantial data on the life history of the species, demographics of the population, and environmental, demographic, and genetic stochasticity (Menges 1991). Of these stochastic factors, environmental stochasticity is thought to have the greatest effect on populations (Menges 1991; Stacey and Taper 1992). While MVPs have been difficult to assess for most species, the long-term monitoring of reintroduced populations may provide an appropriate opportunity to test these models (Pavlik, Chapter Six).

The collection of demographic data requires following individual plants over time and taking repeated measurements. The sampling unit is an individual plant. Individuals are not always easy to identify, especially in clonal species. For this reason, most demographic studies of clonal species have not been based on genetic individuals but have conducted analyses on information collected from ramets (Harper 1977; Panos 1989; Schemske et al. 1994). The planting of individual genets in a reintroduced population provides an opportunity to collect demographic data on a clonal species, at least until individuals grow together. Tagging individuals is the primary method of collecting demographic data. The type of tags and stakes used to mark individual plants will be determined primarily by the environmental conditions in which they will need to survive and the proper level of visibility (visible to the investigator but not easily seen by other humans and animals). Mapping

individuals is also useful to track individuals and can be used as a backup in case individual markers disappear.

Survivorship and fecundity for each age or stage class are the basic attributes to measure in demographic studies. Stage classes, categorized by plant size and reproductive status, are preferred for perennial species because plant size is usually the best predictor of survivorship and fecundity (Harper 1977; Menges 1986, 1991; Guerrant, Chapter Eight). The frequency of sampling these demographic variables will be determined by their temporal variation. To obtain the most accurate assessment of demographic parameters, plants need to be sampled across the range of environmental conditions (soil moisture, light levels, and so on) present in the population.

The emphasis on modeling populations has deterred many investigators from collecting other important life-history data. Detailed measurements of plant condition (or vigor) and observations of other specific life-history components allow a more critical assessment of population status and trends and may also provide insights into the factors that control abundance and distribution. When statistically related to environmental variables or as part of an experimental design, correlation or causality can often be examined.

The two major categories of performance data that can be collected are measures of vegetative condition and reproductive condition.

Measures of vegetative condition	Measures of reproductive condition
Height	Number of flowers
Girth/diameter at breast height	Number of inflorescences
Number of stems	Number of fruit
Number of branches	Number of seeds
Number of leaves	
Number of leaf units (whorls, nodes)	
Leaf size	
Leaf area	
Rosette diameter	

Measuring all of these performance attributes would be extremely time consuming, so variables that are most sensitive to changes in growth and reproduction in the species of interest should be chosen. The selection of the most sensitive variables will differ depending on the life history and growth form of the species. For example, plant height and stem girth is strongly correlated with health and reproduction in most nonrhizomatous herbaceous perennials. Rhizomatous or clonal herbaceous perennials are best monitored by number of stems or cover. The measure used most often for single-stemmed woody species is diameter at breast height. In multistemmed or clonal woody species one would monitor the number of stems, stem volume (Chew and Chew 1965), or cover.

Annuals pose a difficult set of problems. While plant size and reproductive output are relatively good indicators of health, the plasticity of annual species as they respond to variable environmental conditions suggests that individual plant variables may not be as important as measures of areal extent or appropriate habitat. Life-history components such as seed banks, suitable sites for establishment, and the natural processes that form these suitable sites may be especially important for annuals. For all species, reproduction is best measured by flower production and fruit or seed set. Although flower production can also be a valuable reproductive measurement, seed set can be limited by pollinators and seasonal environmental conditions.

It is optimal to measure a large number of attributes at the beginning of a monitoring project and then, through correlation of variables, reduce the number of attributes to those that are the most meaningful. The importance of certain attributes can also change in value over the life history of the species. For example, in woody species, height may be the most important attribute for saplings, while stem diameter is more meaningful for adult trees. When choosing attributes, one should also be aware that some are easier to measure than others. For instance, it is much less time consuming to take one measurement of plant height than it is to count the total number of leaves. Additionally, when interpreting performance data, one should be aware that it may be difficult to separate the effects of management from those of weather or other large-scale factors.

The life history of a plant species is a complex linear sequence of events, from seed germination through seed production to dispersal (Harper 1977). Several diagrammatic models of a species' life history have been presented (Sarukhán and Gadgil 1974; Harper 1977; Whitson and Massey 1981) to illustrate the effect of these factors, all of which are designed to focus attention on the many events that may become limiting factors in the life history of a plant.

If a reintroduced population is declining in condition or abundance then a systematic and comprehensive approach is needed to identify critical or limiting life-history components (Table 10-2). This approach initially involves collecting specific information, including (1) quantitative data on adult mortality and condition, seed production, and recruitment; (2) qualitative data on the occurrence of herbivory, predation, pathogens, and competition; (3) evidence of the species' breeding system (needed for the evaluation of seed production, MVPs, and genetic variation); and (4) quantitative data on climatic conditions from a local weather station. Once these data are collected, then three primary questions can be asked: (1) Are adults declining in vigor? (2) Are flowers and seeds being produced? (3) Are seedlings becoming established?

To analyze data and answer these questions requires knowledge of the range of variation in mortality, seed set, and establishment in naturally occurring populations. This information can be obtained from a reference pop-

TABLE 10-2. A systematic approach to investigating life-history components of reintroduced plants.

---

1. Are adults declining significantly in vigor or increasing in mortality?

    If yes, these factors could be limiting the survival or vigor of plants in the population:

        • Predation, herbivory, or pathogens

        • Competition with other species

        • Environmental factors (such as drought, changes in hydrology, or the absence of fire)

2. Are adequate flowers and seeds being produced?

    If no, these factors could be limiting seed production:

        • Resource availability

        • Inadequate pollination

        • Seed predation

        • Environmental factors (such as drought, changes in hydrology, or the absence of fire)

        • Genetic quality of offspring

3. Are seedlings becoming established in a expected pattern?

    If no, there are several factors that can limit establishment:

        • Seeds are not dispersed to safe sites

        • Safe sites for establishment not present

        • Seed predation after dispersal

        • Climatic/environmental factors not favorable for establishment

        • Herbivory of seedlings

        • Density dependent mortality

        • Pathogens

        • Genetic quality of seedlings

---

ulation, the literature on the species or congeners, or generalized performance criteria. While the first source is the most appropriate, the literature and generalized performance criteria provide significant and unique insights. For example, the percentage of ovules developing into seeds varies considerably among species, ranging from less than one percent to one hundred percent (Bawa and Webb 1984; Stephenson 1981). Seed set differs between annuals and perennials, varies among breeding systems, and also varies over a single season, among years, and among populations. It is evident that this variability makes interpreting seed set data challenging.

For a reintroduced population, one is concerned with the proximate factors that control seed production, such as resource availability, inadequate pollination, seed predation, environmental conditions or events, and the genetic quality of offspring. Many of these factors are influenced by human alteration of the environment and can be managed. If one of these factors is identified as a possible limiting factor, it can be further investigated, and specific management goals can be set. For example, if pollinators are required for seed set, then management efforts may need to address the regional use of pesticides or the effect of development and grazing on foraging, ovipositing, or nesting habitat (Karron 1991). If seed set is low for genetic reasons, then greater genetic diversity can be obtained by planting different genotypes (DeMauro 1993). Genetic causes for low seed set (such as inbreeding,

homozygous deleterious recessives, multigene family traits) are, however, difficult to detect and perhaps harder to mitigate.

The effect and interaction of each of these factors on a reintroduced population will vary from year to year. This is particularly true for environmental conditions and the effects of predation and herbivory. Determining if one of these factors really limits the survivorship of a reintroduced population will require numerous years of monitoring.

2. Monitoring Recruitment of New Individuals into Population

The recruitment and establishment of new individuals is one of the most important measures of success for a reintroduced population (Pavlik, Chapter Six, Gordon forthcoming). It reflects the achievements of many life-history components: adult survivorship and vigor, flower production, pollination, seed set, and seed dispersal. It also reflects the presence of safe sites in the surrounding environment for the establishment of recruits. Yearly recruitment is vital for annual species, while recruitment in long-lived perennials is episodic (Harper 1977). In long-lived perennials the establishment of seedlings is a rare event with thousands of seeds producing few individuals (Harper 1977).

Recruitment can be seen as a function of the number of safe sites in which seeds find the appropriate conditions for germination and establishment (Harper 1977; Primack, Chapter Nine). Safe sites exist at the scale of the seed and are determined by the environmental heterogeneity of soil, soil moisture, nutrients, light, microtopography, and impacts of predators and competitors (Harper 1977). These sites may be continuously available or the result of some irregularly occurring perturbation. The natural disturbance events that create safe sites (such as rodent burrows or downed trees) may already occur in a conservation area or may need to be restored (as with fire, artificial light gaps in even-age forests, and hydrologic regimes). The location of seedlings in a population can tell us a great deal about the conditions and natural processes needed for establishment.

Measuring recruitment requires a different monitoring design than for following already established plants. Instead of following individuals at known locations, areas need to be monitored to detect the occurrence of seedlings. If the area in which seedlings are likely to establish is small, the total area can be monitored. If the potential establishment area is large, or if the species has expanded from the original plantings at later stages in a reintroduction, then the area will require random subsampling. Temporary plots are appropriate to detect spatial changes in a population; their dimensions will depend on seedling size and density, edge effects, and investigator impact.

Defining the area to be sampled for seedling establishment will vary with the seed dispersal characteristics and growth form of the species. Seed dispersal is frequently limited in distance; the majority of seeds fall in the vicinity

of their parents, while a few seeds disperse further (Harper 1977; Howe and Smallwood 1982; Primack, Chapter Nine). This pattern is especially true in wind- and gravity-dispersed species. Other agents can facilitate further dispersal. Ant dispersal, a surprisingly common method of seed dispersal in some habitats, can move seeds up to 70 meters (Handel and Beattie 1990), and vertebrate species can disperse seeds tens of miles from a source plant (Howe and Smallwood 1982). Species dispersed by water also can be transported long distances (Waser 1982). These dispersal modes influence the spatial distribution of seedling establishment. A stratified placement of plots to survey for seedlings is recommended, with one stratum being the area immediately surrounding fruiting plants, and the dimensions of the others defined by the type of dispersal agents and the occurrence of safe sites.

Obtaining accurate establishment data for demographic modeling requires frequent surveys of the sampling universe. Searching for and marking all seedlings is a tedious and massive task, especially in species that have high seedling establishment and mortality (Harper 1977; Howe and Smallwood 1982). Seedling mortality can be related to the distance from its parent that the seedling has established, with those that have dispersed farther having higher survivorship. Causes of mortality close to the parent include density-dependent mortality and herbivory (Howe and Smallwood 1982). Once new recruits have been established, then these individuals can be permanently marked and followed demographically.

A study of a federally listed species, *Hudsonia montana*, illustrates some of the issues involved with the monitoring of recruitment (Frantz and Sutter 1987; Frost 1993). The goal of the study was to test the effect of fire on recruitment and adult growth. A randomized block design was used, with three treatments: burning, clipping, and control. During the two years after the treatments, few seedlings were found, and those that had established had no obvious pattern among the three treatments. After the third growing season, the wettest summer of the three, there was an explosion of seedlings, with the majority found in the burn treatment. Suitable sites for seedling establishment apparently had resulted from fire that had exposed bare mineral soil combined with the appropriate moisture conditions provided by a wet summer. To obtain this information required the restoration of a natural process (fire) and the patience to monitor the population through a range of environmental conditions over time.

3. Monitoring Community Condition and Processes
The ultimate goal in any reintroduction project should be not only the reestablishment of a population but the restoration or maintenance of the surrounding plant community. There are several important biological reasons for restoring community conditions and processes. Restoration preserves

reintroduced species within their natural ecological and evolutionary setting, increases the probability of establishing self-sustaining populations, and may eliminate threats that originally extirpated populations from the site. The goal of reintroducing self-sustaining populations within a restored natural community marks the difference between protective gardening and successful reintroduction (Zedler, Chapter Fourteen).

One primary implication of a community-based reintroduction is that the monitoring and management emphasis expands over time from the population to the community. Monitoring plant communities and community processes is more challenging than monitoring populations (Goldsmith 1991) and is usually more costly and time consuming (Treshow and Allan 1985). The first challenge is to define the sampling universe, which includes all significant influences on the reintroduced population and the community. The immediate universe can be circumscribed by abiotic factors (soil type, topography), remaining natural or restorable vegetation, or a subjectively chosen management unit. The larger area that controls important landscape level processes (such as hydrologic regime, fire, pollinators, or predators) is much harder to determine.

Monitoring communities is also made more difficult by the extensive number of potential variables that could be measured. Among these variables are measures of species composition (floristic lists, richness, diversity, evenness), species abundance (cover, density, frequency), community physiognomy and structure (vertical structure, spatial patterns), and community function (effect of disturbance, productivity, nutrient cycling). Measuring all of these variables would be extremely time consuming and could result in extraneous data that would be difficult to interpret.

Educated selection of community variables is thus necessary. The community variables that are chosen should be those that have the greatest influence on the reintroduced population, are easily measured, or are indicators of critical states of diversity or processes. These include groupings of species, such as keystone species (Soulé and Simberloff 1986; Conway 1989; but see Mills, Soulé, and Doak 1993), indicator species (Landres, Verner, and Thomas 1988), species guilds (Treshow and Allan 1985), dominant species, rare species, and invasive exotic species. Data for these groupings are gathered by the appropriate measure of abundance (Table 10-3). Measures of community structure, such as understory cover or size of canopy gaps, are also often important. The natural processes (fire, grazing, small-scale disturbances, light-gap formation) that have the greatest influence on the community and population should also be monitored. In some communities several natural processes are simultaneously important for species and populations. In tallgrass prairies of the Midwest, species abundance and distribution and community composition are controlled by fire, grazing, and

small-scale disturbances caused by burrowing rodents (Collins and Gibson 1990). In cypress savanna wetlands of the Southeastern United States, both fire and hydrology affect species abundance and community composition (Sharitz and Gibbons 1982).

The monitoring of longleaf pine-wiregrass ecosystems in the Southeastern United States illustrates some of the decisions concerning appropriate variables and sampling methods. These ecosystems are some of the most species-rich communities in the temperate zone (Walker and Peet 1983), especially at the moist end of the hydrologic gradient, and have large numbers of endangered and rare taxa. Once covering over 70 million hectares of the southeast coastal plain, only 2 percent of its original range currently remains (Ware, Frost, and Doerr 1993). The ecosystem is dependent on fire to maintain species richness. With a natural fire regime this ecosystem has an overstory of longleaf pine above a diverse herbaceous layer and sparse shrub and woody understory.

To monitor changes and the effects of fire management in this community, a number of important variables of composition, abundance, structure, and function have been identified (in monitoring projects designed by The Nature Conservancy in Florida, Texas, and North Carolina). These variables include survival and recruitment of longleaf pine, understory structure and pattern, herbaceous species diversity, abundance of selected rare species, and the physical effect of fire. To efficiently measure each of these variables, different sampling designs have been employed. Large plots (1 hectare or more) have been established to follow longleaf survival and recruitment and the dynamics of light gaps that are important to recruitment. Additional designs utilize the line intercept method to measure the cover of understory shrubs and trees, 20-by-20 centimeter plots to measure herbaceous species richness, belt transects to measure fire effects, and a specific sampling design for each selected rare species population. The key to detecting change in this community is based on the choice of several different types of variables within a variable-specific sampling design.

4. Monitoring Genetic Variability in Reintroduced Populations

A central long-term goal for a reintroduced population is to ensure that the evolutionary potential to adapt to variable and changing environments is maintained (Holsinger and Gottlieb 1991; Barrett and Kohn 1991; Simberloff 1988; Gordon forthcoming). One of the measures of this adaptability is the level of genetic variability in the population, the goal being to maintain as much of the original range of genetic diversity as possible (Simberloff 1988) and to encourage the processes that maintain genetic diversity. For small reintroduced populations, two specific factors could lead to a decrease in genetic variation over time. These are changes in allele frequencies due to chance

TABLE 10-3. Methods used to collect measures of abundance in community monitoring (Bonham 1989; Greig-Smith 1983; Kenkel, Juhasz-Nagy, and Podani 1989; Salzer 1993).

Cover Measurements. Defined as the area of ground covered by a species within a specified area. Measures a combination of density and vigor, but does not distinguish if change is a result of condition or density changes. Measures different life forms (shrubs, herbs) in comparable terms. Provides a quantitative measure for species that cannot be effectively measured by density. Sensitive to seasonal differences.

1. Visual estimates of cover in plots

Data:	Ordinal, semiquantitative data
Precision:	Overestimates cover of sparse, dominant, and conspicuous species; underestimates cover of soil and litter; repeatability a problem with different investigators; difficult to use in statistical analyses
Efficiency:	Quick
Communities:	Useful in all community types; most precise at smaller plot sizes in lower vegetation; difficult in tall, dense, species-rich grasslands

2. Point intercept

Data:	Quantitative data, most objective way to measure cover
Precision:	Measurement error from wind moving vegetation, underestimates cover for clumped or sparse species, overestimates cover of soil and litter
Efficiency:	Time-consuming
Communities:	Best use in grasslands

3. Line intercept

Data:	Quantitative data
Precision:	Best if individual plants have discrete edges; underestimates cover of soil and sparse vegetation
Efficiency:	Relatively efficient if plants or plant groups have discrete edges; time consuming if measuring narrow-leaved, overlapping species
Communities:	Useful in all community types, especially with understory vegetation, shrub vegetation; more accurate in communities with different-sized plants

TABLE 10-3. *(Continued).*

Frequency Measurements. Defined as the number of times a species is present in a given number of sample units, expressed as a percent of total number of observations. Data not affected by seasonal difference in growth, only by the seasonal presence of the species. Changes cannot be related to changes in vigor or density. Subject to quadrat size and shape.

1. Nested plots

Precision:	Dependent on size and shape of quadrats; for accurate detection of change, frequencies should be between 30 and 80 percent, with smallest size determined by size needed to reduce the most common species to below 80 percent. Largest size determined by the sparsest species one wants to have occurring above 30 percent. Low observer bias, only need to identify the species and determine if it is in or out of the plot. Need large number of plots (over one hundred) to accurately detect significant change.
Efficiency:	Very efficient, data collected is presence/absence in each size plot
Communities:	Best use in grasslands, shrub grasslands, and herb strata of forests

Density Measurements. Defined as the number of individuals in a given unit of area. A nonsubjective and repeatable measure of abundance if species is easily identified. Difficulties come with definition of individuals. Data dependent on plot size, shape, and number. Best used in low-density populations, when individuals are easily distinguished and species is important enough (rare, dominant) for this type of data collection. Will not detect a change in vigor of individuals least influenced by annual weather variation.

1. Plots

Precision:	High repeatability; low observer bias if species can be identified; edge effects may reduce precision in large plots, but is of little concern in small plots and plots that use quadrat frames; difficulties determining individual plants
Efficiency:	High-density plots time consuming to count and may result in high measurement error
Communities:	All plant communities

2. Distance methods

Precision:	Most methods assume a randomly distributed population (a rarity in nature); distance methods for nonrandom populations have difficult formulas to estimate density
Efficiency:	Efficient
Communities:	All plant communities, best used with woody species

events (random genetic drift) and directional selection as the population adapts to the specific environment (Barrett and Kohn 1991; Karron 1991; Guerrant, Chapter Eight). Inbreeding can also reduce levels of genetic variation, although its influence may be most significant in outcrossing species (Barrett and Kohn 1991). Outbreeding depression may have a stronger effect on fitness for reintroduced populations that use genetic material from several different source populations or subpopulations (Barrett and Kohn 1991).

Monitoring changes in genetic variation may not be a high priority for reintroduced populations that are successfully establishing and expanding. It may be of concern when the population is not vigorously expanding. But even in these situations, information on genetic variation may not be essential for a successful reintroduction. Changes in genetic variation as it is measured may have no relation to changes in genes affecting traits important for long-term survival or the vigor of population (Schmeske et al. forthcoming). Thus, the primary measures of genetic variation, allelic diversity, and levels of heterozygosity within and among populations should be interpreted cautiously. Any interpretation of these measures should be done only in combination with morphological (Schmeske et al. 1994) and life-history and demographic data to judge the success of a reintroduction. The difficulty of interpreting genetic data increases the importance of the design of reintroduced populations (Guerrant, Chapter Eight; Millar, personal communication, Nov. 1993). Establishing genetic conditions that are diverse and locally adapted through the use of local germplasm, collection of propagules that represent a wide array of genotypes, maintenance of equal representation of genotypes through any nursery amplification period, and discouragement of mating among relatives will promote the viability of the population (Millar and Libby 1989).

There may be situations where monitoring genetic variation may be important. These might arise from observed conditions in the population when high adult mortality, adult mortality coupled with low seed set, differential seed set, and poor recruitment suggest genetic impacts. The history of the species may also influence the urgency and timing of monitoring genetic variation. Reintroduced species that have experienced extreme reductions in their total number of populations or the sizes of their populations may be more susceptible to the stresses caused by small population size (Barrett and Kohn 1991).

Additional valid reasons for monitoring the genetic variation of a reintroduced population include (1) evaluating the success of the project by relating to a reference population the levels and patterns of genetic variability of the reintroduced populations; (2) identifying genotypes that have higher reproductive rates, higher growth rates, and increased survival; (3) directing the matings that will produce the greatest increase in overall heterozygosity;

(4) determining when additional genetic input would be necessary to maintain natural levels of genetic variation; and (5) validating theories of genetic variation in small populations (Brisbin and McDonald 1989).

The sampling method and the number of samples needed to assess genetic variation are dependent on population size and the distribution of genetic variation within the population. In small populations each individual can be sampled, while in large populations individuals should be randomly subsampled (Brown and Briggs 1991). Life-history characteristics such as breeding system, seed dispersal, and life form can be used as a guide to the pattern of genetic variation in a population (Hamrick et al. 1991). A greater number of evenly spaced samples should be taken when the species are outcrossers, are wind-pollinated or long-lived perennials, have seeds that are dispersed widely, have high population densities, were historically widespread, or occur in numerous different microhabitats.

Success, Duration, and the Experimental Nature of Monitoring

The ultimate success of a reintroduction project—that is, the self-sustainability of the population—will only be determined after many years of monitoring (see Pavlik, Chapter Six; Guerrant, Chapter Eight). Success has to be evaluated species by species, using taxon-specific information on life history and demography and community-specific information on condition and processes. The most appropriate source for this information is from a reference population occurring in a naturally functioning community. The importance of a reference population for information on life history and demography requires its monitoring whenever a reintroduction is done. If there is no reference population, congeners or experimental populations may provide insight into the standards for success. Whatever their source, monitoring is essential in obtaining the standards by which success is to be measured.

How long does a reintroduced population need to be monitored? Only under the most optimistic situation—when a population has expanded, recruitment is common, and the natural processes in the community are active—will monitoring ever end. In most situations, monitoring will only be reduced in intensity or frequency. The continuation of monitoring is not a sign of failure but a recognition that our natural areas have been altered to the extent that many species are not assured of survival. Intensive monitoring will be necessary for years to examine a population's response to natural processes and management regimes. For communities, decades may be required before a restored community reaches full functional parity with undisturbed sites (Bowler 1990; Langis, Zalejko, and Zedler 1991).

Reintroduction is by nature an experiment. We know little about the dynamics of small populations and how environmental, demographic, and genetic factors control growth, reproduction, and survival. A reintroduction project provides the opportunity to explore which factors influence the persistence of small populations. Approached as an experiment, reintroductions can both help conserve rare species and become a source for increasing conservation knowledge.

ACKNOWLEDGMENTS

There are many colleagues in The Nature Conservancy that deserve thanks, especially Dan Salzer for sharing his expertise in sampling design and statistics; Susan Benjamin, Doria Gordon, Nathan Rudd, Bob Unnasch, and Rick Young, for their enrichment of the conservancy's monitoring workshop; and Susan Bainbridge for helping me with the figures for the final manuscript. Thanks to all the individuals who have taken The Nature Conservancy monitoring workshops and participated in the exercises. A special thanks to Mareah Steketee, who has tolerated my preoccupation with this subject and helped me find the right words.

REFERENCES

Barrett, S. C. H., and J. R. Kohn. 1991. "Genetic and Evolutionary Consequences of Small Population Size in Plants: Implications for Conservation." In *Genetics and Conservation of Rare Plants*. Edited by D. A. Falk and K. E. Holsinger. New York: Oxford University Press.

Bawa, K. S., and C. J. Webb. 1984. "Flower, Fruit and Seed Abortion in Tropical Forest Trees: Implications for the Evolution of Paternal and Maternal Reproductive Patterns." *American Journal of Botany* 71: 736–751.

Bonham, C. D. 1989. *Measurements for Terrestrial Vegetation*. New York: John Wiley and Sons.

Bowler, P. A. 1990. "Coastal Sage Scrub Restoration-I: The Challenge of Mitigation." *Restoration and Management Notes* 8: 78–82.

Brisbin, I. L., Jr., and M. A. McDonald. 1989. "Genetic Patterns and the Conservation of Crocodilians: A Review of Strategies and Options." In *Crocodiles: Proceedings of the 8th Working Meeting of the Crocodile Specialist Group of the Species Survival Committee of the International Union for Conservation of Nature and Natural Resources*. Gland, Switzerland: IUCN.

Brown, A. H. D., and J. D. Briggs. 1991. "Sampling Strategies for Genetic Variation in *ex situ* Collections of Endangered Plant Species." In *Genetics and Conservation of Rare Plants*. Edited by D. A. Falk and K. E. Holsinger. New York: Oxford University Press.

Caswell, H. 1989. *Matrix Population Models*. Sunderland, Mass.: Sinauer Associates.

Chew, R. M., and A. E. Chew. 1965. "The Primary Productivity of a Desert Shrub (*Larrea tridentata*)." *Ecological Monographs* 35: 355–375.

Cochran, W. G. 1977. *Sampling Techniques*. 3d ed. New York: John Wiley and Sons.

Collins S. L., and D. J. Gibson. 1990. "Effects of Fire on Community Structure in Tall-Grass and Mixed-Grass Prairie." In *Fire in North American Tallgrass Prairies*. Edited by S. L. Collins and L. L. Wallace. Norman, Okla.: University of Oklahoma Press.

Conway, W. G. 1989. "The Prospects for Sustaining Species and Their Evolution." In *Conservation for the Twenty-first Century*. Edited by D. Western and M. C. Pearl. New York: Oxford University Press.

Crouse, D. T., L. B. Crowder, and H. Caswell. 1987. "A Stage-Based Population Model of Loggerhead Sea Turtles and Implications for Conservation." *Ecology* 68: 1412–1423.

Dallal, G. E. 1988. *DESIGN: A Supplementary Module for SYSTAT and SYGRAPH*. Evanston, Ill.: SYSTAT.

DeMauro, M. M. 1993. "Development and Implementation of a Recovery Program for the Federally Threatened Lakeside Daisy (*Hymenoxys acaulis* var. *glabra*)." In *Recovery and Restoration of Endangered Species*. Edited by M. Bowles and C. J. Whelan. Cambridge, England: Cambridge University Press.

Eberhardt, L. L., and J. M. Thomas. 1991. "Designing Environmental Field Studies." *Ecological Monographs* 61: 53–73.

Eckblad, J. W. 1991. "How Many Samples Should Be Taken?" *BioScience* 41: 346–348.

Fairweather, P. G. 1991. "Statistical Power and Design Requirements for Environmental Monitoring." *Australian Journal of Marsh and Freshwater Research* 42: 555–567.

Falk, D. A., and P. Olwell. 1992. "Scientific and Policy Considerations in Restoration and Reintroduction of Endangered Species." *Rhodora* 94: 287–315.

Fiedler, P. L. 1991. *Mitigation-Related Transplantation, Relocation, and Reintroduction Projects Involving Endangered and Threatened and Rare Plant Species in California*. Sacramento, Calif.: California Department of Fish and Game, Endangered Plant Program.

Frantz, V., and R. Sutter. 1987. Hudsonia montana: *From Discovery to Recovery*. Raleigh, N.C.: North Carolina Department of Agriculture, Plant Conservation Program.

Frost, C. C. 1993. Hudsonia montana: *Monitoring and Management for 1991*. Raleigh, NC: North Carolina Department of Agriculture, Plant Conservation Program.

Goldsmith, B. 1991. "Vegetation Monitoring." In *Monitoring for Conservation and Ecology*. Edited by F. B. Goldsmith. London: Chapman and Hall.

Gordon, D. R. 1994. "Translocation of Species into Conservation Areas: A Key for Natural Resource Managers." *Natural Areas Journal*.

Green, R. H. 1979. *Sampling Design and Statistical Methods for Environmental Biologists*. New York: J. Wiley and Sons.

Greig-Smith, P. 1983. *Quantitative Plant Ecology*. 3d ed. Berkeley: University of California Press.

Hall, L. A. 1987. "Transplantation of Sensitive Plants as Mitigation for Environmental Impacts." In *Conservation and Management of Rare and Endangered Plants*. Edited by T. Elias. Sacramento: California Native Plant Society.

Hamrick, J. L., M. J. W. Godt, D. A. Murawski, and M. D. Loveless. 1991. "Correlations Between Species Traits and Allozyme Diversity: Implications for Conservation Biology." In *Genetics and Conservation of Rare Plants*. Edited by D. A. Falk and K. E. Holsinger. New York: Oxford University Press.

Handel S. N., and A. J. Beattie. 1990. "Seed Dispersal by Ants." *Scientific American* 263: 76–83.

Harper, J. L. 1977. *Population Biology of Plants*. New York: Academic Press.

Holsinger, K. E., and L. D. Gottlieb. 1991. "Conservation of Rare and Endangered Plants: Principles and Prospects." In *Genetics and Conservation of Rare Plants*. Edited by D. A. Falk and K. E. Holsinger. New York: Oxford University Press.

Howe, H. F., and J. Smallwood. 1982. "Ecology of Seed Dispersal." *Annual Review of Ecology and Systematics* 13: 201–228.

Hurlbert, S. H. 1984. "Pseudoreplication and the Design of Ecological Field Experiments." *Ecology* 54: 187–211.

Karron, J. D. 1991. "Patterns of Genetic Variation and Breeding Systems in Rare Plant Species." In *Genetics and Conservation of Rare Plants*. Edited by D. A. Falk and K. E. Holsinger. New York: Oxford University Press.

Kenkel, N. C., P. Juhasz-Nagy, and J. Podani. 1989. "On Sampling Procedures in Population and Community Ecology." *Vegetatio* 83: 195–207.

Kupper L. L. and K. B. Hafner. 1989. "How Appropriate Are Popular Sample Size Formulas?" *American Statistician* 43: 101–105.

Lande, R. 1988. "Demographic Models of the Northern Spotted Owl." *Oecologia* 75: 601–607.

Landres, P. B., J. Verner, and J. W. Thomas. 1988. "Ecological Uses of Vertebrate Indicator Species: A Critique." *Conservation Biology* 4: 316–328.

Langis, R., M. Zalejko, and J. B. Zedler. 1991. "Nitrogen Assessments in a Constructed and Natural Salt Marsh of San Diego Bay." *Ecological Applications* 1: 40–51.

Likens, G. E. 1983. "A Priority for Ecological Research." *Bulletin of the Ecological Society of America* 64: 234–243.

Likens, G. E., ed. 1989. *Long-Term Studies in Ecology: Approaches and Alternatives*. New York: Springer-Verlag.

MacDonald, L. H. 1991. *Monitoring Guidelines to Evaluate Effects of Forestry Activities on Streams in the Pacific Northwest and Alaska*. Document EPA 910/9-91-001. Seattle, Wash.: EPA.

Magnuson, J. J. 1990. "Long-Term Ecological Research and the Invisible Present." *BioScience* 40: 495-501.

Menges, E. S. 1986. "Predicting the Future of Rare Plant Populations: Demographic Monitoring and Modeling." *Natural Areas Journal* 6: 13–25.

Menges, E. S. 1990. "Population Viability Analysis for an Endangered Plant." *Conservation Biology* 4: 52–62.

Menges, E. S. 1991. "The Application of Minimum Viable Population Theory to Plants." In *Genetics and Conservation of Rare Plants*. Edited by D. A. Falk and K. E. Holsinger. New York: Oxford University Press.

Millar, C. I., and W. J. Libby. 1989. "Disneyland or Native Ecosystem: Genetics and the Restorationist." *Restoration and Management Notes* 7: 18–24.

Mills, L. S., M. E. Soulé, and D. F. Doak. 1993. "The Keystone-Species Concept in Ecology and Conservation." *BioScience* 43: 219–224.

Noon, B. R., and C. M. Biles. 1990. "Mathematical Demography of Spotted Owls in the Pacific Northwest (USA)." *Journal of Wildlife Management* 54: 18–27.

Palmer, M. E. 1987. "A Critical Look at Rare Plant Monitoring in the United States." *Biological Conservation* 39: 113–127.

Panos J. M. 1989. *Variation in the Architecture of White Clover (Trifolium repens L.) in Contrasting Habitats and the Consequences of this Variation on Clonal Dynamics.* Ph.D. diss., University of Connecticut, Storrs.

Pavlik, B. M. 1987. "Autecological Monitoring of Endangered Plants." In *Conservation and Management of Rare and Endangered Plants.* Edited by T. Elias. Sacramento: California Native Plant Society.

Pavlik, B. M. 1993. "Demographic Monitoring and the Recovery of Endangered Plants." In *Recovery and Restoration of Endangered Species.* Edited by M. Bowles and C. J. Whelan. Cambridge, England: Cambridge University Press.

Peterman, R. M. 1990. "The Importance of Reporting Statistical Power: The Forest Decline and Acidic Deposition Example." *Ecology* 71: 2024–2027.

Salzer, D. 1993. *Lectures on Sampling Design.* Portland Oreg.: Oregon Field Office, The Nature Conservancy.

Sarukhán, J., and M. Gadgil. 1974. "Studies on Plant Demography: *Ranunculus repens* L., *R. bulbosus* L., and *R. acris* L. III. A Mathematical Model Incorporating Multiple Modes of Reproduction." *Journal of Ecology* 62: 921–936.

Schemske, D. W., B. C. Husband, M. H. Ruckelshaus, C. Goodwillie, I. M. Parker, and J. G. Bishop. 1994. "Evaluating Approaches to the Conservation of Rare and Endangered Plants." *Ecology* 75: 584–606.

Sharitz, R. R., and J. W. Gibbons. 1982. *The Ecology of Southeastern Shrub Bogs (Pocosins) and Carolina Bays: A Community Profile.* FWS/OBS-82/04. Atlanta Ga.: U.S. Fish and Wildlife.

Simberloff, D. 1988. "The Contribution of Population and Community Biology to Conservation Science." *Annual Review of Ecology and Systematics* 19: 473–512.

Sokal, R. R., and F. J. Rohlf. 1969. *Biometry.* San Francisco: W. H. Freeman and Company.

Soulé, M. E. 1987. *Viable Populations for Conservation.* Cambridge, England: Cambridge University Press.

Soulé, M. E., and D. Simberloff. 1986. "What Do Genetics and Ecology Tell Us About the Design of Nature Reserves?" *Biological Conservation* 35: 19–40.

Spellerberg, I. F. 1991. *Monitoring Ecological Change.* New York: Cambridge University Press.

Stacey, P. B., and M. Taper. 1992. "Environmental Variation and the Persistence of Small Populations." *Ecological Applications* 2: 18–29.

Stephenson, A. G. 1981. "Flower and Fruit Abortion: Proximate Causes and Ultimate Functions." *Annual Review of Ecology and Systematics* 12: 253–279.

Stewart-Oaten, A. 1993. "Evidence and Statistical Summaries in Environmental Assessment." *Trends in Ecology and Evolution* 8: 156–157.

Stuart, A. 1976. *Basic Ideas of Scientific Sampling.* New York: Hafner Press.

Sutter, R. D. 1986. "Monitoring Rare Plant Species and Natural Areas—Ensuring the Protection of Our Investment." *Natural Areas Journal* 6: 3–5.

Sutter, R. D., S. E. Benjamin, N. Murdock, and B. Teague. 1993. "Monitoring the Effectiveness of a Boardwalk at Protecting a Low Heath Bald in the Southern Appalachians." *Natural Areas Journal* 13: 250–255.

Thompson, S. K. 1991. "Adaptive Cluster Sampling: Designs with Primary and Secondary Units." *Biometrics* 47: 1103–1115.

Travis J., and R. Sutter. 1986. "Experimental Designs and Statistical Methods for Demographic Studies of Rare Plants." *Natural Areas Journal* 6: 3–12.

Treshow, M., and J. Allan. 1985. "Uncertainities Associated with the Assessment of Vegetation." *Environmental Management* 9: 471–478.

Usher, M. B. 1991. "Scientific Requirements of a Monitoring Programme." In *Monitoring for Conservation and Ecology*. Edited by F. B.Goldsmith. London: Chapman and Hall.

Waite, S., and M. J. Hutchings. 1991. "The Effects of Different Management Regimes on the Population Dynamics of *Ophrys sphegodes*: Analysis and Description Using Matrix Models." In *Population Ecology of Terrestrial Orchids*. Edited by T. C. E. Wells and J. H. Willems. The Hague, The Netherlands: SPB Academic Publishing.

Walker, J., and R. K. Peet. 1983. "Composition and Species Diversity of Pine-Wiregrass Savannas of the Green Swamp, North Carolina." *Vegetatio* 55: 163–179.

Ware, S., C. Frost, and P. D. Doerr. 1993. "Southern Mixed Hardwood Forest: The Former Longleaf Pine Forest." In *Biodiversity of the Southeastern United States*. Edited by W. H. Martin, S. G. Boyce, and A. C. Echternacht. New York: John Wiley and Sons.

Waser, N. M. 1982. "Patterns of Seed Dispersal and Population Differentiation in *Mimulus guttatus*." *Evolution* 36: 753–761.

White, P. S. and S. P. Bratton. 1980. "After Preservation: Philosophical and Practical Problems of Change." *Biological Conservation* 18: 241–255.

Whitson, P., and J. Massey. 1981. "Information Systems for Use in Studying the Population Status of Threatened and Endangered Plants." In: *Rare Plant Conservation: Geographical Data Organization*. Edited by L. E. Morse and M. S. Henifin. New York: New York Botanical Garden.

CHAPTER 11

FOCUS

Pinus torreyana at the Torrey Pines State Reserve, California

F. Thomas Ledig

Torrey pine is a category 2 species under the Endangered Species Act, which means that it is a taxon "for which information now in the possession of the [Fish and Wildlife] Service indicates that proposing to list [it] as endangered or threatened . . . is possibly appropriate, but for which substantial data on biological vulnerability and threats are not currently known or on file to support the immediate preparation of rules." The California Native Plant Society classifies it as list 1b, 3-2-3. List 1b is for rare plants, and the numerical code means that it is distributed in highly restricted occurrences (3), endangered in a portion of its range (2), and endemic (3).

Natural Distribution and Ecology of Taxon

Torrey pine is among the rarest species of pines in the world, indigenous only to the La Jolla–Del Mar area of coastal San Diego County and on Santa Rosa Island, Santa Barbara County. The mainland population is protected within Torrey Pines State Reserve and the island population within Channel Islands National Park. The two populations are separated by 280 kilometers, and the island is separated from the adjacent mainland by about 50 kilometers. The Torrey pines at the reserve are the only pines on the coast of California south of Santa Barbara, about 330 kilometers away. The nearest pine ecosystem in any direction of the reserve is over 60 kilometers to the east in the Laguna Mountains. The 445-hectare Torrey Pines State Reserve, engulfed by the spread of San Diego and its suburbs, has been called a wilderness island in an urban sea. Santa Rosa Island, by contrast, is undeveloped.

Both island and mainland populations grow on thin soils derived from sandstone or on nearly bare rock on sites exposed to wind and salt spray. The climate is Mediterranean, with almost all precipitation falling during the late autumn and winter months, but is decidedly cooler and more maritime than surrounding areas. Nevertheless, associated vegetation at the Torrey Pines State Reserve contains elements of the Sonoran Desert, such as jojoba (*Simmondsia chinensis*). The seeds of Torrey pine are heavy and nearly wingless, so dispersal is limited.

Little is known of the prehistoric status of Torrey pine. No fossil record exists, which may suggest that the species has always been rare and restricted.

A 1973 census in the Torrey Pines State Reserve counted 3,401 mature trees. The total population (seedlings, saplings, and trees) may be six thousand, according to staff at the reserve. C. C. Parry noted only about one hundred individuals at the mainland site when he discovered the species in 1850 (Lemmon 1888). Thirty-eight years after Parry collected the type specimens, Lemmon enumerated eighty-three trees north of the San Dieguito River and "not above a few hundred individuals" south of the river, in the heart of the present Torrey Pines State Reserve. Preservation of the mainland site and planting of Torrey pines under the direction of Miss Ellen Browning Scripps may have contributed to the present population that numbers in the thousands. Torrey pine's current abundance, relative to that of a century ago, may also reflect cyclical fluctuations related to climatic or other factors.

The Santa Rosa Island population was first reported in 1888 and estimated at about one hundred trees (Lemmon 1888). The Channel Islands National Park staff estimates approximately one thousand individuals on Santa Rosa Island at present.

Although both populations are larger now than they were a century ago, natural recruitment over the past twenty years appears to be relatively low, especially in the Torrey Pines State Reserve (McMaster, Jow, and Jackson, n.d.).

The two populations of Torrey pine differ in a number of characteristics when grown in a common garden: branch elongation, needle length, cone width–length ratio, and number of new branches per year (Haller 1986). The difference in branch elongation shows up as a difference in height and form of the trees as they mature. Needle color of island trees seems to be bluer than that of mainland trees in the common garden (personal observation). Terpene composition of the oleoresin also differs between island and mainland populations (Zavarin et al. 1967). Although the morphological and terpene differences are not great, the lack of intermediates and the uniformity within populations led Haller (1986) to name the island population as a subspecies, *Pinus torreyana* ssp. *insularis*.

Isozymes of Torrey pine were separated by electrophoresis to survey ge-

netic variation in the island and mainland populations (Ledig and Conkle 1983). The results were surprising. Every tree at the Torrey Pines State Reserve was genetically identical to every other at fifty-nine different isozyme loci. If only a single clone were present, the results would be similar. No other tree species is as genetically uniform, although a few approach this level. Furthermore, every Torrey pine sampled on Santa Rosa Island was identical to every other one on the island. However, the island trees differed from the mainland trees at two of the fifty-nine gene loci (a malic dehydrogenase gene and a shikimate dehydrogenase gene), which, based on certain assumptions, corresponds to an estimated 8 percent of their genes. Thus, in Torrey pine all the detected variation is between populations and not within. This contrasts to the situation in nearly all other conifers, where over 80 percent (and usually over 90 percent) of the total genic variation is within populations.

Both the mainland and the island pines must have experienced bottlenecks during which the number of trees was reduced to very low levels, perhaps even to a single tree, for several generations. Under such circumstances, genetic drift could have eliminated variants and fixed the two populations for the alternate alleles that now distinguish them at the malic dehydrogenase and shikimate dehydrogenase loci. The founder effect may account for such a bottleneck on Santa Rosa Island. Torrey pine might have colonized the island about 18,000 years ago when the Santa Barbara Channel separating the prehistoric island Santarosae from the mainland was only 7 kilometers wide. The mainland population may have been drastically reduced in numbers during the xerothermic period, 8,500 to 3,000 years B.P.

Threats

Genetic uniformity leaves a species vulnerable to pests and to environmental change. If an insect or disease organism finds one tree attractive and open to attack, it will find all trees equally susceptible. Or if environmental conditions become unsuitable for one tree, they will be unsuitable for all. The population can crash rapidly. Such crashes are amply documented in genetically uniform crop plants; the Irish potato famine was the result of using a single potato clone throughout Ireland (U.S. Committee on Genetic Vulnerability of Major Crops 1972). Thus, Ledig and Conkle (1983) predicted that Torrey pine was particularly vulnerable because all trees at the Torrey Pines State Reserve were identical as far as could be detected, and all were in close proximity. In such a situation, it seemed prudent to conserve Torrey pine *ex situ*. Accordingly, in 1986 the Institute of Forest Genetics collected seeds from 149 trees for long-term storage.

Conditions that Initiated Reintroduction

The reserve staff noted an unusual amount of Torrey pine mortality due to the California five-spined ips (*Ips paraconfusus* Lanier) in late summer of 1989. Ips usually feed in downed material (dead trees or branches felled by the wind or snow). However, they can kill trees by feeding and breeding in the phloem or inner bark. Ips killed an occasional tree in Torrey Pines State Reserve before 1989, but the ips problem was undoubtedly exacerbated by the prolonged drought in California between 1987 and 1993. The drought reduced the vigor of the trees, which normally can withstand ips attack by "pitching them out"; that is, oleoresins normally flow to the point of attack and drown the beetles. However, when trees are brought under stress from any cause (such as drought, disease, wounding, soil compaction), they lose the ability to resist attack (Furniss and Carolin 1977).

By early May 1991, over 840 trees had been killed by ips and thirty-eight more were infested (Shea and Neustein 1994). This represents more than 25 percent of the mature trees counted in the 1973 census or at least 14 percent of the total 6,000 estimated by Reserve staff. Most of the mortality was in the Parry Grove and the Guy Fleming Grove and some along Razor Point Trail. The line of demarcation between the dead trees and the living, uninfested stand was clear. Virtually every tree between the Pacific Ocean and the line of green trees was killed.

The first concern was to slow or stop the beetles' advance. To accomplish this, Patrick Shea, principal research entomologist of the Pacific Southwest Research Station, used a combination of both aggregation and antiaggregation pheromones (Shea and Neustein 1994). A line of Lindgren funnel traps was placed within the zone of dead trees parallel to the line of green, healthy trees. This line of traps was approximately 800 yards long, with a set of three traps every 60 to 80 yards. The aggregation pheromone of the California five-spined ips was placed in each trap.

In addition to deploying aggregation pheromones in the area of beetle-killed trees, Shea placed antiaggregation pheromones within the boundary of the living, green trees. The line of antiaggregation pheromones paralleled the funnel traps. In effect, this amounted to a double signal to the beetles: the aggregation pheromones pulled the beetles toward the traps and the antiaggregation pheromones pushed them away from the green, uninfested trees.

During the first nine weeks of trapping, Shea removed 131,000 ips from the reserve. Between May 15 and October 24, 1991, a total of 156,520 California five-spined ips were trapped. During this period, only two additional trees were infested and the numbers of beetles that were trapped dropped to very low levels. At this time the pheromone traps were removed. Because reserve staff found fourteen new trees infested with ips in January 1992, the traps were redeployed. From March 18 to December 17, 1992, in excess of 158,000 ips

were trapped, but the weekly average never reached the levels seen in 1991. Only two additional trees were killed during this trapping period.

Traps were deployed again from mid March to late September 1993. During this period, only 16,051 ips were trapped. No additional mortality or infestation of Torrey pine was seen after August 1992. The swift reduction in mortality was attributed to the pheromone trapping program.

Objectives of Reintroduction

With the ips under control, attention turned to restoring the Torrey Pines State Reserve for aesthetic reasons and to increase the numbers of Torrey pine to improve the species' demographic chances for survival. The Institute of Forest Genetics drew upon the conservation collection made in 1986 to provide seed and seedlings for planting. The areas hardest hit by the ips (the Parry Grove, Guy Fleming Grove, and Razor Point) were well represented in the seed inventory. Thus, the California Department of Parks and Recreation was able to restore the groves with the progeny of the dead trees themselves.

Description of Reintroduction Project

The seed collection of 1986 was made to ensure that Torrey pine was not lost should a catastrophe eliminate the *in situ* reserves. Staff from the Institute of Forest Genetics collected cones from 149 trees at the Torrey Pines State Reserve. These trees were distributed within the Parry Grove and the Guy Fleming Grove (areas where Torrey pines reached their greatest density), on Razor Point and in the Extension Area. The cones were taken to the Institute of Forest Genetics at Placerville and opened, and the seeds were extracted. After drying, the seeds were put into freezers. We anticipated that cold, dry pine seeds could be stored for decades, perhaps even a century, and still be germinated when needed. The seed inventory totalled 29,512 seeds.

After the ips were controlled, seeds were sown directly in the groves and protected by screenwire. Unfortunately, seed quality had deteriorated in storage and germination was very low (less than 2 percent). The fault probably lay with the seed storage facilities, which needed upgrading, and not necessarily with any peculiarity of Torrey pine seeds. An additional 4,000 seeds were sown in 1994 in an experiment to test the effect of soil type, openings, and slope aspect on germination and survival. Results will be monitored.

In late 1992, seeds were sown in containers in the greenhouse at the Institute of Forest Genetics, Placerville, to grow seedlings for outplanting. In February 1994, 513 containerized seedlings were lifted and delivered to the Torrey

Pines State Reserve. The reserve's staff and volunteers planted each seedling near a beetle-killed tree and protected it with plastic sleeves to prevent animal damage. Planting conditions were favorable because rains came later than normal. Seedling survival was excellent, better than 98 percent as of May 1994, and restoration seems assured.

Funding

The Torrey pine seed collections for *ex situ* conservation were, in a sense, bootlegged on research funds by the USDA Forest Service. Costs of cone collection, seed extraction, and storage were about $6,000. Approximate cost of producing and delivering seedlings to the Torrey Pines State Reserve was an additional $2,000. Sowing within the Torrey Pines State Reserve was accomplished by the California Department of Parks and Recreation in cooperation with San Diego State University. The seedlings were planted by volunteers under the direction of the California Department of Parks and Recreation. The Department budgeted $30,000, but this included a strong research and monitoring component, including palynological studies of the Los Peñasquitos Marsh in the Torrey Pines State Reserve in an attempt to chart past fluctuations in the pine population. We guess that actual planting and sowing costs probably would be under $3,000.

Partnerships

The California Department of Parks and Recreation, the USDA Forest Service's Pacific Southwest Research Station, and San Diego State University all played roles in the restoration.

REFERENCES

Furniss, R. L., and V. M. Carolin. 1977. *Western Forest Insects*. U.S. Forest Service, miscellaneous publication no. 1339.

Haller, J. R. 1986. "Taxonomy and Relationships of the Mainland and Island Populations of *Pinus torreyana* (Pinaceae)." *Systematic Botany* 11: 39–50.

Ledig, F. T., and M. T. Conkle. 1983. "Gene Diversity and Genetic Structure in a Narrow Endemic, Torrey Pine (*Pinus torreyana* Parry ex Carr.)." *Evolution* 37: 79–85.

Lemmon, J. G. 1888. *Second Biennial Report of the California State Board of Forestry, for the Years 1887–88, to Governor R. W. Waterman*. Sacramento.

McMaster, G. S., W. M. Jow, and A. C. Jackson. n.d. *Final Report on the Basic Demographic Study of the Torrey Pine (Pinus torreyana) for the CA Department of Parks and Recreation, Southern Region*. San Diego State University.

Shea, P. J., and M. Neustein. 1994. Protection of a Rare Stand of Torrey Pine from *Ips paraconfusus*. In Proceedings of an Informal Conference on Application of Semio-chemicals for the Manage. of Bark Beetle Infestations, Annals of the Entomological Society of America Meeting, 12–16 Dec., 1993. Indianapolis, Ind.: U.S. Forest Service General Technical Report.

U.S. Committee on Genetic Vulnerability of Major Crops. 1972. *Genetic Vulnerability of Major Crops*. National Academy of Science, Washington, D.C.

Zavarin, E., W. Hathaway, T. Reichert, and Y. B. Linhart. 1967. "Chemotaxonomic Study of *Pinus torreyana* Parry Turpentine." *Phytochemistry* 6: 1019–1023.

Reintroduction
in a Mitigation Context

As we noted in the Introduction, the most controversial application of reintroduction and restoration is in the context of compensatory mitigation. Mitigation can take many forms, but the practices generally have in common the destruction of an existing population or parcel of habitat—the act that is to be "mitigated" or softened in impact—in exchange for habitat elsewhere. In some cases, the site "elsewhere" is, like the location to be destroyed, existing habitat of very high quality. Mitigation tradeoffs of this kind are used to enhance the security of the second location, by acquisition, transfer of development rights to a conservation group, or establishment of a management trust fund. In other, more radical cases, the tradeoff is for *new* habitat or populations to be created, using the techniques of ecological restoration and reintroduction. It is to this latter circumstance that this section pertains most directly, especially where rare or endangered species of plants are concerned.

For an endangered species in this situation, two kinds of questions pertain. First, how valuable is each existing population to the ecological function and evolutionary survival of the species? In other words, can we afford the potential, partial loss of what now exists? If each population is unique and irreplaceable, then mitigation tradeoffs involving destruction of an existing site must almost automatically be proscribed. Such questions can be best answered by ecologists and population biologists, who evaluate the number, size, and distribution of populations, the genetic variation and gene flow within and among sites, the persistence and stability of existing populations, the evidence of metapopulation dynamics, and the prospects for continued survival in the present habitat.

The second set of questions concerns the use of reintroduction or community-level restoration to create new or relocated populations for the species. Do we have the ability to establish a new population (or new habitat) where one has not existed previously? Once established, will the new population or community function ecologically as did the original? In other words, does the proposed new population or community meet the standards of sustainability developed elsewhere in this book? Such questions are extremely difficult to answer in general form, and few empirical data exist to support the contention for individual species.

The crux of the mitigation issue thus lies in the relationship between the existing populations or habitat and the proposed new ones—or, more formally, in the relationship between conservation and restoration. Resources and imperatives will be strategically displaced away from protecting existing habitat and toward more politically expedient solutions involving relocating populations. The "no-net-loss" approach to wetlands embodies this thinking in policy. The pivotal word is *net*, which reveals the underlying assumption that acres of existing wetland can be replaced by an equal number of acres

elsewhere, with no "net" loss of biological or cultural values. The same reasoning can be applied to populations of threatened species, but the question is, Are these assumptions valid?

Mitigation practices continue to grow, despite substantial uncertainty about the feasibility, reliability, and ethics of translocation and creation of endangered species populations. Recovering a damaged wetland or establishing a new population of a threatened species may take decades before "success" can be declared with any confidence. However, few developers—public or private—are likely to post a fifty-year preconstruction bond before they break ground for a new shopping center or highway, and few regulatory agencies are willing to risk the political heat that might accompany such a stipulation. In the absence of such guarantees, nature bears most of the risk. The destruction of habitat is instantaneous and occurs up front. The promised replacement may take decades to become established, if it works out at all.

One school of thought argues that compensatory mitigation is (and should be) a legitimate conservation tool. It is unrealistic, the argument goes, to believe that conservationists will actually stop development in more than a handful of cases. To pretend otherwise is to be intentionally (or even hypocritically) blind to the realities of the world in which we live and the society of which we are all a part. Not everything or every place can be saved; under the circumstances, the best strategy is to adopt a position of constructive engagement, playing for increased security for at least some areas by sacrificing other (presumably more expendable) ones. In this way, conservationists can capture some of the momentum—and some of the resources—of the development community and transform these into permanent conservation gains.

The contrary argument asserts that accepting the practice of compensatory mitigation effectively concedes defeat for biodiversity. Mitigation tradeoffs could potentially be applied anywhere if the compensation ratio is high enough; consequently, no natural area would be safe from proposals for replacement. Moreover, mitigation is based entirely on the unproven premise that the populations, communities, and landscapes created by people have the same biological diversity, structure, function, and long-term prospects for survival as do those that occur naturally. In addition, critics point out, the time frames for development and ecosystem recovery differ by several orders of magnitude; populations, communities, and ecosystems may take decades or centuries to equilibrate locally after disturbance, whereas mitigation politics have a time scale of a few years at best. Ultimately, mitigation can guarantee no security even for the replacement areas. In the meantime, the destruction of existing habitat is instantaneous and irreversible.

The chapters in this section address these and other questions concerning uses of mitigation involving endangered plants. Berg opens Chapter Twelve

with an overview, noting that good alternatives frequently exist to mitigation-related loss of populations. In many cases the best approach is to envision and seek out pathways that will avoid mitigation altogether and to direct energy toward larger conservation objectives.

Two chapters in this section examine mitigation-related reintroduction in parts of the country where it is used frequently in relation to land development. In Chapter Thirteen, Howald summarizes the use of mitigation with endangered species in California, evaluating the outcomes of forty-one mitigation-related rare plant translocations statewide. She also reviews fifteen translocation attempts in San Diego County, ten projects involving vernal pools in the Santa Rosa Plain of Sonoma County, and a case study of the Santa Cruz tarplant (*Holocarpha macredenia*). California has some of the nation's strictest environmental laws, such as the California Environmental Quality Act, that establish avoidance of impacts to the original population as the preferred alternative. Nonetheless, Howald cites evidence that many attempted translocations have been performed without adequate baseline information, critical review, performance standards, realistic evaluation of costs, or recognition of the uncertainty of the outcome.

One of the most important aspects of mitigation is how replacement systems function ecologically. As Zedler demonstrates in Chapter Fourteen, coarse compositional parity can be established in dominant species with original systems. Full ecological functional equivalency, however, can be a more elusive and difficult goal. Her research in the San Diego Bay has focused on determining the extent to which constructed tidal marshes are comparable to natural systems in nutrient cycling, soil organic content, biomass, invertebrate diversity and abundance, and other functional attributes. In addition, Zedler and her colleagues have evaluated the extent to which the constructed wetlands function as nesting habitat for the light-footed clapper rail (*Rallus longirostris levipes*), the species for which the mitigation was originally required. Their research has found that, in such systems, mitigation has yet to produce wetlands that are functionally equivalent to natural marshes or that provide suitable rail habitat. Moreover, their observations suggest that progress toward such parity is likely to take many decades, if in fact it is achieved at all.

One overlooked opportunity for rare plant conservation may be on corporate lands, including land previously used for mining, timber cutting, solid waste storage, and other industrial uses. In Chapter Fifteen, Klatt and Niemann provide a corporate perspective on endangered species management, including the use of such lands as receptor sites for reintroduction. Collectively, corporate lands represent millions of acres that conservationists can ill afford to ignore altogether. While many difficult issues would need to be worked out—such as guarantees for long-term conservation, security, and

management—these lands are part of the landscape as a whole and represent a potentially important opportunity for conservation.

The section ends with a thoughtful essay on mitigation strategies by Michael Bean (Chapter Sixteen) exploring the origins of mitigation policy and practice in wetlands and its gradual extrapolation to endangered plants. He also reviews innovative legal approaches such as mitigation banks, which would operate in a manner comparable to the pollution credit systems now in wide use.

In our "focus" illustration for this section (Chapter Seventeen), Gann and Gerson examine the use of mitigation-related reintroduction in Florida, in which some habitats that are richest in rare species—such as the sand scrub uplands, pine rocklands, and oak hammocks—have been particularly fragmented by commercial land development. Ironically, statutory protection of some Florida wetlands has actually shifted development pressure toward sensitive upland ecosystems.

Only long experience with compensatory mitigation will reveal if it is part of the solution or part of the problem. We are sure, however, that it is not too early to begin thinking critically about the issue, even as scores of reintroduction and restoration programs are already underway across the landscape.

Rare Plant Mitigation: A Policy Perspective

Ken S. Berg

Translocation of rare, threatened, or endangered plants as mitigation for development impacts is one of the most complicated and controversial issues in plant conservation. Complex interactions between political, legal, social, and biological factors make it difficult to effectively mitigate the adverse impacts of development and conserve biodiversity.

Because plants receive limited protection under the U.S. Endangered Species Act (ESA), most attempts to translocate de facto rare, threatened, or endangered plants are not regulated, and therefore official records are not maintained (see McDonald Chapter Four). Consequently, it is difficult to objectively assess the effectiveness of mitigation-related translocations in the United States. One of the few studies available (Fiedler 1991) found that the success rate was less than 25 percent for thirty-eight mitigation-related translocations in California.

The subject of translocation and reintroduction in a mitigation context can be confusing because the terms *mitigation, translocation, reintroduction,* and *introduction* are used broadly and inconsistently.

Translocation is the general term for moving plants (individuals, propagules, or populations) from one place to another, regardless of source or destination. The destination determines the more specific cases of introduction and reintroduction. *Reintroduction* involves moving plants back into a site where the species was known to occur or, more generally, into an area within the species' historical range. *Introduction* is moving plants into a site where the species has not been previously known to occur or into an area outside the species' historical range. *Transplantation* is often synonymous with *translocation* but can be used for the more specific case where plants are moved from one wild site to another.

Mitigation is a general term used to describe a wide variety of actions taken to avoid, reduce, or compensate for adverse impacts of development. In discussions and in practice, there is often little distinction made between the

different types of mitigation measures, even though they vary considerably in their methods and effectiveness. In preferred order, mitigation means and measures are actions taken to (1) avoid, (2) minimize, (3) rectify, (4) reduce or eliminate over time, or (5) compensate for adverse environmental impacts (U.S. Fish and Wildlife Service 1981).

The most effective way to mitigate for adverse impacts to rare plants is to select a development site and design the project so that both direct and indirect impacts to the rare plant habitat are avoided or minimized. When this is infeasible, a more experimental approach, such as translocation or habitat restoration, is often used in an attempt to compensate for unavoidable adverse impacts.

One major reason mitigation projects are so controversial is that the type of mitigation used, along the spectrum from avoidance to compensation, is more often determined by negotiation rather than by applying principled standards in a consistent manner (see Bean, Chapter Sixteen). Because we lack an effective means of guiding development away from significant natural areas, mitigation efforts often amount to making the best of a bad situation. Difficult negotiations are often followed by litigation when one or more parties believe that regulatory procedures or property rights have been violated. Negotiating skills and legal maneuvering may have more influence on the final outcome than the value of the natural resources (Howald, Chapter Thirteen; Tyson 1993; Yaffee 1991).

Mitigation translocations often fail because the strategic and biological considerations necessary for success cannot be adequately addressed within develop scenarios and schedules. Usually, limitations of time, money, or commitment prevent the detailed planning, careful implementation, and sustained population monitoring and habitat maintenance necessary for a successful translocation (Howald, Chapter Thirteen; Gann and Gerson, Chapter Seventeen; Pavlik, Chapter Six; Sutter, Chapter Ten; Barlow 1993; Howald 1993; Hall 1987). The broader temporal, spatial, and biological scale issues that are important for a strategic approach to plant conservation are rarely considered in politically driven land-use decisions (Falk and Olwell 1992).

In response, plant conservation and botanical organizations have adopted policies to discourage the use of translocation as mitigation for impacts to rare plants (New England Wild Flower Society 1992; California Native Plant Society 1988). The Center for Plant Conservation (Falk and Olwell 1992) has suggested as a national standard that "mitigation projects should never entail the destruction of existing significant and irreplaceable natural areas. Reintroduction and restoration should, in other words, be employed to heal damage that has already been caused, not to rationalize further destruction" (p. 301).

In spite of these concerns, mitigation-related translocations continue because conservationists lack the legal authority or political support to require or negotiate the use of alternate mitigation measures. Inadequate legal protection for plants results in ineffective mitigation, since in many cases no mitigation is required (McDonald, Chapter Four; Berg 1993; Campbell 1993; Howald 1993).

More effective policies on the use of translocation as mitigation are needed. However, better mitigation policies will only be effective if current barriers to their successful implementation are overcome. Experience in California and Florida show that current rare plant mitigation practices are ineffective because of legal, political, technical, and financial constraints (Howald, Chapter Thirteen; Zedler, Chapter Fourteen; and Gann and Gerson, Chapter Seventeen).

This chapter surveys some of the practical impediments to achieving better mitigation and some current national efforts to improve the context in which it occurs.

Impediments to Effective Mitigation

Declining plant diversity is a symptom of inappropriate land use. Solving the plant extinction crisis requires a fundamental change in land use patterns. The Endangered Species Act, considered by many to be the premier environmental law, can only limit conservationists' reactions to the inevitable impacts of increasing population and resource consumption (Grumbine 1992); it can do little to guide overall land use.

Efforts to improve plant mitigation must focus on removing the political, institutional, technical, educational, and philosophical barriers that impede biodiversity conservation (Noss and Cooperrider 1994). Plant conservationists must increase collaborative efforts to overcome inadequate laws, increase available funding and scientific information, and bring continuity to currently fragmented programs (Bean, Chapter Sixteen; Howald 1993; Keystone 1991; Campbell 1987, 1993).

Political and Institutional Barriers
The Endangered Species Act provides a poor safety net for declining plant species. Since only about half of the 430 federally listed plants occur on federal lands, most of the fifty-seven hundred "at risk" plant species in the U.S. do not receive protection under the ESA (Morse and Kutner 1993).

The ESA provides full protection to listed plants only when they occur on federal land or when they will be affected by a federally funded or permitted action. It also provides limited protection to plants on private lands but does

not necessarily protect their habitats if they are harmed by actions undertaken in knowing violation of a state law. Consequently, standards for plant mitigation are lower than standards for wetlands and endangered animals (Bean, Chapter Sixteen; McDonald, Chapter Four).

The biggest barriers to effective mitigation are its reactive nature and its lack of strategic context. In 1979, Rupert Cutler, then Assistant Secretary for Agriculture, recognized that "almost by definition, mitigation is, an afterthought, . . . an attempt to compensate for a mistake, and at least in part, a failure." Cutler suggested we move project planning to a more holistic approach, where habitat needs become an integral part of the project design from the beginning (Cutler 1979).

Efforts to avoid impacts are often unsuccessful because the mitigation negotiations begin long after the development site has been selected and the project has been designed. Without a means either to compensate landowners for conserving habitat or to allow development to proceed at an alternative site, translocation is likely to be the selected mitigation measure, regardless of its biological viability (Bean, Chapter Sixteen; Tyson 1993; Yaffee 1991).

Because plant conservation and recovery programs are poorly funded, conservationists spend most of their efforts reacting to proposed impacts rather than developing conservation strategies. Without rangewide habitat conservation and species recovery strategies as a guide, individual mitigation decisions lack direction and are haphazard, ineffective, and often detrimental (Galeano-Popp 1993).

In 1990, Fish and Wildlife Service recovery funding for plants was only 8 percent of the agency's total recovery budget (Campbell 1993). Consequently, as of September 1994, fewer than 200 of the 480 federally listed plants in the United States had final recovery plans. Unfortunately, many of the existing plans lack sufficient details on the species' habitat conservation requirements to help guide land use planning (Cook and Dixon 1989; U.S. General Accounting Office 1988). In many cases recovery goals are set at levels that risk extinction rather than ensure survival, due to political, social, and economic consideration (Tear et al. 1993).

Technical Barriers

Plant conservation programs are impeded by a lack of critical information about the distribution, abundance, and ecology of rare plants in the United States. Large areas in some states still lack basic floristic and vegetation inventories (Bureau of Land Management 1992). In California, Nevada, and Utah, for example, the rate of discovery of new species in recent years indicates that many more species have yet to be discovered and classified (Shultz 1993; Clemmer 1991; Shevock and Taylor 1987).

Much more basic information on the conservation needs of rare plants is needed to make effective management decisions (Messick 1987). Efforts must focus on gathering the information necessary to protect and restore remaining habitat areas before limited resources are spent on experimental translocations that may hinder species' survival.

Unintelligent tinkering with endangered plants is extremely risky, because most species are critically endangered by the time they are federally listed. Of 332 taxa that were listed or proposed between 1985 and 1991, the median number of populations was 4 and the median total population size was 120 individuals. For 39 of these plants, the total population size was 10 or fewer individuals at the time of listing (Wilcove, McMillan, and Winston 1993).

Educational Barriers

Increased public education is critical to gaining support for the difficult trade-offs inherent in adopting more effective plant mitigation programs.

To a layperson, the idea of moving an endangered plant out of harm's way may seem reasonable and desirable. Most gardeners have probably been successful doing this in their yards with cultivated plants. But transferring this approach from a few individuals of a hardy cultivar to an entire population of a rare species involves an increase in complexity that exceeds our technical capabilities.

Consequently, scientists and conservationists argue that nature is too complex for simplistic translocation approaches and that conservation programs must emphasize protecting large expanses of existing natural areas (Zedler, Chapter Fourteen; Roberts 1993; Sutherland and Gibson 1988; Turner 1988; Steinhart 1987). This must be better communicated to decision makers and the public so that mitigation efforts shift toward avoiding impacts rather than trying to compensate for them.

Current National Efforts to Improve Policies and Programs

Conservationists must work together to gain broader public support for more effective endangered species policies. But policies are only effective when people and funds are committed to implement them. In this era of declining government budgets and staffing, it is increasingly important that conservationists establish clear priorities, share expertise, and minimize duplication of effort.

In 1994, new policies and efforts were developed to increase collaboration and improve implementation of the Endangered Species Act. These efforts emphasize increased communication and coordination, development of regional and national priorities and strategies, and sharing of expertise and technology.

Improving Endangered Species Implementation

In January 1994, to provide a better safety net for declining species, the U.S. Fish and Wildlife Service, National Marine Fisheries Service, Bureau of Land Management, U.S. Forest Service, and National Park Service entered into a memorandum of understanding to establish a framework for improved coordination and cooperation in the conservation of candidate species. These efforts will emphasize implementing rangewide conservation strategies to protect candidate species and their habitats to the extent that federal listing is not necessary. Early coordination between federal agencies has shown that candidate species can be protected before they become critically endangered and conservation options become more limited.

In June 1994, the Fish and Wildlife Service and National Marine Fisheries Service issued six joint policy directives to improve the ESA's effectiveness while making it easier for Americans to understand the law and its require-ments. A key element of this initiative was the formation of the Federal Inter-agency Working Group, whose function is to identify, develop, and imple-ment reforms that streamline and improve overall performance of the ESA.

In September 1994, the Fish and Wildlife Service, National Marine Fish-eries Service, Bureau of Land Management, National Park Service, U.S. Forest Service, and several other federal agencies agreed to improve their ef-forts to conserve listed species and implement actions identified in recovery plans. They also agreed (1) to determine whether their respective planning processes effectively help conserve threatened and endangered species; (2) to resolve problems identified with interagency consultations under section 7(a)(2) of the ESA, the process that determines mitigation for activities when listed species are affected by federal actions or occur on federal lands; and (3) to reward the performance of employees who successfully conserve or recover listed species and the ecosystems upon which they depend.

These common goals will be accomplished through national and regional interagency recovery teams and working groups and will be implemented with appropriate involvement of the states, tribal and local governments, and the public.

Policy proclamations from Washington, D.C., can only be effective when people inside and outside the agencies are committed, authorized, and funded to implement them. Continued public support and scrutiny is needed to ensure that the new policies are implemented at the field level.

Toward a Coordinated National Strategy

In May 1994, to help build additional public support for plant conservation and to develop and implement regional and national plant conservation strate-gies, seven federal agencies launched the North American Native Plant Con-servation Initiative.

Under this initiative, the Agricultural Research Service, Bureau of Land Management, National Biological Survey, National Park Service, Soil Conservation Service, U.S. Fish and Wildlife Service, and U.S. Forest Service have joined together to establish common priorities, improve communication and coordination, share expertise, encourage conservation-oriented research, export solutions across agency jurisdictions, and provide increased support and connectivity to many collaborative conservation efforts taking place in the field. Key plant conservation organizations such as the Botanical Society of America, Center for Plant Conservation, Garden Club of America, Society for Ecological Restoration, The Nature Conservancy, and others have joined the initiative as formal cooperators to lend additional expertise and capability. By September 1995, the initiative had grown to include the Department of Defense, the Office of Surface Mining and Reclamation, and a total of forty-five nonfederal cooperators.

One goal of the initiative is to provide the botanical component to the landscape-level or ecosystem-management conservation planning efforts that are taking shape, such as the Northwest Forest Plan, the Eastside Plan for the Pacific Northwest, the Great Plains Initiative, the Everglades Ecosystem Plan, and others.

To be effective, these ecosystem management efforts should aggregate individual plant and animal conservation and recovery strategies, conduct additional inventories and analyses to identify priority habitat areas, coordinate habitat acquisition and management efforts of federal, state, and local governments, guide public and private development into less environmentally sensitive sites, and provide incentives and compensation for landowners who participate (Keystone 1991; Yaffee 1991).

Plant conservationists must take an active role in these ecosystem-management efforts to ensure that endangered plants receive adequate attention. For plants to receive effective protection in a landscape-scale conservation plan, their needs must be incorporated at the earliest stages. My experience with multispecies conservation planning efforts has been that the needs of listed vertebrates (such as the California gnatcatcher, desert tortoise, and spotted owl) drive the planning process, and that plants are a secondary priority, if they are included at all. This usually results in inadequate protection for plants, since their distribution and conservation requirements are not identical to animals in the same ecosystem.

A New Political Landscape

The November 1994 federal election changed the short-term politics of national environmental policies and programs by placing in both the House

and Senate a relatively conservative majority that opposes many of President Clinton's programs and initiatives. This adversarial relationship between Congress and the administration profoundly influences all aspects of national biodiversity conservation policy, including efforts to: (1) implement the existing ESA, (2) reauthorize a strong ESA, (3) ensure adequate funding for federal agencies and programs, and (4) implement and reauthorize other environmental laws and programs, such as the Clean Water Act and National Environmental Policy Act (Benenson 1995a; Cushman 1995).

In the first six months of the 104th Congress, this power struggle between Congress and the administration resulted in several efforts to establish environmental policy through legislative riders on appropriations bills. In addition, House and Senate versions of fiscal year 1996 appropriations for endangered species conservation and land acquisition programs make significant budget reductions. These actions foreshadow the obstacles that federal biodiversity conservation efforts face in the near term.

In March 1995, Senator Kay Bailey Hutchison (R-TX) led a successful effort to enact a moratorium on final endangered species listings and critical habitat designations. The moratorium passed as an amendment to HR 889, a department of defense supplemental appropriations bill that President Clinton signed into law on April 10, 1995. The moratorium was created by rescinding $1,500,000 from the Fish and Wildlife Service fiscal year 1995 budget and prohibiting any remaining funds from being used for these activities. The moratorium will expire October 1, 1995 or sooner if the ESA is reauthorized.

One of the most contentious clashes between Congress and the administration involved forestry policy on federal lands. On July 27, 1995, President Clinton signed Public Law 104-19, commonly referred to as the "1995 rescissions bill." This act provided supplemental funds for disaster relief, made cuts in some fiscal year 1995 appropriations, and included several environmental policy riders, including a controversial one directing logging on federal lands.

The president had vetoed the first version of the bill seven weeks earlier, and included criticism of the logging rider in his veto message. However, a slightly modified version was retained in the bill that the president signed into law. The rider directs "salvage logging" and other timber harvest activities on U.S. Forest Service and Bureau of Land Management lands, including areas covered by the president's Forest Plan for the Pacific Northwest. The controversy is over concerns that the legislative language effectively suspends existing environmental and endangered species protections in order to expedite logging (Devroy 1995). By late August 1995, the administration's interpretation was that "P.L. 104-19 gives us the discretion to apply current environmental standards to the timber salvage program, and we will do so." (Clinton

1995). This interpretation may lead Congress to further direct and clarify expedited logging procedures on federal lands.

The rescissions bill process exemplifies the antagonism between the administration and Congress that is likely to dominate activities during the 104th Congress (Devroy 1995). Additional efforts to legislate environmental policy on appropriations bills and to reduce the federal budget for endangered species programs followed.

Both the House and Senate versions of the 1996 Interior and Related Agencies Appropriations bill include an extension of the moratorium on final endangered species listings and critical habitat designations until October 1996 or until the ESA is reauthorized, whichever is sooner. These bills also make substantial cuts in endangered species funding for the Fish and Wildlife Service, Forest Service, and Bureau of Land Management. If averages between the House and Senate figures emerge from conference, then budgets for the Forest Service and Bureau of Land Management endangered species programs will be reduced by about 7 percent, and the Fish and Wildlife Service endangered species budget will be cut 18 percent.

The Interior appropriations bills also propose significant cuts in funds for Land and Water Conservation Fund acquisitions (House: 78 percent reduction; Senate: 46 percent reduction) (U.S. Congress 1995).

With a moratorium on future listings in place and spending cuts proposed, Congress has turned to reauthorizing the underlying statute. The upcoming ESA reauthorization debate will focus on private landowner prohibitions and incentives, the definition of take under Section 9, streamlining the consultation process, and expanding participation in recovery (Beedy 1995).

In March 1995, the Clinton Administration released a ten-point plan to guide congressional reauthorization and preempt drastic weakening of the ESA. This package includes the six 1994 joint-policy directives the Fish and Wildlife Service and National Marine Fisheries Service issued to strengthen science, streamline Section 7 consultations, minimize social and economic impacts, provide certainty to landowners, and prevent species endangerment. It also includes proposed regulatory exemptions for private landowners and suggested legislative changes in the act, which would give an increased role to states and create incentives for private landowners (Department of Interior 1995; Lehman 1995).

In May 1995, Senator Slade Gorton (R-WA) introduced S 768, a bill intended to amend the ESA. This bill would make much of the act discretionary and give greater authority to the secretary of interior to establish conservation objectives for listed species, which could fall far short of the current ESA goal of recovery in the wild for all listed species. The bill received negative reviews from environmentalists, who contended that it was written by

lobbyists for the timber industry and would give too much authority to the secretary of interior, a political appointee.

In July 1995, the Keystone Center released "The Keystone Dialogue on Incentives for Private Landowners to Protect Endangered Species," a compendium of incentive options compiled by individuals from environmental, industry, and government interests. These ideas for the encouragement of greater private landowner participation in endangered species habitat conservation are likely to be given significant attention, especially in light of the Supreme Court's affirmation that habitat on private lands can be regulated by the Fish and Wildlife Service.

On June 29, 1995, the Supreme Court upheld a key tenet of the Endangered Species Act when it ruled that the 1975 Fish and Wildlife Service regulation, which prohibits habitat modification when it harms listed species, is legal. The 6-3 decision in the case of *Babbitt v. Sweet Home* allows the continued protection of endangered species habitat on private lands by the FWS. Environmentalists and congressional advocates of the law cheered the court decision, while opponents of the current ESA felt it was a clear indication that the law should be changed to protect private-property rights (Benenson 1995b).

Conclusion

Ultimately, we need a coordinated process to identify and protect significant natural areas and to guide development into the least environmentally sensitive places (Noss and Cooperrider 1994; Keystone 1991; Yaffee 1991). We must shift away from piecemeal, reactive efforts that attempt to mitigate development impacts and move toward broader ecosystem-management and restoration programs that incorporate endangered species conservation as a primary objective (Scheuer 1993; Keystone 1991; Yaffee 1991). These species conservation objectives must be based on an improved understanding of the biological, ecological, and recovery needs of rare plants and natural plant communities. New land conservation tools and business mechanisms are needed to successfully implement these ecosystem-management plans.

A stronger Endangered Species Act is also needed to give plants greater legal protection. The evolution of social values, especially regarding the rights of nature and the interplay between private property rights and social responsibility, must lead to a declaration that endangered plants are more than real estate and to legal protection for plant communities and endangered ecosystems (Keeler-Wolf 1993; Noss 1991; Rolston 1991; Nash 1989).

The November 1994 election dramatically underscored the need for increased public support. Biodiversity cannot be conserved in the long term

without broad-based citizen cooperation. Stronger laws can work only if they are supported by innovative programs to increase incentives and options for private landowners to conserve natural habitats. Land exchanges, transfer of development rights, conservation tax credits, cooperative watershed management, and mitigation banking are land conservation methods that must be more fully utilized to ensure equitable treatment of landowners and continued public support (Bean, Chapter Sixteen; Klatt and Niemann, Chapter Fifteen; Grenell 1987; Carlton 1986; Jahn 1979).

Creative solutions to future mitigation negotiations are needed to overcome the social, legal, and political biases that perpetuate the treatment of endangered plants as second-class citizens. Without creativity and innovation, translocations will continue to be used ". . . to rationalize further destruction" (Falk and Olwell 1992, p. 301) and endangered species lists will be mere "chronicles of extinction" (Reffalt 1991, p. 77).

REFERENCES

Barlow, P. 1993. "Mitigating Impacts to Rare Plants in Pipeline Construction." In *Proceedings of the Southwestern Rare and Endangered Plant Conference. Miscellaneous Publication* No. 2. Edited by R. Sivinski and K. Lightfoot. Santa Fe: New Mexico Forestry and Resources Conservation Division.

Beedy, E. C. 1995. "Ten Ways to Fix the Endangered Species Act." *Endangered Species Update* 12 (6): 12.

Benenson, B. 1995a. "GOP Sets the 104th Congress on New Regulatory Course." *Congressional Quarterly Special Report.* June 17, 1995.

Benenson, B. 1995b. "Court Upholds Law Protecting Species on Private Property." *Congressional Quarterly Special Report.* July 1, 1995.

Berg, K. S. 1993. Endangered plants as second class citizens. In *Proceedings of the Symposium Interface Between Ecology and Land Development in California.* Edited by J. E. Keeley. Los Angeles: Southern California Academy of Sciences.

Bureau of Land Management. 1992. "Rare Plants and Natural Plant Communities: A Strategy for the Future." *Fish and Wildlife 2000.* Washington, D.C.: Department of Interior.

California Native Plant Society. 1988. Policy and Guidelines for the Mitigation of Rare Plants. Sacramento: California Native Plant Society.

Campbell, F. T. 1987. "The Potential for Permanent Plant Protection." In *Conservation and Management of Rare and Endangered Plants.* Edited by T. Elias. Sacramento: California Native Plant Society.

Campbell, F. T. 1993. "Plants: Competing for the Light." In *Proceedings of the Southwestern Rare and Endangered Plant Conference. Miscellaneous Publication No. 2.* Edited by R. Sivinski and K. Lightfoot. Santa Fe: New Mexico Forestry and Resources Conservation Division.

Carlton, R. L. 1986. "Property Rights and Incentives in the Preservation of Species." In *The*

Preservation of Species: The Value of Biological Diversity. Edited by B. G. Norton. Princeton, N.J.: Princeton University Press.

Clemmer, G. 1991. "Stacked Deck." *Biodiversity News* 4 (1): 1–7.

Clinton, W. J. 1995. Memorandum from the White House to Secretaries of Interior, Agriculture, and Commerce and Administrator of the Environmental Protection Agency on Implementing Timber-Related Provisions to Public Law 104-19, dated August 1, 1995.

Cook, R. E., and P. Dixon. 1989. A Review of Recovery Plans for Threatened and Endangered Plant Species. A report for the World Wildlife Fund. Washington, D.C.: World Wildlife Fund.

Cushman, J. 1995. "GOP's Plan for Environment Is Facing a Big Test in Congress." *The New York Times.* July 17, 1995.

Cutler, M. R. 1979. "The Need to Move from Mitigation to Multi-Objective Planning." In *The Mitigation Symposium: A National Workshop on Mitigating Losses of Fish and Wildlife Habitats.* General Technical Report RM-65. Fort Collins, Colo.: USDA Forest Service.

Department of Interior. 1995. Administration proposes Endangered Species Act exemptions for small landowners; "Guideposts for Reform" would give more authority to states. News Release. March 6, 1995. Six pages with attachment.

Devroy, A. 1995. "President Approves Rescissions Measure." *The Washington Post.* July 28, 1995.

Falk, D. A., and P. Olwell. 1992. "Scientific and Policy Considerations in Restoration and Reintroduction of Endangered Species." *Rhodora* 94 (879): 287–315.

Fiedler, P. 1991. *Mitigation-Related Transplantation, Relocation and Reintroduction Projects Involving Endangered, Threatened, and Rare Plant Species in California.* Sacramento: California Department of Fish and Game.

Galeano-Popp, R. 1993. "Rare Plant Protection Strategies: A Manager's Perspective." In *Proceedings of the Southwestern Rare and Endangered Plant Conference. Miscellaneous Publication No. 2.* Edited by R. Sivinski and K. Lightfoot. Santa Fe: New Mexico Forestry and Resources Conservation Division.

Grenell, P. 1987. "Innovative Programmatic Approaches to Resource Conservation." In *Conservation and Management of Rare and Endangered Plants.* Edited by T. Elias. Sacramento: California Native Plant Society.

Grumbine, R. E. 1992. *Ghost Bears: Exploring the Biodiversity Crisis.* Washington, D.C.: Island Press.

Hall, L. A. 1987. "Transplantation of Sensitive Plants As Mitigation for Environmental Impacts." In *Conservation and Management of Rare and Endangered Plants.* Edited by T. Elias. Sacramento: California Native Plant Society.

Howald, A. M. 1993. "Finding Effective Approaches to Endangered Plant Mitigation." In *Proceedings of the Symposium Interface Between Ecology and Land Development in California.* Edited by J. E. Keeley. Los Angeles: Southern California Academy of Sciences.

Jahn, L. R. 1979. "Summary of the Symposium and Recommendations for Improving Mitigation in the Future." In *The Mitigation Symposium: A National Workshop on Mitigating Losses of Fish and Wildlife Habitats.* General Technical Report RM-65. Fort Collins, Colo.: USDA Forest Service.

Keeler-Wolf, T. 1993. "Rare Community Conservation in California." In *Proceedings of the Symposium Interface Between Ecology and Land Development in California.* Edited by J.E. Keeley. Los Angeles: Southern California Academy of Sciences.

Keystone Dialogue Group. 1991. Biological Diversity on Federal Lands: Report of a Keystone Policy Dialogue. Keystone, Colo.: The Keystone Center.

Lehman, W. 1995. "Relief for Private Landowners." *Endangered Species Bulletin* 4 (4): 16–17.

Messick, T. 1987. "Research Needs for Rare Plant Conservation in California." In *Conservation and Management of Rare and Endangered Plants.* Edited by T. Elias. Sacramento: California Native Plant Society.

Morse, L. E., and L. S. Kutner. 1993. "Plants." *Biodiversity Network News* 7 (1): 5.

Nash, R. F. 1989. *The Rights of Nature: A History of Environmental Ethics.* Madison: University of Wisconsin Press.

New England Wild Flower Society. 1992. "New England Plant Conservation Program." *Wild Flower Notes* 7 (1): 1–79.

Noss, R. F. 1991. "From Endangered Species to Biodiversity." In *Balancing on the Brink of Extinction: The Endangered Species Act and Lessons for the Future.* Edited by K. A. Kohm. Washington, D.C.: Island Press.

Noss, R. F., and A. Cooperrider. 1994. *Saving Nature's Legacy: Protecting and Restoring Biodiversity.* Washington, D.C.: Island Press.

Reffalt, W. 1991. "The Endangered Species Lists: Chronicles of Extinction?" In *Balancing on the Brink of Extinction: The Endangered Species Act and Lessons for the Future.* Edited by K. A. Kohm. Washington, D.C.: Island Press.

Roberts, L. 1993. "Wetlands Trading Is a Loser's Game, Say Ecologists." *Science* 260: 1890–1892.

Rolston, H. III. 1991. "Life in Jeopardy on Private Property." In *Balancing on the Brink of Extinction: The Endangered Species Act and Lessons for the Future.* Edited by K. A. Kohm. Washington, D.C.: Island Press.

Scheuer, J. H. 1993. "Biodiversity: Beyond Noah's Ark." *Conservation Biology* 7 (1): 206–207.

Schultz, L. 1993. "Patterns of Endemism in the Utah Flora." In *Proceedings of the Southwestern Rare and Endangered Plant Conference. Miscellaneous Publication* No. 2. Edited by R. Sivinski and K. Lightfoot. Santa Fe: New Mexico Forestry and Resources Conservation Division.

Shevock, J., and D. Taylor. 1987. "Plant Exploration in California: The Frontier Is Still Here." In *Conservation and Management of Rare and Endangered Plants.* Edited by T. Elias. Sacramento: California Native Plant Society.

Steinhart, P. 1987. "Mitigation Isn't." *Audubon* 89: 8–11.

Sutherland, W., and C. Gibson. 1988. "Habitats to Order (Man-Made Habitats Are No Substitute for the Real Thing)." *New Scientist* 117: 70.

Tear, T. H., J. M. Scott, P. H. Hayward, and B. Griffith. 1993. "Status and Prospects for Success of the Endangered Species Act: A Look at Recovery Plans. *Science* 262: 976–977.

Turner, T. 1988. "The Myth of Mitigation." *Sierra* 73 (1): 31–33.

Tyson, W. 1993. "Reconciling Mitigation, Preservation, and Ecosystem Restoration." In *Proceedings of the Symposium Interface Between Ecology and Land Development in*

California. Edited by J. E. Keeley. Los Angeles: Southern California Academy of Sciences.

U.S. Congress. 1995. "Department of the Interior and Related Agencies Appropriations Bill, 1996." *Committee Report 104-125.* July 28, 1995. 168 pages.

U.S. Fish and Wildlife Service. 1981. "U.S. Fish and Wildlife Service Mitigation Policy." *Federal Register* 46 (15): 7644–7663.

U.S. General Accounting Office. 1988. Endangered Species Management Improvements Could Enhance Recovery Programs. Report to the Chairman, Subcommittee on Fisheries and Wildlife Conservation and the Environment, Committee on Merchant Marine and Fisheries, House of Representatives. GAO/RCED-89-5.

Wilcove, D. S., J. McMillan, and K. C. Winston. 1993. "What Exactly Is an Endangered Species? An Analysis of the U.S. Endangered Species List: 1985–1991." *Conservation Biology* 7 (1): 87–93.

Yaffee, S. C. 1991. "Avoiding Endangered Species/Development Conflicts Through Interagency Consultation." In *Balancing on the Brink of Extinction: The Endangered Species Act and Lessons for the Future.* Edited by K. A. Kohm. Washington, D.C.: Island Press.

Translocation As a Mitigation Strategy: Lessons from California

Ann M. Howald

———————————⇒•⇐———————————

One can spend a lifetime in California and never see all of its natural diversity: The low desert's ocotillo woodlands and palm oases; the South Coast, where sage-covered slopes rise above the surf; the Central Valley, home to remnant tule marshes and riparian forests; the Sierra Nevada's oak-covered foothills and hidden groves of giant sequoias; the Great Basin's western fringe, with its ancient salt lakes and stands of pinyon pine; and the North Coast, where stately redwoods shade ferny canyons and trillium hollows. These landscapes and the promises they hold have brought millions to the Golden State; they are the basis for California's mystique, for its history, and for its current crisis in endangered plant protection.

In California, land development has resulted in the permanent loss of thousands of acres of natural habitat and scores of populations of endangered plants. Translocation has been proposed as a method of reducing these losses. If translocation can be implemented successfully (that is, if an endangered plant population can be removed from one site and established as a self-sustaining entity in another natural site), then it could provide benefits far beyond those of mitigating development impacts. For example, translocation could be used to restore endangered plants to habitats where they have been lost as a result of reversible environmental changes, such as fire and drought. It could also be used to enhance the biological diversity of natural preserves and other protected wildlands. Establishing additional populations of endangered plants in natural sites could give those that have been reduced to only one population, such as marsh sandwort (*Arenaria paludicola*) and showy Indian clover (*Trifolium amoenum*), an important hedge against extinction. However, as previous efforts have made clear, translocations are risky. Our ecological ignorance of most endangered plants forces guesswork into

translocation projects. Impatience, economic constraints, and the desire for success, have resulted in less-than-ideal outcomes. In California, the use of translocation to mitigate endangered plant losses remains highly controversial.

How the "California Dream" Has Imperiled Rare Plants

The California flora boasts extremely high levels of diversity and endemism. As Raven (1977) has noted, the California Floristic Province "has more plant species than the central and eastern United States and adjacent Canada, an area more than ten times its size. (p. 111)" The 1978 findings of Raven and Axelrod that 30 percent of California's plant species are endemic to the state have scarcely needed revision even after new taxonomic treatments (Hickman 1993) and recent plant discoveries (Shevock, Ertter, and Taylor 1993; Shevock and Taylor 1987). The new Jepson Manual (Hickman 1993) describes 6,013 native California taxa, of which 31 percent are endemic.

In California, rare plants are found in every region, in nearly every habitat. Figure 13-1 shows the distribution of the more than nine thousand rare plant populations currently mapped by the California Natural Diversity Data Base (CNDDB), a computerized inventory that tracks the status of California's rare plants, animals, and natural communities.

Many of California's rare plants have become endangered due to direct loss of habitat from development, a result of the state's burgeoning population. Through the 1980s the state's population grew by about 2 percent per year, so that by 1990, one in every ten Americans was a Californian (Hoffman 1993). California's population is projected to increase to nearly 35 million by the year 2000, putting further strain on its rare plants. In addition to development, California's flora is imperiled by other human activities that have increased with population growth: agriculture, grazing, mining, off-road vehicle use, and the spread of invasive weeds (CDFG 1992a).

Significant losses of rare plants and their habitats already have occurred in California (Table 13-1). The California Department of Fish and Game (CDFG) and others have estimated habitat losses as high as 94 percent for interior wetlands (Jones and Stokes Associates 1987) and greater than 97 percent for vernal pools in the San Diego area (Bauder 1986). Estimates of the number of California plant species that have become extinct range from twenty-six in the Jepson Manual (Hickman 1993) to thirty-seven in the fifth edition of the California Native Plant Society's Inventory of Rare and Endangered Vascular Plants (Skinner and Pavlik 1993). The Center for Plant Conservation's California Endangered Plant Task Force has identified 143 California plant taxa that are at risk of extinction in the next five to ten years if

FIGURE 13-1.   Known locations of rare plants in California.

TABLE 13.1.   Losses of habitats, endangered plant species, and endangered plant populations in California.

Resource type	Amount lost
Habitat Type	
Central valley riparian woodland[1]	89 percent
Coastal wetlands	80 percent
Interior wetlands	94 percent
Central valley vernal pools	66 percent
San Diego county vernal pools[2]	97 percent
Endangered plant species	
Extinct[3]	26–37 species
At risk of extinction in 5–10 years[4]	143 species
*Endangered plant populations*5	
Extirpated	256 populations
Probably extirpated	254 populations

1. Estimated losses since 1800 (Jones and Stokes Associates 1987).
2. Estimated losses through 1990 (Bauder 1986).
3. Hickman 1993; Skinner and Pavlik 1993.
4. Center for Plant Conservation 1993.
5. California Natural Diversity Data Base 1993.

actions are not taken to reverse their declines (Center for Plant Conservation 1993). The California Natural Diversity Data Base (1993a) has data to show that more than 250 populations of endangered plants have been extirpated for certain, and another 250 or so are thought to be extirpated.

In response to these threats, the U.S. Fish and Wildlife Service (FWS) and the CDFG already have listed 220 California plant taxa as endangered, threatened, or rare. The service is actively pursuing the listing of another 126 candidates that appear to be eligible for listing (FWS 1993). There is compelling evidence that many other California plants meet the criteria for state or federal listing. For example, the fifth edition of the California Native Plant Society's Inventory lists 841 California taxa that are considered "rare and endangered" (Skinner and Pavlik 1993).

Why Translocation?

Resource agencies such as the CDFG and the FWS are engaged in an ongoing search to identify the most effective methods of mitigating endangered

plant losses (Howald 1993). Translocation is frequently used, although it is not the preferred mitigation method of resource agencies, according to pertinent environmental legislation and agency policies. Both the California Environmental Quality Act (CEQA Section 15370) and the National Environmental Policy Act (NEPA) regulations (40CFR 1508.20) identify the following as mitigation:

- *Avoiding* an impact altogether by not taking a certain action or parts of an action

- *Minimizing* impacts by limiting the degree or magnitude of the action and its implementation

- *Rectifying* the impact by repairing, rehabilitating, or restoring the affected environment

- *Reducing* or eliminating the impact over time by preservation and maintenance operations during the life of the action

- *Compensating* for the impact by replacing or providing substitute resources or environments

Avoidance: The First Priority

NEPA regulations stipulate that avoidance is to be given first priority in reducing project impacts. Many agencies, such as the U.S. Army Corps of Engineers, the U.S. Environmental Protection Agency, the FWS, and other federal agencies responsible for protecting wetlands and endangered species follow this guidance in their reviews of development projects that could harm wetlands and endangered plants. The Sacramento Endangered Species Office of the FWS regularly recommends avoidance rather than compensation, because endangered species' populations and their habitats can't be duplicated, and the long-term viability of these species is seriously threatened as a result of past habitat losses. The CDFG, as the state's trustee agency for natural resources, regularly requests avoidance as its preferred means of mitigation through comments it provides on projects undergoing CEQA review. The CDFG requires developers to demonstrate that avoidance is not feasible before approving translocation projects for state-listed plants (Howald and Wickenheiser 1990).

In spite of resource agency preferences for avoidance, CEQA "lead" agencies, which in many cases are local (county and city) government agencies, frequently do not require project applicants to avoid endangered plant populations. (In California, local governments exercise final control over most development projects through zoning and permitting.) In many cases where impacts to endangered plants are deemed significant under CEQA and

therefore require mitigation, lead agencies often turn to translocation as a form of compensation, even though compensation has the lowest priority of the five mitigation alternatives outlined in CEQA and NEPA. (Another form of compensation is protection through acquiring existing natural habitat, sometimes accompanied by restoration.) Translocations typically have consisted of efforts to relocate threatened populations to new sites, sometimes including attempts to "create" the special habitats they require (Howald 1993). Since 1983, CDFG records show that more than sixty development projects have used translocation to mitigate adverse impacts to endangered plants (Fiedler 1991). Another option lead agencies exercise is to allow the project to proceed without mitigation by adopting "overriding considerations," which permit economic or other needs to take precedence over the need to avoid environmental damage.

In relatively few cases do state and federal agencies have the authority to require avoidance. Examples include (1) Army Corps–regulated projects proposed for wetlands that contain federally listed species where the FWS has classified the site as "Resource Category 1," meaning that the site's natural resources are considered unique and irreplaceable; and (2) projects for which a state or federal biological assessment (under provisions of Federal or California Endangered Species Acts) has determined that the project is likely to cause the extinction of a listed species.

"No-Net-Loss" and Endangered Plant Mitigation

The use of translocation to mitigate endangered plant impacts also can be viewed as an outgrowth of agency policies on wetlands. The Army Corps (Clean Water Act, Section 404) and the CDFG (Fish and Game Code, Wetlands Resources Policy) have policies that call for a "no-net-loss" approach to wetland mitigation. This approach requires that loss of wetlands be mitigated by replacement at a 1 to 1 or greater ratio through the creation of new wetlands, preferably the same kind as those lost. Implicit in the strategy is the assumption that the methodology exists to create naturally functioning wetlands — an assumption that has been challenged in California by those who have studied wetlands created for mitigation purposes (Zedler, Chapter Fourteen). Zedler (1991) has stated that "mitigation policies must be put on hold until constructed wetlands are proved capable of attracting and sustaining the full complement of native species. (p. 35)"

In California, the no-net-loss approach has been applied to endangered plants of both wetlands and uplands, although no policy specifically calls for this approach. Mitigation plans propose translocation to "replace" endangered plant populations on a more or less 1:1 basis. They require creation of a population of the same size in both area and numbers of individuals, often without regard for the carrying capacity of the translocation site or the biology of the translocated species.

The Complexity and Difficulty of Translocation

Support for translocation as a viable method of mitigating endangered plant losses has emerged from the favorable results of a few successful experimental translocations and reintroductions. However, the successes only underscore the difficulty of such efforts (and obscure the failures of other efforts). They suggest that it may be possible, under highly controlled and thoroughly monitored circumstances, to establish an endangered plant in a carefully selected new location or even in habitat that has been created expressly to support it. These experimental attempts include a vernal pool creation project in San Diego County (Zedler et al. 1993); the translocation of *Holocarpha macradenia* (Santa Cruz tarplant) in Contra Costa and Alameda counties (Havlik 1989); the introduction of *Oenothera deltoides* ssp. *howellii* (Antioch Dunes evening primrose) to three sites, including dunes at Brannon Island State Recreation Area (Matthews 1990); and the recovery of *Amsinckia grandiflora* (large-flowered fiddleneck) through reintroduction to sites within its historical range (Pavlik 1990).

As Zedler et al. (1993) report, in 1986 forty artificial basins were created in a chaparral environment in San Diego County to mitigate for the loss of *Pogogyne abramsii* (San Diego mesa mint), a state and federally endangered vernal pool endemic. Since their creation, the artificial pools have been monitored annually and compared with nearby natural vernal pools. The authors conclude that this project has resulted in "some degree of success" because many of the vernal pool plants, including the endangered *Pogogyne abramsii*, have become established and have reproduced in the created pools. Their data suggest it is possible to increase the biological diversity of a specific type of coastal chaparral by creating vernal pools within the chaparral matrix. They caution, however, that their monitoring period of five years is relatively short, and that their results may not be applicable in other habitats.

In the case of an ongoing seven-year recovery program for *Amsinckia grandiflora*, Pavlik (1991a, 1992; Chapter Six) has found that three or four out of seven experimental populations may have the potential to survive over the long term, although their survival probably will depend upon long-term management.

Special Problems of the Mitigation Context

Translocations of endangered plants to mitigate impacts to natural populations are burdened with unique problems. The same unanswered biological questions that impede research-based recovery efforts likewise hinder the design of mitigation-related translocations. At least as problematic as biological uncertainties (and often more so) is the need to fit a complicated biological experiment into a framework dictated by the scheduling

requirements and financial constraints that accompany land-development projects (Howald 1993).

Fahselt (1988) discusses several objections to the use of transplantation as a conservation technique. Many of his points also are relevant to the use of translocation as a mitigation strategy. He notes that transplanting endangered plants into natural sites may result in degradation of a natural ecosystem that is itself rare. He points out the substantial costs of transplantation projects and suggests that the funds could be better spent protecting the natural ecosystems upon which endangered plants depend. He warns that transplantation advocates may undermine other protection efforts by giving the impression that moving plants from inconvenient sites is an easy and acceptable solution.

Translocating an endangered plant successfully is far from easy, especially when conducted as a form of mitigation. Imagine all that can go wrong in this realistic, although hypothetical, example. A mitigation-related translocation, poorly planned and hastily implemented, is installed at a site selected for reasons of expediency and budget rather than biological suitability. California's notoriously unpredictable rainy season begins with an early deluge that washes away both soil and seedlings. Even after a frantic round of ad hoc remediation, first-year results don't meet performance standards. Pressure to fix what is wrong, speed up the process, and create a "successful" outcome results in actions that clearly go beyond what is natural. Situations like this lead to biological crisis management, wasted time and money, and ultimate failure. They can force actions that do not benefit the species, the environment, or the success of the mitigation effort. Research projects certainly are not immune to pressures of this type, but the need to achieve a specific result within a narrowly defined time frame generally is lacking. Failure in research—rather than a biological and financial catastrophe—is a potential and acceptable outcome.

Lack of Baseline Information

A translocation that stands a good chance of success relies upon a comprehensive knowledge of the ecological and reproductive requirements of the species. One notable example is the experimental reintroductions of *Amsinckia grandiflora* by Pavlik (1991b; Chapter Six). For many plants, especially endangered plants, such information is unknown or is limited to what is derived from short-term field observations of one or a few populations. Many plants now recognized as endangered are found in unusual habitats (such as vernal pools and serpentine areas) and are therefore thought to require unusual soil types, reduced competition, specific pollinators or hosts, or other specialized environmental factors. Understanding these requirements and applying the findings to translocation efforts may require a research program

of several years' duration—a commitment of time and money that can be difficult to obtain in the context of mitigation.

Time constraints imposed by a project's lead agency and by the developer (neither of whom want responsibility for mitigation efforts that extend beyond the construction phase of a project) mean that preproject research to fill baseline data gaps and small-scale experiments to identify effective techniques may be viewed as unrelated to mitigation and as unnecessary extra costs, though these activities could increase the possibilities for successful mitigation.

Lack of Critical Review for Methods and Procedures

Although translocations of endangered plants have been taking place in California for more than ten years, there has not been a comprehensive review of translocation methods. Such a review would provide a basis for selecting the most appropriate methods and for further testing of those that appear most promising. Although no single method or procedure is likely to be shown the most effective in all cases, the translocation design process would be greatly improved if a comparative analysis of methods were available. Planting methods have ranged from direct transplants of whole plants (Hall 1987) and broadcasting of seeds into unmodified habitat (Havlik 1989) to precision-planting of seeds in relocatable plots subjected to a variety of pretreatments (Pavlik 1990). These methods have produced widely varying results. Site selection procedures have resulted in the use of botanical gardens and open space areas chosen without regard to the ecological requirements of the species (Hall 1987), while in other cases sites have been chosen because they contained habitats that appeared to mimic the natural habitat of the species (Havlik 1989; Pavlik and Heisler 1988). Clearly, the choice of methods can strongly influence a translocation's outcome.

Critical review of methods has been stymied by a lack of project documentation. Many projects, especially earlier efforts that could offer valuable opportunities for review of long-term results, were not documented thoroughly enough to permit critical review. Hall (1987) found that in five out of the fifteen mitigation projects she reviewed, documentation was incomplete or inaccurate, or the written records had been lost or misplaced. When available, written documentation of mitigation projects often is in the form of unpublished reports that are not easy to obtain (Fiedler 1991).

No Widely Accepted Performance or Success Standards

The issue of determining how to set standards, both for *performance* (approved methods of carrying out a project) and *success* (criteria by which the outcome of the project will be judged), is unquestionably the most difficult task that agencies face in overseeing mitigation-related translocations. Agencies with mandated responsibility for setting these standards have so far

been unable to act, possibly because the studies, analyses, and critical reviews needed to support standards haven't been conducted. As a result, mitigation projects involving translocations and habitat creation have been implemented and evaluated without the benefit of agreed-upon goals for performance and success. A project's "success" or lack thereof has most often been defined and determined by the organization that implemented and monitored it. The question remains open as to whether success standards for mitigation projects should be different from those used for reintroductions conducted for research purposes.

As Pavlik discusses in Chapter Six, deciding what constitutes success in a research-based reintroduction is not a simple matter. He notes that, in this context, success must not imply finality, perfection, or victory, but rather is unpredictable, with no clear endpoint. He has identified a framework for developing a definition of success that could be project-specific yet encourage consistency in design and measurement so that projects on very different plants and habitats could be evaluated and compared. His approach uses four components of a definition of success: abundance, extent (distribution), resilience, and persistence. For each component, he suggests several measurable factors that could be used in crafting a definition. If applied to mitigation-related translocations, Pavlik's approach would be a significant departure from past practices that have ranged from no success criteria to project-specific success criteria that used relatively simple measures, such as estimated abundance and distribution of the "target" species. Most determinations of success have been made after fewer than five years of monitoring, although it may take decades for a translocated population to be subjected to the full range of California's climatic variation.

Cost Factors

Mitigation-related translocations can be expensive, especially if they are done correctly. Although reliable data on these costs are difficult to obtain, the limited information available suggests significant costs are involved. Mitigation funding becomes a particularly significant issue when a small-scale development project encounters the need for a large-scale mitigation effort, such as creating an off-site habitat. In much of California, high real-estate values mean that finding and acquiring a suitable mitigation site represents a significant investment. Developers regularly claim that mitigation costs make their projects financially infeasible (Dunn 1993), and elected officials, particularly those with pro-growth agendas, are reluctant to impose hefty mitigation fees, especially on developers of small-scale projects.

Because mitigation projects are often implemented by companies with a profit motive, mitigation costs are subject to the economics of the marketplace, where competition results in developers' cutting project costs to achieve the lowest bid. Such cost-cutting can compromise the viability of the

mitigation effort if it leads to inadequate planning, design, implementation, monitoring, or maintenance or to selection of a mitigation site that is too small to accommodate the project. This is especially problematic in the absence of clear regulatory guidelines for endangered species mitigation.

Long-term management costs also can be significant. The Center for Natural Lands Management (CNLM) is currently completing an analysis of long-term management costs for the U.S. Environmental Protection Agency (CNLM 1994). The center has data to show that, when overall costs are considered, building in areas of high biological sensitivity is not cost effective. Their study indicates that it is more economical to direct development to areas of lower biological sensitivity than to pay the costs of finding, purchasing, and maintaining over the long term an off-site mitigation area.

Uncertain Outcomes

Experiments with highly complex biological systems are subject to inherent risks. In spite of the best efforts, failure is always a possibility. When mitigation-related translocations fail, intense dissatisfaction may be felt by developers who have spent a lot of money, by agencies who have invested scarce resources to provide oversight and guidance, and by conservationists who have trusted others to protect and conserve irreplaceable natural resources. Although the risks can scarcely be eliminated, the value of these projects would be increased if they were viewed as experimental and thus not likely to provide full mitigation. They would also be more valuable if thorough, long-term monitoring were required, so that even if the translocation is itself a failure the effort is not, because it has produced information that will guide future decisions.

So far, little progress has been made in modifying the mitigation process in California so that it actually delivers what CEQA intended—reduction of impacts to a level of insignificance. In 1989, legislation (Assembly Bill 3180) that requires mitigation monitoring took effect, but its intent seems to be to determine whether mitigation measures were carried out at all rather than to discern success in reducing project impacts. Strengthening any of CEQA's provisions to protect endangered species doesn't appear likely in the near future. In 1993 the California legislature adopted three bills (Assembly Bill 1888, Senate Bills 659 and 919) that weaken the environmental review process by reducing review requirements, shortening review periods, and limiting resource agencies' ability to comment, among other measures.

Mitigation-Related Translocations in California

Critical evaluation of mitigation-related translocations in California has been hindered by the lack of generally accepted definitions of translocation success (Pavlik, Chapter Six) and by inadequate documentation and monitoring.

Monitoring of most projects has lasted less than five years and has lacked rigor (Sutter, Chapter Ten). Hall (1987) reviewed fifteen mitigation-related translocations in San Diego County that involved ten endangered plant species (Table 13-2). She found that only four projects were 100 percent successful (although "success" was generally undefined), and eleven projects had success levels of 50 percent or less. In her study, the aspect of mitigation that was most often neglected was monitoring. To improve success rates she recommended that mitigation-related translocation projects be subject to peer review and bonding requirements and that longer maintenance and monitoring periods be required.

A more recent questionnaire-based survey commissioned by the California Department of Fish and Game (Fiedler 1991) identified forty-five mitigation-related translocations involving thirty-two species that were initiated between 1983 and 1989 (Table 13-3). These translocations were conducted to mitigate the impacts of forty development projects. This survey included some of the same projects that Hall reviewed. The success levels shown in the table are based mainly on the evaluations of those who conducted the projects. Few of the projects utilized specific, measurable criteria for success and, although more recently initiated projects have placed greater emphasis on monitoring and documentation, most have relied on monitoring methodologies that are qualitative, unrepeatable, or unrepresentative. In addition, monitoring has been carried out over too few years to allow reliable evaluation of success.

Case Studies

The following examples illustrate the outcomes of mitigation-related translocations in California. These two examples summarize the findings from a group of projects involving endangered plants found in vernal pools in Sonoma County and a project concerned with *Holocarpha macradenia* (Santa Cruz tarplant), an annual of coastal grasslands. While certain aspects are unique, other mitigation-related translocations in California have experienced similar problems.

Santa Rosa Plain (Sonoma County) Vernal Pools

Vernal pools are seasonal wetlands that fill with water during the winter rainy season and dry out as summer approaches (Zedler 1987). They are found in several parts of California, including the Santa Rosa Plain in Sonoma County, north of San Francisco Bay. They support numerous species of endemic plants, invertebrates, and amphibians and are visited regularly by waterfowl, shorebirds, and other wildlife (Zedler 1987). The Army Corps considers them to be wetlands regulated under Section 404 of the Clean Water

TABLE 13-2. Evaluation of mitigation-related translocation, San Diego County, California, 1986 (Hall 1987).

Case no/species	Success (%)	Technique	Environment	Documentation	Maintenance	Monitoring	Total
1. *Acanthomintha ilicifolia*	0	1	1	0	0	0	2
2. *Acanthomintha ilicifolia*	100	1	1	1	1	1	5
3. *Acanthomintha ilicifolia*	100	1	1	1	1	1	5
4. *Ambrosia pumila*	0	0	0	1	0	0	1
5. *Dudleya attenuata orcuttii*	0	1	0	0	0	0	1
6. *Brodiaea filifolia*	0	1	0	0	0	0	1
7. *Brodiaea filifolia*	100	1	1	0	1	1	4
8. *Baccharus vanessae*	10	1	0	1	0	1	3
9. *Baccharus vanessae*	10	1	1	1	1	1	5
10. *Arctostaphylos glandulosa crassifolia*	50	1	1	1	1	0	4
11. *Monardella linoides viminea*	10	1	0	1	0	0	2
12. *Monardella linoides viminea*	50	1	1	1	0	1	4
13. *Ferocactus viridescens*	100	1	1	1	0	1	4
14. *Opuntia parryi serpentina*	0	0	0	1	0	0	1
15. *Hemizonia conjugens*	0	1	1	0	0	0	2

1 = Satisfactory completion

0 = Deficiency adversely affected project outcome

TABLE 13-3. Mitigation-related translocations in California, 1983-1989.

Species name	Year initiated	Project name/proponent	Success
Acanthomintha ilicifolia	1988	1. Westview/Pardee Company	Ongoing
A. ilicifolia	1988	2. Palos Vista/Shea Homes	Ongoing
A. ilicifolia	1989	3. Sabre Springs/Pardee Company	Ongoing
A. ilicifolia	1985	4 . Indian Hill 5. Las Brisas 6. Spyglass/Unknown	Limited
Antennaria flagellaris	1983	7. None/Bureau of Land Management	Not successful
Blennosperma bakeri	1989	8. Montclair Park/Christopherson Homes	Not successful
B. bakeri	1989	9. San Miguel Estates I/Cobblestone Development Corporation	Ongoing
Brodiaea filifolia	1988	10. San Marcos College Area Specific Plan/Baldwin Company	Ongoing
Brodiaea insignis	1989	11. Kaweah Reservoir Dam Expansion/Calif. Dept. of Water Resources	Planning stage
Calochortus greenei	1989	12. None/Siskiyou County	Not successful
Chorizanthe howellii	1989	13. None/Univ. of Calif., Davis	Ongoing
Cirsium occidentale var. *compactum*	1986	14. Little Pico Bridge Replacement and Piedras Blancas Shoulder Widening/Calif. Dept. of Transportation	Partial
Croton wigginsii	Unknown	15. None/Bureau of Land Management	Not successful
Eriastrum densifolium ssp. *sanctorum*	1988	16. Santa Ana Woollystar Relocation Project/Calif. Dept. of Transportation	Not successful
Eriophyllum mohavense	1988	17. Luz SEGS VII/Calif. Energy Commission	Not successful
Eryngium aristulatum var. *parishii*	1986	18. Del Mar Mesa Vernal Pools/Calif. Dept. of Transportation	Partial
Erysimum capitatum var. *angustatum*	1989	19. Vaca Dixon-Contra Costa 230-kv Reconductoring Project/Pacific Gas and Electric Co.	Partial
Erysimum menziesii	1989	None/Univ. of Calif., Davis	Ongoing
E. menziesii	1987	20. Spanish Bay Golf Course/ Unnamed	Not successful
E. menziesii	1988	21. None/Unnamed timber company	Ongoing
Erysimum teretifolium	Unknown	22. Olympia Quarry Revegetation/ Lone Star Industries	Planning stage
Gilia tenuifolia ssp. *arenaria*	1987	Spanish Bay Golf Course/Unnamed	Not successful
Hemizonia increscens ssp. *villosa*	1988	23. Gaviota Interim Marine Terminal/Texaco	Ongoing
Hemizonia minthornii	1988	24. Twin Lakes Tank No. 2 /Las Virgenes Municipal Water District	Not successful

TABLE 13-3. (*Continued*)

Species name	Year initiated	Project name/proponent	Success
H. minthornii	Unknown	25. Woolsey Canyon Development/ Chateau Builders	Planning stage
Holocarpha macradenia	1986	26. Hilltop Commons Development/ Nylen Company	Successful
Lasthenia burkei	1984	27. Airport Blvd. Business Park/ Unnamed	Successful
L. burkei	1986	28. Sonoma Co. Airport Expansion/ Sonoma Co. Airport	Successful
L. burkei	1989	San Miguel Estates I/Cobblestone Development Corporation	Ongoing
L. burkei	1988	29. Area 31 Waste Water Storage Pond/Sonoma Co. Public Works	Partial
Lilaeopsis masonii	1989	30. Baker Slough Bank Revetment Calif. Dept. of Water Resources	Ongoing
L. masonii	Unknown	31. None/Calif. Dept. of Parks and Recreation	Planning stage
Lupinus milo-bakeri	1985	32. None/Calif. Dept. of Transportation	Not successful
Lupinus tidestromii var. *tidestromii*	1985	Spanish Bay Golf Course/Unnamed	Unknown
Mahonia nevinii	1988	33. Vesting Tentative Tract No. 23267/RANPAC Corporation	Unknown
Monardella linoudes ssp. *viminea*	1983	34. None/Calif. Dept. of Transportation	Not successful
Oenothera deltoudes ssp. *howellii*	1989	Vaca Dixon-Contra Costa 230-kv Reconductoring Project/Pacific Gas and Electric Co.	Partial
Opuntia basilaris var. *treleasei*	1983	35. Kern River Cogeneration Power Plant/Calif. Energy Commission	Successful
O. basilaris var. *treleasei*	1989	36. Sycamore Cogeneration Project/ Sycamore Cogeneration Company	Unknown
Orcuttia viscuda	Unknown	37. Sunrise/Douglas Wetland Creation Program/Unnamed	Ongoing
Pentachaeta lyonii	1988	38. Lake Sherwood Golf Course/ Unnamed	Not successful
Pogogyne abramsii	1986	Del Mar Mesa Vernal Pools/Calif. Dept. of Transportation	Partial
Pseudobahia peirsonii	Unknown	39. Round Mountain Flood Control Project/Fresno Co. Metropolitan Flood Control District	Planning stage
Sedum albomarginatum	1986	40. Feather River Canyon Storm Damage Repair/Calif. Dept. of Transportation	Not successful
Sidalcea pedata	1988	41. *Sidalcea pedata* Transplantation Project/K-Mart Corporation	Successful

Sources: Fiedler 1991; CDFG file information

Act. The vernal pools of the Santa Rosa Plain support populations of three state and federally endangered plants: *Blennosperma bakeri* (Sonoma sunshine), *Limnanthes vinculans* (Sebastopol meadowfoam) and *Lasthenia burkei* (Burke's goldfields) (CNDDB 1993b; Waaland et al. 1989).

The Santa Rosa Plain is one of many areas in California where rapid expansion of urban centers has brought development into direct conflict with endangered plant protection (Snyder 1993; Benfell 1993). Mitigation typically has consisted of attempts to "create" new vernal pool habitat and relocate endangered plant populations into created habitat (Patterson 1990a). As with other translocation projects, critical evaluation of these projects has been hin-

TABLE 13-4.   Santa Rosa Plain (Sonoma County) mitigation

Project name	Year initiated	Translocated species	Species at mitigation site before mitigation	Size of mitigation site (ha)
Airport Business Park[1]	1984	*Lasthenia burkei*	None	≈0.5
Sonoma County Airport[2]	1986	*Lasthenia burkei*	*Lasthenia burkei*	n/a
Sonoma County Wastewater[3]	1988	*Lasthenia burkei*	*Lasthenia burkei*	4.8
Montclair Park[4]	1989	*Blennosperma bakeri*	None	0.4
San Miguel Rancho/San Miguel Estates I (Alton Lane Mitigation Site)[5]	1989	*Lasthenia burkei* *Blennosperma bakeri*	*Lasthenia burkei* *Blennosperma bakeri*	11.6
Northpoint Village[6]	1991	*Limnanthes vinculans*	*Limnanthes vinculans*	3.1
San Miguel Estates II (Alton Lane Mitigation Site)[7]	1992	*Lasthenia burkei* *Blennosperma bakeri*	*Lasthenia burkei* *Blennosperma bakeri*	11.6
TMD-Brown (Alton Lane Mitigation Site)[8]	1992	*Lasthenia burkei* *Blennosperma bakeri*	*Blennosperma bakeri*	2.4
Graynor[9]	1992	*Lasthenia burkei*	*Lasthenia burkei*	0.4
Santa Rosa Air Center[10]	Proposed	*Limnanthes vinculans*	*Limnanthes vinculans*	20

Sources: 1. Patterson 1990a   2. Patterson 1990a   3. Western Ecological services 1993   4. Patterson 1990b
9. Patterson 1992c   10. Patterson 1992a

dered by inadequate documentation and lack of accepted success criteria. Even in cases where project-specific success criteria have been applied, short monitoring periods and qualitative monitoring techniques have made it difficult to evaluate the results of individual projects or compare one project to another. Table 13-4 summarizes key information from nine projects utilizing seven mitigation sites, and one proposed project. This table will be referred to throughout this section.

Vernal pool creation has been attempted in other regions of California where natural vernal pool habitats have different species composition, substrate, hardpan type, and surrounding community from those found within

projects involving endangered plant translocation and vernal pool creation.

Area of created pools (ha)	Monitoring period (years)	Success criteria	Comments
0.1	3	No. of plants	Regrading and supplemental seeding required.
0	3	None	Unquantified numbers of seeds were spread in existing swales and ditches.
1.8	5	Created wetlands meet corps criteria; no. of plants	Two of three created pools supported *Lasthenia* in 1992.
0.2	5	No. of plants; area of wetland	CDFG deemed failure in 1993; created pools failed to hold water after regrading; located in Sonoma Valley.
2.2	5	No. of plants; area of wetland	Wetland created by excavating basins and berming swales; natural wetlands present at mitigation site.
0.2	5	Water-ponding capacity; no. of plants; plant height; flowers/plant	"Test" pools created without agency authorization or approval of mitigation plan; reshaping, regrading, reseeding proposed as remediation.
21.0	5	No. of plants; area of wetland	Uplands disturbed when pools excavated after beginning of rainy season.
1.3	5	Water-ponding capacity; cover of wetlands plants; no. of endangered plants	Uplands disturbed when pools excaved after beginning of rainy season; endangered plant seeds *from mitigation site* used to inoculate created pools.
0.2	5	Water-ponding capacity; no. of plants; area of wetlands	Seeds from another site used to enhance initial seeding effort.
4.4	5	No. of plants; area of wetland	Proposed mitigation would result in 44 percent wetland, 56 percent upland.

5. Patterson 1990a; 1992b   6. Patterson 1990b   7. Patterson 1993   8. Monk and Associates 1993

the Santa Rosa Plain (Ferren and Pritchett 1988; Zedler et al. 1993). Some of these projects have included strong research components (Ferren and Pritchett 1988; Zedler et al. 1993), but even these have been carried out to mitigate the loss of natural vernal pool habitat.

The construction methods used to create vernal pools are generally similar from project to project (Ferren and Pritchett 1988; Patterson 1990a). As described by Patterson (1990a), earth-moving equipment is used to create shallow depressions in poorly drained soils or to berm swales to increase the size and depth of the area of inundation. Grading and scraping are carried out during the dry season in order to avoid destruction of the seedlings of nearby native upland plants by movements of heavy equipment. Seeds and other materials (called inoculum) are removed from the source site by raking or vacuuming and are distributed, usually by hand-sowing, in the created depressions. Sometimes the uppermost layer of soil is collected from the source site and respread in the created basins. The intention is that the created basins, like natural vernal pools, will fill with water during the rainy season and later will support populations of plants and invertebrates derived from transported soil and inoculum, and local amphibians and other organisms will colonize the new breeding site. Some practitioners (Stromberg 1994) have employed extensive soil testing and elevation surveys to achieve desired ponding depths and hydrologic flow patterns, but these methods have not been used on the Santa Rosa Plain.

ESTABLISHING STANDARDS FOR SUCCESS

In most cases, success criteria applied to Santa Rosa Plain vernal pool creation projects have been developed by those conducting the project, and they have focused on the "target" endangered species and the hectares of new wetland created (see Table 13-4). None of these projects has been evaluated for success on the basis of more complex measures associated with ecosystem functions, such as providing food-chain support, breeding sites, and habitat for nontarget vernal pool endemics, as has been suggested by Ferren and Gevirtz (1990). Nor have there been useful comparisons of created pools with control or "reference" vernal pools. Preoccupation with growing the same number of individuals of an endangered plant in the created habitat as had been counted (usually during one season) in the sacrificed natural habitat has led to the practice of reseeding mitigation sites repeatedly (Patterson 1990b, 1991a) in an attempt to grow more plants and reach "success" levels, without considering that the carrying capacity of a newly created vernal pool may be far lower (or far higher) than that of the natural site. Focus on numbers of plants and hectares of wetlands as success criteria may derive from the developers' and consultants' perceptions that the only requirement of mitiga-

tion is to replace what is specifically protected by law—state and federally listed plants and jurisdictional wetlands.

In California, no agency or other entity has yet proposed statewide standards for defining success either for vernal pool creation or for translocation of endangered plants. The difficulty of this task can't be overestimated. The functional requirements of complex wetlands that support endangered species aren't readily translatable into regulations. Yet a way must be found to evaluate mitigation success. As a first step, a task force of state and federal resource agency representatives is developing standards for vernal pool creation projects in Sacramento County that will be implemented by the Army Corps. Pavlik (Chapter Six) has identified the critical components of a biological definition of success for reintroductions of vascular plants of different life forms (such as annual herbs, perennial herbs, shrubs, and trees), but so far these ideas have not been translated into standards. Kentula et al. (1992) have outlined a strategy to improve decision making in wetland restoration and creation. This strategy, called the Wetland Research Program (WRP) Approach, could be used as a guide in developing performance, success, and monitoring standards for vernal pool creation projects. The WRP Approach emphasizes using natural sites as controls in evaluating created and restored sites.

Agency Attempts to "Regulate" Mitigation Projects

Attempts by the CDFG and other resource agencies to "regulate" the Santa Rosa projects illustrate the difficulties in keeping up with mitigation methods when there are no accepted standards of success or performance. Without accepted standards, each project requires the overseeing agency to defend anew its requests for monitoring, maintenance, and long-term protection—a process that cannot be completed adequately for all projects at current agency staffing levels. As a result, projects in Santa Rosa have proceeded without consistent requirements for implementation or results. In response to the lack of consistency in goals, methods, monitoring, long-term protection, and other aspects of mitigation-related translocation projects, in 1990 the CDFG adopted translocation guidelines (Howald and Wickenheiser 1990) that have served as a first step in regulating mitigation-related translocations of endangered plants. These guidelines call for

- A legally binding *mitigation agreement* that commits the project proponent to complete all aspects of the mitigation program

- A written *mitigation plan* that spells out in detail the technical components of the mitigation program

- Project-specific *performance criteria* that must be approved by the CDFG

- A *monitoring* period of at least five years

- *Performance secured* through a letter of credit or other negotiable security

- Long-term *habitat protection and management* that is funded through an endowment fund

Even though these guidelines lack the force of law, and their use is generally limited to projects under CEQA involving state or federally listed plant species, they could be reasonably effective if the staff needed to implement them were available. In California, the reality is that unmet agency staffing needs leave some mitigation projects with far less oversight than is needed to ensure that implementation and monitoring are proceeding according to plan.

THE NEED FOR BETTER DOCUMENTATION AND QUANTITATIVE MONITORING METHODS

Most of the mitigation plans and monitoring reports for the Santa Rosa Plain projects have lacked specificity in one or more key elements, such as project design, implementation schedule, monitoring methods, or long-term management. For example, most project designs fail to provide detailed specifications for the size and shape of the created vernal pools. Although some include sketches of the proposed location and size of the created basins, none of the reports includes either preproject topographic maps showing the "before construction" topography or "as-built" maps showing the actual results of construction. This level of detail, recommended for wetlands creation by Kentula et al. (1992), is needed to evaluate whether vernal pool creation projects replicate natural vernal pool habitat. Too often these projects create fewer, larger, and deeper pools than the natural pools that were destroyed. As an example, an insufficiently described phase of vernal pool creation at the Alton Lane mitigation site resulted in created basins that were significantly larger and deeper than any existing local vernal pools. The project was initiated after the rainy season had begun (although the dry season is the recommended period for construction) (Patterson 1992b) and without informing either of the agencies overseeing the project. By the time the CDFG and the FWS became aware that the project was being installed, it was too late to force meaningful changes in design.

Many plans also lack quantitative, repeatable monitoring methods, as described by Patterson (1991a):

> [Typically the full description consists of] multiple visits during the growing season to record visual observations of water depth, [and] water quality and to check the progress of the pool vegetation. Two

visits in spring (April to May) are used to record estimates of endangered plant numbers and vigor (height and flowers per plant), to document the vegetation character (floral composition and dominance, cover, richness, vigor) and overall pool ecology (use by birds, amphibians, insects, etc.). Colony numbers are estimated based on and extrapolated from small calibration plots (one square foot to several square meters) whereby a specific count of stems is made within the designated "quadrat" and the extent of plants at similar density is measured by pacing and visual gauging across homogeneous areas.

As this example illustrates, monitoring methods are not quantitative, and data analysis procedures are not provided. The monitoring results typically consist of a narrative description of pool vegetation and rough estimates of numbers of endangered plants present in each natural and created vernal pool. For mitigation sites that include pre-existing populations of endangered plants, the assumption is made that all plants growing in created habitat are a direct result of translocation. Also, in swales that contained endangered plants prior to mitigation-related enlargement, estimates of plant numbers attempt to separate plants resulting from mitigation from naturally occurring plants. However, the method for differentiating "mitigation plants" from "nonmitigation plants" is not described (Patterson 1991a). Estimates of cover, numbers of flowers per plant, plant height, and other measures, when given, are not supported by sampling data or statistical analysis (Patterson 1991a, 1992b). Estimates of the amount of wetland created lack verification that could be provided by conducting a wetland delineation according to the accepted federal method (Environmental Laboratory 1987).

The primary problem with the lack of rigor that characterizes most mitigation monitoring efforts on the Santa Rosa Plain is that it limits the potential for an impartial analysis by someone other than those who conducted the project. This has important implications when it comes to deciding whether the project proponent should be released from further responsibility for mitigation and whether there is a need for remediation. Also, comparisons of results of different projects cannot be made effectively without reliable data. Development of monitoring standards would improve the quality of monitoring and analysis of monitoring results. A more recent Santa Rosa project (Monk and Associates 1993) was required to provide cover estimates for "target" native plants in created vernal pools, but the monitoring of translocated endangered plants still does not go beyond estimates of population size and general observations of whether the plants are "fertile." (see Table 13-4.) In Chapter Ten, Sutter describes elements of a monitoring program that could be

adapted to these mitigation projects. As new standards are proposed, resource agencies have a responsibility to inform the consultant community prior to instituting them so that those responsible for designing and monitoring mitigation projects have the opportunity to become properly trained and experienced.

FINDING SUITABLE MITIGATION SITES

Within the Santa Rosa Plain, as elsewhere in California, it has been difficult to find acceptable mitigation sites—ones that are large enough to permit long-term retention of habitat values, are close to the impact site (to minimize gene pool manipulation), have environmental conditions likely to support the target species, and are protectable over the long-term through purchase by the project proponent. Private land is sought for mitigation purposes because in California most public land is not available for this use. Policies of some agencies that manage public lands (such as the California Department of Parks and Recreation) do not permit their use to mitigate the impacts of private developments.

The scarcity of good mitigation sites within the Santa Rosa Plain and the desire to protect some existing vernal pool habitat as a component of mitigation has led to the use of sites that already contain natural vernal pools, including some with existing endangered plant populations (see Table 10-4). After mitigation is implemented, these sites contain both created and natural pools, in close proximity. For example, natural pools at the Alton Lane mitigation site are literally a stone's throw from pools that were created in 1989 (Patterson 1990a) and 1992. This site also includes natural swales that were bermed to increase the area of ponded water, thus significantly modifying the hydrology of these features.

Academic and agency scientists have expressed concerns over mitigation projects that have modified existing natural wetlands and have added created pools to sites that already contain natural vernal pools. One result of this practice is the overconversion of upland to wetland. At the Alton Lane mitigation site, the mitigation plan called for 4.2 hectares of wetland to be created on an 11.6-hectare site that already contained at least 1.0 hectare of wetland, resulting in a postcreation ratio of at least 5.2 hectares (45 percent) of wetland to 6.4 hectares (55 percent) of upland (Patterson 1992b).

For a project at the Santa Rosa Air Center, proposed mitigation is to create 4.4 hectares of vernal pool wetlands on a 20-hectare site that already contains 4.4 hectares of wetland, for a final ratio of 8.8 hectares of wetland (44 percent) to 11.2 hectares of upland (56 percent) (Patterson 1992a). These ratios of wetland to upland far exceed the 20 to 25 percent wetland component found at typical natural vernal pool sites within the Santa Rosa Plain (Waaland and Dixon, n.d.). Healthy uplands are an important part of the vernal pool

ecosystem. They provide essential breeding sites for native bees that are "specialist" pollinators of vernal pool plants, and they provide habitat for other species with critical roles in vernal pool ecosystem function (Thorpe 1990; Fiedler and Laven, Chapter Seven).

Results of Unauthorized
Movements of Endangered Plants

Although it is the intent of the CDFG and the USFWS to minimize the distances that endangered plant populations are moved during translocations and to decrease the manipulation of gene pools that can result from population movements, these goals are not always achieved. In Sonoma County two cases have been documented of unauthorized movements of state and federally listed endangered plants outside of their recorded natural ranges as a result of mitigation-related translocations. In the Montclair Park mitigation project, an attempt to create vernal pools to provide replacement habitat for *Blennosperma bakeri* resulted in the inadvertent movement of a second endangered plant, *Lasthenia burkei*, to a watershed where the latter species was previously unknown (Patterson 1990b; CNDDB 1993b). The explanation was that *Lasthenia* seeds were probably transported to the new site on the shoes of the consulting biologist after he had collected seeds of that species elsewhere for another project (Patterson 1990b). In 1993 the CDFG determined that the Montclair Park mitigation project was a failure because the created pools failed to pond water and failed to support sustainable populations of *Blennosperma bakeri*. The City of Sonoma plans to convert the site to another use, leaving unresolved the question of what to do with the out-of-range *Lasthenia* plants, which have reappeared in low numbers each year since 1990.

Another endangered plant, *Limnanthes vinculans*, was the subject in 1987 and 1992 of unauthorized translocations that moved plants north of the species' recorded natural range (Patterson 1989, 1994). In the 1987 incident, seeds collected without authorization from the CDFG's Laguna de Santa Rosa Ecological Reserve were introduced to a receptor site near the Sonoma County Airport, approximately five miles north of the northern limit of its recorded range (Waaland et al. 1989). The purpose was to determine if the airport site could be used as a mitigation area for this species (Patterson 1989). In this case, monitoring in 1988 did not detect the presence of *Limnanthes vinculans* at the airport site, and the experiment was terminated. In 1992 *Limnanthes vinculans* seed collected from a site not affected by development was added to the inoculum for newly constructed pools at the Alton Lane mitigation site without informing regulatory agencies or modifying the original mitigation agreement (Patterson 1994). The intention was to increase the biological diversity of the mitigation site. The consultant was not aware that these activities required special authorization.

These examples highlight the need for standards that would identify a basis for determining the ecogeographic and genetic limits that should be placed on movements of endangered plants in the course of a translocation attempt. Also needed are guidelines for dealing with inadvertant movements out of recorded natural range. One obvious concern is the confusion about authenticity that may result if out-of-range movements are not thoroughly documented.

Even when populations are moved minimal distances, concerns remain over genetic mixing of previously disjunct vernal pool plant populations. Several translocation receptor sites within the Santa Rosa Plain also contain natural vernal pools with existing populations of the translocated species. At present, no studies have been completed investigating the effects of bringing into contact through translocation populations of vernal pool endemics that were previously disjunct. The amount of interpopulation genetic diversity present in these plants is unknown, although an ongoing study is using electrophoresis to investigate this in *Blennosperma bakeri*. Bauder (1993) presents evidence that differences in selection pressure may have fostered genetic differentiation within vernal pool species from Southern California, and that these differences are important to the species' longevity. A debated but unresolved question is whether the mixing of gene pools of endangered plant populations from different microsites creates a beneficial effect by increasing heterozygosity or results in reduced fitness. This is an issue of special concern for the California flora, with its many narrowly endemic species and genera that contain many similar species with incomplete reproductive isolation (Raven and Axelrod 1978; Guerrant, Chapter Eight).

Another genetic concern is raised by the techniques used to collect seeds from source sites. Typically, seeds are collected by raking or vacuuming the current year's seed crop from the pool surface. Most vernal pool plants are annuals and, as Vivrette (1993) has noted, gene pools of annual plants in California often possess a genetic reservoir, in the form of a soil seed bank. These seed banks contain gene combinations derived from many years of varying environmental conditions. Collecting one or even several years' aboveground seed crops captures only a fraction of the real genetic diversity of a population and therefore puts the "new" translocated population at a disadvantage in dealing with future environmental variation.

The Outcome of "No-Net-Loss" Policies
No net benefit to the endangered plants of the Santa Rosa Plain has been achieved through the application of "no-net-loss" wetland policies. CDFG and Army Corps mitigation policies sometimes have the effect of minimizing on-site wetland loss through avoidance, but this often results in "postage

stamp" preserves of less than half a hectare, surrounded by houses. There is virtually no chance that these tiny preserves will protect natural resources over the long term. Lacking buffers and without the ability to preserve natural hydrology or maintain biological relationships with former uplands, these sites become degraded rapidly from trash-dumping, vandalism, and invasion by exotics (Howald 1993). The "no-net-loss" requirement to create new wetlands if any wetlands are destroyed, no matter how small or degraded, means that there is no incentive to protect and leave as is large parcels with relatively undisturbed habitat for endangered plants. Any habitat that is acquired, even that which contains natural vernal pools, will be used to create additional pools. This means that all sites that are protected through mitigation are changed in the process; none remains untouched. Many of these mitigation sites end up compromised or degraded, having lost natural functions and values.

The conservation community has been reluctant to accept an alternative approach to "no-net-loss"—for example, one that is willing to trade the loss of some smaller, less protectable sites for the acquisition and permanent protection of other large, high-quality natural sites. This reluctance may derive from an understandable unwillingness to accept the unmitigated loss of a population and its habitat. Unfortunately, when translocation efforts end in failure, the result is not only the complete loss of endangered plant populations but the continued lack of protection for the very few natural populations that remain (see Bean, Chapter Sixteen). A recent review of CEQA's effectiveness in protecting endangered plants notes that agencies "are attempting to move in the direction of obtaining off-site intact occupied habitat" for endangered plants, "acknowledging that some loss may be offset by the reduced fragmentation and isolation of populations, and the opportunity to protect communities or ecosystems rather than species" (Dennis 1994, pp. 10–11).

LONG-TERM MANAGEMENT CONCERNS

Within the Santa Rosa Plain and the nearby Sonoma Valley there are at least seven existing mitigation sites for vernal pools and endangered plants; preliminary plans have been developed for an additional site (see Table 13-4). Several others are planned. The long-term management requirements for most of these sites have not been spelled out, nor is there a comprehensive management plan for the group, although it is likely that even the most successfully implemented projects will require a consistent level of long-term oversight and maintenance. The proliferation of mitigation sites without long-term protection was one factor that led to the development of CDFG's translocation guidelines (Howald and Wickenheiser 1990). CDFG records show that, since 1980, at least fifty-five endangered plant mitigation sites have

been established. Since the CDFG guidelines took effect, translocation projects requiring CDFG approval have included provisions for long-term site protection and maintenance.

According to the CDFG guidelines, long-term site protection consists of transferring the property in fee title or placing a conservation easement on the property, with the easement held (or the property owned) by an appropriate agency or conservation group. The Legal Advisor's Office of the CDFG has crafted language for a conservation easement specifically for this purpose. Long-term maintenance is ensured through the transfer of an endowment fund to the agency or conservation group that assumes long-term maintenance responsibilities. The amount of the endowment fund is calculated by estimating the average annual cost of maintaining the site, including expenses for fencing, exotics control, trash removal, and biological monitoring. The endowment fund does not cover expenses associated with education or public use. At CDFG, endowment fund monies are pooled and invested in secure interest-producing instruments such as certificates of deposit. Accumulated interest can be used to fund approved maintenance and protection activities. To date, about 4.4 million dollars have been deposited in this management fund from all mitigation projects, including those for plants and wildlife. In 1993 this fund generated approximately $200,000 for site monitoring and maintenance.

One option for the Santa Rosa Plain sites is management by the Center for Natural Lands Management, a private nonprofit corporation that provides long-term management for mitigation and compensation lands throughout the state. Recognizing California's growing need for management services for an expanding network of mitigation sites, the center's objectives are "to manage natural resource conservation lands in perpetuity . . . to protect and manage for biological diversity, to ensure intelligent and biologically sound mitigation projects . . . to promote through public education and awareness the values of resource conservation and maintaining diversity, and to encourage public involvement and volunteerism in resource conservation" (CNLM, 1993).

The center currently manages sixteen mitigation sites with endangered plants, including two preserves in the southern San Joaquin Valley that comprise thirty-one habitat management units supporting seven species of listed plants.

Santa Cruz Tarplant

Santa Cruz tarplant (*Holocarpha macradenia*) is a summer-blooming annual of coastal grasslands, with a historic distribution that included a northern component in Alameda and Contra Costa counties (part of the "East Bay" region of the San Francisco Bay area) and a southern component in Santa Cruz

and Monterey counties. By the early 1980s urban and industrial development in the East Bay had resulted in the loss of all known populations of the tarplant in the northern part of the species' range. A population in the City of Pinole, thought at the time to be the largest in existence, was lost when a shopping center was built on the site (Rae 1981).

Havlik (1989) led a last-ditch effort to preserve some representation of Santa Cruz tarplant in the northern part of its range. He salvaged seeds and plants from three populations slated for destruction. The salvaged plants died, but the seeds were used in a series of twenty-two translocations on nearby public land, in habitat that appeared similar in soil type and plant composition to natural tarplant sites (Havlik 1987). Table 13-5 shows the results of annual estimates of population size conducted since 1982 (except 1989). Although these data must be interpreted loosely due to varying census methods and other factors, they indicate that through 1988 most translocation sites supported sizeable populations of tarplant. In 1990, with California well into a seven-year drought, the numbers plummeted (CNDDB 1990, 1991) and remained low until 1993, when some sites showed modest increases (CNDDB 1993c), possibly accounted for by the presence of soil seed banks. These results illustrate that the final outcome of a translocation may not be known for many years. Long-term monitoring is required if we are to find out how these introduced populations fare over time as they are exposed to the full range of California's local climatic variation.

In 1988 a remnant population of Santa Cruz tarplant was found on open land adjacent to the existing Pinole shopping center—land proposed at that time for a shopping center addition. CEQA review of the proposed project resulted in a Negative Declaration, an environmental document prepared when a project is found to result in no significant environmental impacts (or, as in this case, a project whose significant environmental impacts the developer agrees to mitigate "up front" as a part of the project). CDFG biologists sought avoidance of the population through an on-site preserve, but the developer's position that establishing a preserve would make the project financially infeasible was accepted by the CDFG's director.

In November of 1988 the developer signed an agreement committing to mitigate impacts to Santa Cruz tarplant "to the satisfaction of the California Department of Fish and Game" (CDFG 1988), although no specific mitigation plan had been submitted. Over the next four and a half years the CDFG attempted to secure the implementation of an acceptable translocation effort conducted according to CDFG guidelines. This attempt ended in June of 1993 when, after the original developer sold the project, the CDFG's director signed an agreement releasing the new owner from the responsibility of conducting the translocation and committing the CDFG to carry out the mitigation using funds provided by the original developer (CDFG 1993). The

TABLE 13-5.   Census data for Santa Cruz tarplant (Holocarpha macradenia) translocations in Wildcat Canyon, Contra Costa County, California.

Pop no.	1982	1983	1984	1985	1986	1987	1988	1990	1991	1992	1993
1	100	175	1	47	30	250	130	nd	nd	0	0
2	600	1,200	900	2,600	3,000	1,750	5,100	1	1	5	463
3	100	130	25	125	50	300	450	112	15	2	35
4	300	400	400	550	250	100	375	1	nd	1	4
5	60	75	0	50	15	0	0	nd	nd	nd	0
6		300	125	700	700	500	800	nd	nd	nd	4,051
7		130	125	175	100	70	125	0	nd	0	0
8		115	85	34	15	0	15	nd	nd	nd	0
9			250	15	1,200d	200	600	nd	nd	nd	nd
10			0	100	100	100	60	nd	nd	0	16
11			400	1,600	275	1,600	4,000	3	3	18	133
12			750	600	1,050	1,200	800	8	nd		047
13			30	750	600	400	700	nd	nd	nd	11
14			2	7	10	nd	nd	nd	nd	nd	nd
15					750	250	300	1	nd	0	5
16					750	700	1,500	10	nd	0	160
17					30	325	500	12	nd	1	0
18					250	400	1,000	12	12	0	20
19					0	0	0	0	nd	0	0
20*					400	400	600	nd	nd	1,500	nd
21†					20	1	nd	nd	nd	nd	nd
22†						850	3,000	nd	nd	nd	nd

nd = no data   *Located on Sobrante Ridge
† Located in the San Pablo Reservoir watershed
Sources: Havlik 1989; CNDDB 1990, 1991, 1992, 1993.

Pinole population of tarplant was destroyed in July of 1993 during construction of a shopping center addition.

This example teaches many lessons. Some deal with the straightforward problems that can arise in any well-intentioned mitigation effort. More important are the seemingly intractable conflicts that arise when a resource agency tries to oversee the solution of a complicated biological problem—one whose satisfactory resolution depends upon the cooperation of a second party that acts in an increasingly recalcitrant manner. Both types are discussed in the sections that follow. They show that the search for more effective means of resolving such situations must lead to recognition of the need for stronger legal protection for plants, alternative acceptable strategies to translocation, and national, agency-sanctioned guidelines for carrying out mitigation-related translocations.

Finding a Suitable Mitigation Site

The Pinole mitigation project provides an all-too-common example of the practical difficulties involved in finding a suitable mitigation site in an essentially urban landscape where property values are high and willing sellers are few. Although the CDFG provided the developer with biological selection criteria and an example of how these could be used (Pavlik and Heisler 1988), a systematic search for a suitable mitigation site was never conducted. One privately owned candidate site was located, but it was owned by another development company unwilling to sell it. As an alternative, the CDFG and the developer engaged in a lengthy negotiation with a local park district to secure permission to use some of their land, which already contained populations of Santa Cruz tarplant introduced by Havlik (1989). In the end this effort was unsuccessful, primarily due to opposition from conservation groups favoring avoidance, who opposed using public land to mitigate the impacts of private development.

Storage and Treatment of Seeds

While the search for a mitigation site was going on, and the details of the mitigation plan were being negotiated, seeds were collected each year from the remnant Pinole population (discovered in 1988). Seeds were collected by hand-picking individual seed heads as they matured. Harvested seed was stored in a trash can in the basement of the environmental consulting company hired to plan the mitigation program until alternate arrangements were requested by CDFG. Any loss of seed viability that may have resulted from these storage conditions was undocumented. Even though essentially all the seed produced by the remnant population was collected each year, new seedlings of this annual plant appeared each spring; in 1993 the population was estimated to be larger (more than five thousand individuals) than at any time since it was discovered in 1988, providing evidence of an extensive seed bank composed of seeds that break dormancy under a variety of conditions. The presence of a soil seed bank may explain some of the observed fluctuations in population size recorded in Table 13-5.

An early draft mitigation plan (LSA Associates 1990) proposed to increase or "bulk" the number of seeds available for the translocation by growing the plants in cultivation, although the seed thus produced would not have had the benefit of selection as it occurs under natural conditions.

If we assume that the success of translocation is largely dependent upon the viability of the propagules used to establish the new population, then proper treatment and storage of an irreplaceable seed source is of fundamental importance. Although the CDFG guidelines call for collection, storage, and treatment procedures that protect seed viability, suitable storage facilities are sometimes not readily available, and resource agency understaffing makes it

often impossible to provide the level of oversight needed to assure that proper procedures are used.

BALANCING DEVELOPER AND AGENCY GOALS

Mitigation efforts proceed most smoothly when CEQA lead agencies, developers, consultants, and regulatory agencies understand each other's goals and limitations and work cooperatively to reach an endpoint that satisfies the needs of all parties. In California, CEQA lead agencies are responsible for keeping the CEQA environmental review process on track and for providing adequate opportunity for public contribution to the review process. Developers must concentrate on acquiring the permits required for their project, proceeding as quickly as possible and keeping an eye on the bottom line. Consultants must provide the biological and technical expertise to design and implement mitigation programs. Agencies that manage natural resources must focus on long-term success in maintaining the health of the environment. An attitude of cooperation and respect for each other's divergent goals is essential if effective mitigation is to be achieved.

The saga of the Pinole mitigation project illustrates what happens when cooperation is lacking. In this case, attempts to reach a solution were on several occasions scuttled when the developer, who viewed CDFG's requests for information or additional mitigation plan specificity as unreasonable, used political pressure to avoid taking the steps necessary to develop a satisfactory mitigation plan. No one wins in such situations. A great deal of time and money is wasted, and, in the end, the environment suffers. The primary lesson is that we need stronger legislation to protect endangered plants and their habitats and more options for mitigating unavoidable impacts.

Conclusions and Recommendations

The California experience with using translocation to mitigate impacts to endangered plants shows that, although there are potential benefits, the current process has a number of problems with no quick or easy solutions. The "science" of translocation has its problems: lack of basic information about endangered plants, lack of accepted methods, difficulties in finding and obtaining a mitigation site, and the inherent risks of biological experiments. There are also problems with the regulation of translocation projects, such as lack of established standards for performance and success. There are problems with the framework imposed by the requirements of development projects and problems with the long-term management of mitigation sites. One

response is to throw out the process, but before we advocate that approach, we need to consider the alternatives.

At present, protection for endangered plants in California derives from a patchwork of overlapping and sometimes contradictory laws, policies, and guidelines of local, state, and federal agencies. Only under rare circumstances do endangered plants enjoy the level of legal protection that is required to render them truly secure (Bean, Chapter Sixteen; Berg, Chapter Twelve). Until that situation changes, translocation will remain one of a limited number of mitigation options. Given this scenario, for now it is in the best interests of endangered plant conservation to ensure that translocations, when performed, are carried out in a way that gives endangered plants the greatest possibility for long-term persistence and resilience and provides data to improve future decision making.

More Mitigation Choices Are Needed

At the same time, we must increase the choices for mitigating endangered plant impacts. Mitigation through translocation is too experimental and risky to stand alone. In cases where avoidance is not an option, we need to consider alternatives to the "no-net-loss" approach. In the long run, we may protect an endangered plant more effectively if we allow the loss of some small, degraded populations with little potential for long-term survival in return for permanent protection of larger, more pristine sites. Currently, we are losing some of the best sites through failed translocations without gaining any additional protection for the few good sites that remain.

One way to gain permanent protection for large important sites is to use mitigation banking to consolidate smaller mitigation "debts" toward the purchase and protection of existing high-quality sites. So far, mitigation banking in California has been used mainly in conjunction with wetland creation schemes. One project that seeks to go beyond this, protecting existing alkali wetlands and endangered species, has been delayed for several years, while agency reviewers grapple with the fact that current mitigation banking guidelines do not encompass projects of this type.

Selecting which sites to protect through mitigation banking or other mechanisms could be accomplished through regional habitat conservation planning. A Southern California pilot project of the CDFG's Natural Communities Conservation Program is using a regional evaluation process to prevent the listing of the California gnatcatcher and to protect coastal sage scrub sites and their endangered species (CDFG 1992b). In 1993 federal funding was obtained to prepare a regional conservation plan for the Santa Rosa Plain's vernal pool ecosystem. The plan will identify areas that must be

protected and those that can be developed and will streamline the permit process for developers who comply with the goals of the plan.

Stronger Legal Protection Needed for Endangered Plants

Although current challenges to the Federal Endangered Species Act threaten to be the strongest put forth so far, we need to move in the direction of greater protection for endangered plants. Plants have had the reputation of being "second-class citizens" (Berg 1993) in the world of endangered species conservation since the state and federal ESAs were enacted. In California there is a particular need to eliminate exemptions for agriculture (including grazing) and to strengthen the protection plants receive under CEQA and CESA. (see Bean, Chapter Sixteen).

Effective State and Federal Guidelines Needed for Translocations

Improving the quality of mitigation-related translocations requires the implementation of effective state and federal guidelines that cover all aspects of the process. Established guidelines would put regulatory agencies in a stronger position, would reduce the ad hoc nature of current regulation, and would offer developers the security of knowing exactly what is expected of them. The CDFG's guidelines are a start, but to be genuinely effective, comprehensive guidelines should be adopted by all agencies that are involved with mitigation of endangered plants and should address broader questions (for example, as discussed by Falk and Olwell [1992] the need to ensure that mitigation efforts do not contribute to the degradation of valuable natural habitat). In addition, agency staffing levels must be increased if new guidelines are to have a chance of being implemented successfully.

Groups such as the Center for Plant Conservation, native plant societies, botanical gardens, research institutions and conservation groups (IUCN 1992) all have valuable expertise to contribute to the process of developing recognized standards. Part Five presents the guidelines that arose from workshops sponsored by the Center for Plant Conservation during the 1993 *Restoring Diversity* conference.

Recovery Plans Should Guide Process

In the final analysis, the role of mitigation-related translocations in the conservation of an endangered plant should be dictated by the goals of an agency-approved comprehensive recovery program for the species. However, in California only about twenty plant species have approved recovery plans, all of them prepared for federally listed plants. The CDFG is trying to address this lack of recovery planning with a new tool, the Species Management Data Base, a species-specific computerized database that will summarize the infor-

mation base for the species, list both ongoing and recommended actions, and identify funding sources and experts who can help guide the recovery program. Intensive workshops are being used as a fast-track method to develop priorities and ideas for accomplishing recovery goals.

Recovery planning must develop realistic methods to reduce the level of threat to endangered species and their habitats. For some species, translocation may be the only feasible method for dealing with certain categories of threat and for recovery of species that have diminished appreciably in abundance and distribution within their natural recorded ranges. But the appropriateness of translocation should be determined as a part of an overall program of conservation developed through regional land use planning, not as a last-resort response to imminent destruction.

ACKNOWLEDGMENTS
I wish to thank the following people for reviewing the manuscript and for many useful discussions: Caitlin Bean, Ken Berg, Barbara Dean, Don Falk, Jan Knight, David Leland, Constance Millar, Sandra Morey, Peggy Olwell, Charles Patterson, Ruth Pratt, Barbara Youngblood, and Carl Wilcox. For helping with special information requests, my thanks to Scott Flint, Dawn LaBarbera, Tom Lupo, Sherry Teresa, and Nanci Williams.

REFERENCES
Bauder, E. T. 1986. *San Diego Vernal Pools: Recent and Projected Losses, Their Condition, and Threats to Their Existence, 1970–1990*. Final report to California Department of Fish and Game, Sacramento.

———. 1993. "Genetic Diversity: Esoteric or Essential?" In Proceedings of the Symposium *Interface Between Ecology and Land Development in California*. Edited by J. E. Keeley. Los Angeles: Southern California Academy of Sciences.

Benfell, C. "Wetlands Rule May Hike Cost of Housing." *Santa Rosa Press Democrat*, 19 March, 1993.

Berg, K. S. 1993. "Plants As Second-Class Citizens." In Proceedings of the Symposium *Interface Between Ecology and Land Development in California*. Edited by J. E. Keeley. Los Angeles: Southern California Academy of Sciences.

CDFG (California Department of Fish and Game). 1988. Agreement Between Gemtel and the Department of Fish and Game to Mitigate Impacts to *Holocarpha macradenia*. Public document.

———. 1992a. *1991 Annual Report on the Status of California State-Listed Threatened and Endangered Animals and Plants*. Sacramento: State of California, Resources Agency.

———. 1992b. *Southern California Coastal Sage Scrub Natural Community Conservation Planning Process Guidelines*. Unpublished report. Sacramento: State of California, Resources Agency.

———. 1993. California Endangered Species Act Memorandum of Understanding by and Between Pinole Redevelopment Agency and California Department of Fish and Game Regarding Pinole Vista IV Shopping Center Development. Unpublished Memorandum, reference No. 9310.

CNDDB (California Natural Diversity Data Base). 1990. Field survey forms for 1990 surveys of *Holocarpha macradenia* at Wildcat Canyon Regional Park, Contra Costa County, California Department of Fish and Game, Natural Heritage Division, Sacramento.

———. 1991. Field survey forms for 1991 surveys of *Holocarpha macradenia* at Wildcat Canyon Regional Park, Contra Costa County, California Department of Fish and Game, Natural Heritage Division, Sacramento.

———. 1992. Field survey forms for 1992 surveys of *Holocarpha macradenia* at Wildcat Canyon Regional Park, Contra Costa County, California Department of Fish and Game, Natural Heritage Division, Sacramento.

———. 1993a. Analysis of RareFind data base. Unpublished data. California Department of Fish and Game, Natural Heritage Division, Sacramento.

———. 1993b. RareFind records for *Blennosperma bakeri, Lasthenia burkei,* and *Limnanthes vinculans.* Unpublished records. California Department of Fish and Game, Natural Heritage Division, Sacramento.

———. 1993c. Field survey forms for 1993 surveys of *Holocarpha macradenia* at Wildcat Canyon Regional Park, Contra Costa County, California Department of Fish and Game, Natural Heritage Division, Sacramento.

Center for Natural Lands Management. 1993. "Statement of Qualifications." Unpublished document.

———. 1994. "Habitat Management Cost Analysis." Unpublished document. Prepared for Environmental Protection Agency, Sacramento.

CPC (Center for Plant Conservation). 1993. List of California Plants in Danger of Extinction Within the Next Ten Years. Unpublished document. Prepared by the California Task Force.

Dennis, N. B. 1994. "Does CEQA Protect Rare Plants?" *Fremontia* 22 (1): 3–13.

Dunn, J. 1993. "Endangered Builders." *Sonoma Business* 18 (6): 41–44.

Environmental Laboratory. 1987. "Corps of Engineers Wetlands Delineation Manual." Technical Report Y-87-1. Vicksburg, Miss.: U.S. Army Engineer Waterways Experiment Station.

Fahselt, D. 1988. "The Dangers of Transplantation As a Conservation Technique." *Natural Areas Journal* 8 (4): 238–244.

Falk, D. A., and P. Olwell. 1992. "Scientific and Policy Considerations in Restoration and Reintroduction of Endangered Species." *Rhodora* 94: 287–315.

Ferren, W. R., Jr., and D. A. Pritchett. 1988. "Enhancement, Restoration, and Creation of Vernal Pools at Del Sol Open Space and Pool Reserve, Santa Barbara County, California." The Herbarium. Department of Biological Sciences, University of California, Santa Barbara. Environmental Report No. 13.

———. and E. M. Gevirtz. 1990. "Restoration and Creation of Vernal Pools: Cookbook Recipes or Complex Science?" In *Vernal Pool Plants: Their Habitat and Biology.* Edited by D. H. Ikeda and R. A. Schlising. Studies from the Herbarium, California State University, Chico. No. 8.

Fiedler, P. L. 1991. "Mitigation-Related Transplantation, Relocation and Reintroduction Projects Involving Endangered, Threatened and Rare Plant Species in California." Unpublished report. Sacramento: California Department of Fish and Game, Endangered Plant Program.

Hall, L. A. 1987. "Transplantation of Sensitive Plants As Mitigation for Environmental Impacts." In *Conservation and Management of Rare and Endangered Plants*. Edited by T. S. Elias. Sacramento: California Native Plant Society.

Havlik, N. A. 1987. "The 1986 Santa Cruz Tarweed Relocation Project." In *Conservation and Management of Rare and Endangered Plants*. Edited by T. S. Elias. Sacramento: California Native Plant Society.

———. 1989. "Final Report of the Santa Cruz Tarweed Local Preservation Project, Contra Costa and Alameda Counties, California." Unpublished report. Sacramento: California Department of Fish and Game, Endangered Plant Program.

Hickman, J. C., ed. 1993. *The Jepson Manual*. Berkeley: University of California Press.

Hoffman, M. S., ed. 1993. *The World Almanac and Book of Facts, 1993*. New York: Pharos Books.

Howald, A. M. 1993. "Finding Effective Approaches to Endangered Plant Mitigation." In Proceedings of the Symposium *Interface Between Ecology and Land Development in California*. Edited by J. E. Keeley. Los Angeles: Southern California Academy of Sciences.

Howald, A. M., and L. P. Wickenheiser. 1990. "Mitigation Plan Annotated Outline for Endangered Plants of California." Unpublished report. Sacramento: California Department of Fish and Game.

International Union for the Conservation of Nature and Natural Resources. 1992. "Draft Guidelines for Reintroductions." Species Survival Commission, Reintroduction Specialist Group. Royal Botanic Gardens, Kew, U.K.

Jones and Stokes Associates. 1987. *Sliding Towards Extinction: The State of California's Natural Heritage, 1987*. Unpublished report. San Francisco: The California Nature Conservancy.

Kentula, M. E., R. P. Brooks, S. E. Gwin, C. C. Holland, A. D. Sherman, and J. C. Sifneos. 1992. *An Approach to Improving Decision Making in Wetland Restoration and Creation*. Edited by A. J. Hairston. Corvallis, Ore.: U.S. Environmental Protection Agency, Environmental Research Laboratory.

LSA Associates. 1990. *Draft Mitigation Plan, Santa Cruz Tarplant Relocation, Pinole Vista IV*. Unpublished report.

Matthews, J. R. ed. 1990. *Official World Wildlife Fund Guide to Endangered Species of North America*. Vol. I. Washington, D.C.: Beacham.

Monk and Associates. 1993. "TMD-Brown Wetlands Monitoring Report, Waltzer Meadows Subdivision, Santa Rosa, CA." Unpublished report prepared for TMD-Brown Builders, Santa Rosa, Calif.

Patterson, C. A. 1989. Letter to Mr. Robert Becker, Sonoma County Airport, Calif., 25 February.

———. 1990a. "Vernal Pool Creation and Rare Plant Translocation, Sonoma County, California." Unpublished report. .

———. 1990b. "Vernal Pool Creation and Sensitive Plant Translocation at the Montclair Park Development Site, Annual Monitoring Report, Year 1, Sonoma County,

California." Unpublished report prepared for Christopherson Homes, Santa Rosa, California

———. 1991a. "Vernal Wetland and Sensitive Plant Monitoring Report, Year 2, Alton Road Mitigation Site, Sonoma County, California." Unpublished report prepared for Cobblestone Development, Santa Rosa, California.

———. 1991b. "Mitigation Plan for Sebastopol Meadowfoam (*Limnanthes vinculans*), Northpoint Village, Sonoma County, California." Unpublished report prepared for Pine Creek Properties, Santa Rosa, California.

———. 1992a. "Mitigation Plan for Sebastopol Meadowfoam (*Limnanthes vinculans*) at the Santa Rosa Air Center and Madera Parcels, Sonoma County, California." Unpublished report prepared for Madera Avenue Land Associates and Santa Rosa Associates II, San Carlos, California.

———. 1992b. "Vernal Wetland and Sensitive Plant Monitoring Report, Year 3 - 1992, Alton Lane Mitigation Site, Sonoma County, California." Unpublished report prepared for Cobblestone Development, Santa Rosa, California.

———. 1992c. "Mitigation Plan for Burke's Goldfields (*Lasthenia burkei* at the Skylane Business Park (Graynor Property)." Unpublished report prepared for Mr. Rand Graynor, Petaluma, California.

———. 1993. Unpublished maps and tables documenting 1993 census of Alton Lane mitigation site. Prepared for California Department of Fish and Game, Yountville, California.

———. 1994. "Vernal Wetland and Sensitive Plant Monitoring Report, Year 4–1993, Alton Lane Mitigation Site, Sonoma County, California." Unpublished report prepared for Cobblestone Development, Santa Rosa, California.

Pavlik, B. M. 1990. "Reintroduction of *Amsinckia grandiflora* to Stewartville." Unpublished report. Sacramento: California Department of Fish and Game, Endangered Plant Program.

———. 1991a. "Management of Reintroduced and Natural Populations of *Amsinckia grandiflora*." Unpublished report. Sacramento: California Department of Fish and Game, Endangered Plant Program.

———. 1991b. "Reintroduction of *Amsinckia grandiflora* to Three Sites Across Its Historic Range." Unpublished report. Sacramento: California Department of Fish and Game, Endangered Plant Program.

———. 1992. "Inching Towards Recovery: Evaluating the Performance of *Amsinckia grandiflora* Populations Under Different Management Regimes." Unpublished report. Sacramento: California Department of Fish and Game, Endangered Plant Program.

Pavlik, B. M., and K. Heisler. 1988. "Habitat Characterization and Selection of Potential Sites for Establishment of New Populations of *Amsinckia grandiflora*." Unpublished report. Sacramento: California Department of Fish and Game, Endangered Plant Program.

Rael, S. P. 1981. "Efforts Made to Save Endangered Tar Plant." *Outdoor California* 42: 16.

Raven, P. H. 1977. "The California Flora." In *Terrestrial Vegetation of California*. Edited by M. G. Barbour and J. Major. New York: John Wiley and Sons.

Raven, P. H., and D. I. Axelrod. 1978. *Origin and Relationships of the California Flora*. University of California Publications in Botany, Vol. 72. Berkeley: University of California Press.

Shevock, J. R., B. Ertter, and D. W. Taylor. 1993. "*Neviusia cliftonii* (Rosaceae: Kerrieae), an Intriguing New Relict Species from California." *Novon* 2: 285–289.

Shevock, J. R., and D. W. Taylor. 1987. "Plant Exploration in California, the Frontier is Still Here." In *Conservation and Management of Rare and Endangered Plants.* Edited by T. S. Elias. Sacramento: California Native Plant Society.

Skinner, M. W., and B. Pavlik. 1993. *California Native Plant Society Inventory of Rare and Endangered Vascular Plants of California.* Fifth Edition. Sacramento: California Native Plant Society.

Snyder, G. "Hope for Survival of Rare Vernal Pools." *San Francisco Chronicle*, 17 July, 1993.

Stromberg, L. P. 1994. "Physical Site Conditions and Vernal Pool Compensation, Sun City, Roseville, Agency Field Workshop." Unpublished report.

Thorpe, R. W. 1990. "Vernal Pool Flowers and Host-Specific Bees." In *Vernal Pool Plants, Their Habitat and Biology.* Edited by D. H. Ikeda and R. A. Schlising. Studies from the Herbarium, No. 8. Chico: California State University.

FWS (U.S. Fish and Wildlife Service). 1993. Listing packages in progress, plant species. Unpublished list. Sacramento: Endangered Species Office.

Vivrette, N. 1993. "Low Seed Viability in Rare and Endangered Species: A Warning to Vegetation Managers." Paper presented at the Southern California Academy of Sciences symposium Interface Between Ecology and Land Development in California. Abstract only. Los Angeles, California. May 1–2 1992.

Waaland, M., J. Vilms, and R. Thompson. 1989. "Santa Rosa Plains Endangered Plant Protection Program Report." Unpublished report. Sonoma County Planning Department and California Department of Fish and Game.

Waaland, M., and C. Dixon. n.d. "Characteristics of Vernal Pools at Three Sites in the Santa Rosa Plains." Unpublished data.

Western Ecological Services. 1993. "Fourth Year Monitoring Report, County Service Area 31, Airport Wastewater Storage Pond, Wetland Mitigation Project." Unpublished report prepared for County of Sonoma, Santa Rosa, California.

Zedler, J. B. 1991. "The Challenge of Protecting Endangered Species Habitat Along the Southern California Coast." *Coastal Management* 19: 35–53.

Zedler, P. H. 1987. *The Ecology of Southern California Vernal Pools: A Community Profile.* U.S. Fish and Wildlife Service Biological Report 85(7.11).

Zedler, P. H., C. K. Frazier and C. Black. 1993. "Habitat Creation as a Strategy in Ecosystem Preservation: An Example from Vernal Pools in San Diego County." Proceedings of the symposium *Interface Between Ecology and Land Development in California.* Edited by J. E. Keeley. Los Angeles: Southern California Academy of Sciences.

Ecological Function and Sustainability in Created Wetlands

Joy B. Zedler

───────◆━◉━◆───────

Major questions have arisen over the quality of habitats that have been constructed in mitigation projects, e.g., are they more like gardens than self-sustaining ecosystems? This chapter summarizes research that evaluates the functional equivalency of an excavated salt marsh (nine years old in 1993) with an adjacent reference wetland, within the perspective of our broader understanding of wetland functioning in Southern California.

Before considering whether the outcome of mitigation is a garden or a self-sustaining ecosystem, the terminology should be clarified. *Mitigation* is the lessening of environmental impacts through a variety of techniques such as avoidance, minimizing damages, and compensating for negative impacts through restoration, enhancement, or replacement measures. In wetlands and other habitats used by endangered species, restoration of degraded wetlands is widely viewed as the preferred form of compensatory action. *Gardens* are manipulated systems. The site is prepared and requires continual maintenance (that is, the soils are amended, the desired species are planted, and so on). In wetland gardens, tidal flows may be under control of tide gates; erosion and sedimentation are likely problems; fertilizers may be needed; predators and herbivores require continual control. *Functionally equivalent habitats* are often mentioned as desirable but not defined legally. In this chapter, functional equivalence is viewed in relation to reference wetlands. When a constructed wetland can on longer be distinguished from natural wetlands in the same region and with the same physiography and hydrology, either through examination of structural or functional attributes, then they are considered functionally equivalent. *Self-sustaining ecosystems* do not require maintenance and are able to support populations of organisms and functional attributes without human intervention. This does not mean they are stable or at equilibrium. More likely, they are dynamic systems with seasonal and

interannual variability. Self-sustaining ecosystems are resilient and have mechanisms that allow recovery from disturbances.

I am skeptical about the ability of mitigation actions to provide self-sustaining ecosystems—I have developed this view from working in the Southern California coastal zone. Statewide, California has lost 91 percent of its wetland area in the past two centuries (Dahl 1990). This rate of wetland loss is the highest of the United States and greatly exceeds the average for the conterminous states (53 percent) (Dahl 1990). California has thirty-two species of wetland-dependent plants and animals that are endangered with extinction; this number also tops the nation's list, in a tie with Tennessee (Feierabend 1992). A link between habitat loss and potential extinction is likely.

While drainage of Central Valley wetlands for agriculture was responsible for most of California's wetland loss, the Southern California coast has also experienced a substantial decrease in the area of tidal salt marshes. Several lagoons that were once open to tidal flow now close more frequently due to filling for roadways, which constrict tidal flows and reduce *tidal prisms*—the volume of water that can flow in and out with the tide. These irregularly tidal and nontidal systems support fewer plant and animal species than fully tidal wetlands (PERL 1990). Three tidal systems in the San Diego area remain fully tidal, but filling and dredging have eliminated 85 percent of their salt marshes and intertidal flats (Table 14-1). Associated, shallow subtidal habitats for eelgrass have also been depleted. The wetland areas that remain are further degraded by urban development and altered water quality and quantity.

Several attempts to restore these habitats have been undertaken in Southern California, but only one project (in San Diego Bay) has been studied in any detail. This case study has been widely cited as an example of how mitigation sites fail to achieve functional equivalency with natural wetlands and fail to support endangered species.

Lessons from Southern California Tidal Marsh Construction Efforts

In San Diego Bay, there are three constructed cordgrass (*Spartina foliosa*) marshes. One is on a dredge spoil island, and two are on filled wetlands that have been excavated, although not to their historical contours. The island was planted with cordgrass in 1984; the older of the two excavated sites received transplants in 1985, and the newer one in 1991. At each site, habitat was designed to attract the light-footed clapper rail (*Rallus longirostris levipes*) and to support nesting of this endangered species. None of the sites has met this ex-

TABLE 14-1. Historical habitat changes at Tijuana Estuary, San Diego Bay, and Mission Bay (Macdonald 1990). Shallow habitats (including the shallow subtidal area) have all declined. At the same time, there has been an increase in deepwater habitats (over 6-foot mean lower low water [MLLW]) due to dredging of channels for ships (San Diego Bay) and recreational boats (Mission Bay).

Dates compared:	Historical:	Recent:	Area	
Date of maps, photos:	1856, 1902 (acres)	1984–1987 (acres)	Change (acres)	Percent Change
Intertidal				
Salt marsh	4760	630		
Mudflats, sandflats	6186	1005		
Total intertidal loss			– 9311	– 85.1
Shallow subtidal				
0 to – 6 ft. MLLW	7672	2404	– 5268	– 68.7
Deep subtidal				
6 to –18 ft. MLLW	2431	5727		
Below –18 ft. MLLW	2286	4268		
Total deep subtidal gain			+ 5278	+ 52.8
Salt ponds	0	1252	+ 1252	

pectation, although cordgrass grows over large areas of each constructed marsh. The 1985 and 1991 sites were specifically designed to mitigate damages to clapper rail habitat under an Endangered Species Act Section 7 consultation (FWS 1988).

Clapper rails prefer to nest in tall cordgrass (Jorgensen 1975). The birds build a nest that can float when the high tide inundates the lower salt marsh. Tall vegetation supports the nest and allows the birds to weave a canopy overhead, so the eggs and chicks are not obvious to aerial predators (raptors). In an ongoing research effort, we have compared cordgrass marshes with and without nesting rails.

Detailed studies of the 1985 site suggested why it is not functionally equivalent to nearby natural marsh remnants. We began documenting structural and functional attributes of this sites when it was four years old (Langis, Zalejko, and Zedler 1991; Zedler 1991; Zedler and Langis 1991; Zedler 1993; Zedler and Powell 1993) and theses (Swift 1988; Zalejko 1989; Rutherford 1989). This initial work documented significant differences between constructed and reference wetlands for eleven attributes, half of which were related to the texture of the fill material (Table 14-2). The artificial soils were sandy (mostly old dredge spoil deposits), and they failed to supply nitrogen, which is known to be a limiting factor for cordgrass growth in the region's tidal marshes (Covin and Zedler 1988). Impaired growth was initially measured as lower mean height and lower foliar nitrogen concentrations (Table 14-2).

TABLE 14-2. Comparison of the "best" areas of cordgrass (based on cover data) in the constructed marsh with natural cordgrass stands in the adjacent reference wetland. Sampling locations were at the same intertidal elevation and had a common source of tidal water.

Summary data sets	Percent*
Organic matter content	51
Sediment nitrogen (inorganic N)	45
Sediment nitrogen (total Kjeldahl N)	52
Pore-water nitrogen (inorganic N)	17
Nitrogen fixation (surface cm)	51
N fixation (rhizosphere)	110
Biomass of vascular plants	42
Foliar nitrogen concentration	84
Height of vascular plants	65
Epibenthic invertebrate density	36
Epibenthic invertebrate species richness (no. of spp.)	78
Average of comparisons	**57 percent**

*Relative to value in adjacent reference wetland

More recent studies revealed that canopy architecture (height distributions) differs substantially between the transplanted and natural marshes (Zedler 1993), and that the coarse soil retains little nitrogen (Gibson, Zedler, and Langis, 1994).

We then turned our attention to developing methods for correcting the shortcomings of transplanted marshes. In our first experiment, we added inorganic and organic nitrogen to condition the soil prior to transplantation. These treatments increased the biomass of cordgrass transplants but did not produce plants that were as tall as in natural marshes (Gibson, Zedler, and Langis, 1994). Litterbag experiments indicated that decomposition is extremely rapid and that nitrogen is readily lost from the sandy soil, even when applied as organic amendments (such as with straw and alfalfa). More recent experiments indicate that biweekly fertilization may be essential to stimulate growth to equivalent heights.

While nitrogen amendments may increase plant heights, they may not be a panacea. Natural salt marshes are complex, and their overall functioning is unlikely to be duplicated simply by adding nitrogen to sandy soils. Even though we now know how to grow tall plants in small experimental plots, the widespread addition of nitrogen may have unexpected consequences for whole-ecosystem functioning. For example, macroalgal blooms may be triggered (Fong, Zedler, and Donohoe 1993), with impacts on the epibenthos and infauna, which are foods for the clapper rail. Furthermore, we do not yet understand the long-term effects of nitrogen addition on herbivorous insects.

Several insect grazers occur in these marshes, and at least one, the native scale insect (*Haliaspis spartina*) has damaged plants in all three of San Diego Bay's artificial marshes. Kathy Boyer has assessed scale insect damage in both our 1990 and 1993 experiments (Boyer and Zedler, forthcoming), but it is not yet clear how nitrogen, scale insects, and predators on scale insects (such as *Coleomegilla fuscilabris*, a native beetle) interact to control plant heights. The second- and third-order effects of nitrogen fertilization require further study, especially in the long term.

Even if nitrogen addition is the "solution" to providing equivalent canopy architecture, we have much to learn. For example, to predict whether nitrogen will need to be added every year, we need to understand how the natural marshes function in nutrient uptake and retention. I have posed two alternative hypotheses: The plants may take up small quantities of nitrogen from tidal waters and require decades to centuries for nutrient pools to accumulate; or the marsh ecosystem may respond to infrequent pulses of nitrogen from flood flows, with rapid uptake and sequestering. The first is supported by the work of Winfield (1980), who documented removal of inorganic nitrogen as the tide waters passed over the salt marsh. To operate by itself, this mechanism would require efficient storage (little leakage) and internal recycling. Alternatively, the rapid removal of nitrogen during the region's rare flood events would require a mechanism for luxury uptake, perhaps by macroalgae. (This hypothesis is supported by the experiments of Fong, Zedler, and Donohoe [1993].) When we understand how the wetlands accumulate and retain nitrogen, we will be able to develop long-term recommendations concerning fertilizer addition.

A Broader View of Functional Equivalency

It is very difficult to measure wetland functions, since repeated sampling and complicated analyses are usually required (Table 14-3). Like a bank account balance, a one-time assessment of soil nitrogen concentration is only a status report; it does not tell us whether there are adequate reserves; whether the amount is increasing, decreasing, or stable; whether it comes from streamflows, tidal flows, or nitrogen-fixation; or whether it is being sequestered by cordgrass and algae or lost through denitrification or leaching. To be able to manage either a depauperate financial situation or an impaired wetland, one must understand the causes and effects of the situation. Low supplies of available nitrogen (or cash) could mean that plants are stressed (impending bankruptcy) or that they have absorbed all available nutrients to produce a luxuriant canopy (efficient investing).

Species composition (the presence of plants and animals) is a simple structural attribute that can be determined quantitatively from surveys. The processes

TABLE 14-3 Examples of structural and functional attributes of coastal wetlands. Structural attributes are measurable at one point in time, while functions are processes that occur through time. Many functions may be responsible for each structural feature of an ecosystem, and individual structural attributes may be responsible for several functions.

Structure	Examples of functions
Nutrient status	Nitrogen supply (import, fixation)
	Storage, recycling, and export
Plant community composition	Maintenance of gene pool
(species occurrence and abundance)	Resistance to invasive exotics
	Gap regeneration/succession
	Self-sustaining populations
	Pollination and sexual reproduction
	Dispersal and establishment
Canopy architecture (cover,	Provides nesting habitat
height distribution, density)	Protection from predators
	Provides refuges for predators
Food base (biomass, foliar nitrogen)	Primary productivity
	Food web support
Food web (animal community	Feedback mechanisms that control
composition)	herbivory, predation, parasitism
	Recovery from outbreaks (resilience)
Corridors to adjacent habitats	High-tide refuges
	Ability to reestablish (resilience)
Channels and tidal creeks	Nursery function
	Food support for older fishes,
	shellfish, and wildlife
Large intertidal flats	Attracts migratory shorebirds
	Feeds birds, provides resting habitat

that determine species presence and abundance are much more complicated. They include the environmental tolerances and habitat requirements of each species, various competitive interactions, predator–prey interactions, and facilitative processes (such as shading and mycorrhizal associations). In constructed wetlands, it may be decades before species composition stabilizes enough to conclude that populations are self-sustaining and that the desired community is as likely to persist as a natural marsh, because "stability" depends on a network of interacting processes with complex feedback mechanisms. Stability per se is unlikely in coastal wetlands, where floods and other catastrophic events can change salinities and introduce or eliminate species. It is also unlikely where sea level rise is accelerating. Without an understanding of how natural marsh populations are sustained, we are less able to manage constructed wetlands for desired species and communities.

A second basic ecosystem attribute is food web structure. To create habitat

suitable for clapper rails, one must know not just what they eat at one point in time but whether food habits shift as birds mature or preferred foods decline. To ensure suitable nesting habitat, a constructed wetland must provide food support for all life stages.

Food web diagrams are commonly displayed in textbooks, yet few are based on real data. Feeding habits have rarely been quantified for whole ecosystems, so most of the arrows that depict who eats whom must be derived from casual observations or extrapolated from similar species or habitats. For some species, the only clues may be the shapes of their mouth parts. Even if we clearly understood food webs in natural habitats, we might lack specific information that is critical for managing constructed sites. The scale insects that became pests in our transplanted cordgrass marshes are a good example. These native insects do little damage to the vegetation of natural wetlands. They become a problem only in an artificial situation, perhaps because their native predator is too rare. Such complex predator–prey control mechanisms are not obvious in natural systems and may only be elucidated when the control organism is removed. The predatory beetle that eats scale insects on cordgrass may have been too rare in artificial marshes because the short vegetation did not provide the necessary high-tide refuge.

In other parts of the world, habitat restoration is being undertaken for fish and shellfish support. The restoration of tidal flows to some twenty-five hundred acres of former mangrove and salt marsh habitat is being planned near Newcastle, New South Wales, Australia (Zedler, Nelson, and Adam 1995). To maximize fish support functions, we need to know whether seaweeds, seagrasses, mangroves, or salt marsh plants are the main components of the food base. In Texas, Minello and Zimmerman (1992) have shown that transplanted marshes can develop vigorous stands of *Spartina alterniflora* and attract fishes, although invertebrate densities were generally low. In comparison with reference marshes, amphipods in constructed marshes were 22 to 40 percent as dense, large macrofauna were 30 to 38 percent as abundant, and decapod crustaceans were 20 to 26 percent as abundant. Data on feeding and growth patterns are still needed to determine whether these marshes can sustain fish, not just attract them.

Shorebirds that migrate along the Pacific Highway make use of natural, degraded, and constructed mudflats (B. Kus, unpublished data, 1993). At San Diego Bay, the excavated tidal flats attracted large numbers of species but in very low densities compared to reference sites. Whether the low densities of birds were due to the lower densities of epibenthic invertebrates found by Rutherford (1989) is unknown. In North Carolina, Moy and Levin (1991) suggested that lower invertebrate densities result from low organic matter concentrations in constructed marsh substrates. From the standpoint of food-chain support, constructed marshes appear to be functionally impaired.

The work of Landin, Clairain, and Newling (1989) suggests that a four-teen-year-old riverine wetland that was built of dredge spoil in Virginia is successful in many functional aspects. Their study considered vegetation, benthos, fish populations, mammals, and birds and found high plant productivity, greater wildlife densities, and more species than at reference sites. Riverine wetlands may include species that are more resilient to perturbations (due to frequent flooding, scouring, and accretion) than those of coastal salt marshes. It is also likely that the potential for functional equivalency is greater where a larger proportion of the natural wetland resource still remains and where individual sites are not as degraded and isolated as in San Diego Bay (NRC 1992).

The most challenging restoration goal may be the reconstruction of habitat for endangered species. Many of these species became endangered because of habitat loss and degradation. As habitats diminish, it is probably the species with very specific requirements that decline most rapidly and run the risk of extinction. Such species pose major challenges to restoration efforts. It seems reasonable to expect generalists and opportunistic species to colonize newly constructed habitats much more readily than endangered species. This is why weeds are a problem in restoration sites. Our efforts to re-establish the salt marsh bird's beak (*Cordylanthus maritimus* subspecies *maritimus*) to San Diego Bay marshes support this claim. This endangered annual plant must set seeds to sustain the population. While that may not seem difficult, it does limit the resilience of the population, in part because native pollinators are no longer plentiful (Parsons 1994), seed production is often low, granivores damage many of the seeds that are produced, and good seeds do not remain viable indefinitely. Seed germination and seedling establishment seem to require moisture from rainfall (which is sporadic) rather than from tides (which are dependable). Furthermore, the plant is a hemiparasite. It prefers selected host plants but does not grow well when the host canopy casts too much shade (Fink and Zedler 1989). It is no wonder that such a "particular" plant is endangered. We have approached this reintroduction project cautiously, beginning with seeding remnants of natural marsh. Only when these patches have proven self-sustaining and have produced surplus seeds will we attempt to introduce it to constructed high marshes, where the risk of failure is high due to their sandier soil, low organic matter, low nutrient content, and low soil moisture–retention capability.

Southern California's mitigation sites are never in pristine settings. Rather, they are places that have been filled, diked, or otherwise modified. They are small and isolated by urbanization, with adjacent freeways, railroads, and power lines. Night lighting and vertical barriers may impair bird access and use. They receive runoff year round from irrigation—not a natural hydrologic feature in this Mediterranean-type climate. They are downstream of dammed

rivers, so flood pulses can never be natural. Some have permanent tidal ob-
structions that attenuate the tidal range. They are plagued by noise from cars,
planes, and helicopters. Because of these permanent disturbances, they can
never match the functional capacity of a pristine ecosystem. In mitigation
agreements, it is not required that constructed marshes match the condition
of pristine ecosystems; nor is functional equivalency with disturbed wetland
remnants required. To date, the criteria that have been specified are much
less demanding.

Requirements for Successful Mitigation

The two excavated marshes in San Diego Bay have fairly strict criteria for
compliance with the Endangered Species Act (Section 7 consultation, FWS
1988). That is, they must offer more than the presence of the appropriate vegetation; a suitable canopy and food base must be sustained for three consecutive years. However, the endangered birds are not required to be present. In
all, there must be seven potential clapper rail home ranges (2 to 4 acres), each
with at least 15 percent of the area in cordgrass (and 90 to 100 m² having specific height and cover, with evidence that it is self-sustaining) and 15 percent
of the area in high marsh (to serve as a high-tide refuge). Nine years after
highway widening jeopardized the clapper rail's native habitat, there is still
no assurance that these criteria can be met. The seven areas were selected in
1994 and the first assessment of cordgrass height and cover undertaken in
1995.

Mitigators and planners should recognize that compliance with even
simple criteria may take more than a decade, and that assessment efforts will
continue throughout that time period. Given the problems at San Diego Bay,
it seems reasonable to recommend much stricter criteria for successful replacement of natural wetlands—particularly those that support endangered
species—to ensure that damages are fully compensated. Southern California
faces a dilemma in that the pressure to develop remaining wetlands is high,
while the potential for successful compensation is low.

Future Research Needs

Larger questions must be answered before mitigation measures proceed in
Southern California's coastal wetlands. Can mitigation sites ever become functionally equivalent to reference wetlands? Are they more susceptible to exotic
species invasions? What level of functional equivalency can they attain in
comparison to pristine wetlands? What are the spatial requirements of the

region's endangered species? To what extent can they be met by constructed wetlands? What types of buffers and corridors must be adjacent to constructed wetlands for biodiversity to be sustained?

Conclusions and Recommendations

Functional equivalency with reference sites is a relevant criterion for successful mitigation, especially for losses to endangered species habitat. Until we understand how the structure of constructed sites is controlled (that is, until we understand the processes that control nutrient cycles, vegetation composition, animal uses, and food web dynamics), we cannot predict that biodiversity will be sustained.

The San Diego Bay mitigation sites indicate that functional equivalency is not easily achieved. The criteria for compliance with the Endangered Species Act Section 7 Consultation are less stringent than functional equivalency, yet they are not fulfilled, even at the 1985 site.

The experiments with nitrogen addition suggest that taller cordgrass can be grown, but whether the result will be a garden or a self-sustaining ecosystem is still unclear. We will consider them gardens until there is evidence of long-term sustainability.

Studies of constructed wetlands in San Diego Bay indicate that some controls on structure (such as insect outbreaks) may not be recognizable in natural systems. Hence, an adaptive research program will need to accompany projects that seek to provide mitigation for damages to endangered species habitat.

The 1985 mitigation marsh has proven valuable in identifying problems and functional inequivalence. Without the cooperation of the mitigating agencies (California Department of Transportation and U.S. Army Corps of Engineers), it would not have been possible to develop corrective measures.

To maintain regional biodiversity, it would be prudent to cease further damages to coastal wetlands in Southern California. Other regions with larger proportions of habitat remaining should heed the lesson that our experience offers: It is not easy to repair damages once habitats have declined to the point of endangering several species. Meanwhile, we should continue to improve restoration methods through experimentation and scientific study, and we should apply these new techniques to expand the region's tidal marsh resource through means other than mitigation. Tijuana Estuary provides such an opportunity. A plan already exists to restore about 500 acres, including the removal of sediments from about 400 acres of salt marsh that are no longer tidal. Implementing such plans would be a significant step toward

achieving the NRC (1992) call for 10 million acres of wetland restoration by the year 2010.

ACKNOWLEDGMENTS

I thank Don Falk for encouraging this submission and Tom Kwak, Bruce Nyden, and anonymous reviewers for editorial assistance. Without the work of my students and associates at the Pacific Estuarine Research Laboratory (especially Patrice Ashfield, Kathy Boyer, John Cantilli, Kevin Gibson, Barbara Kus, René Langis, Sue Rutherford, Kendra Swift, and Margaret Zalejko), this chapter would not have been possible. Our research has been supported by the U.S. Fish and Wildlife Service, the California Department of Transportation, NOAA and NOAA's Coastal Ocean Program (Estuarine Habitat Program), the National Sea Grant College Program, the Department of Commerce, under grant number NA89AA-D-SG138, projects numbered R/NP-1-20H, R/NP-1-21E, and R/CZ/106, the California Sea Grant College Program, and in part by the California State Resources Agency. The U.S. government is authorized to reproduce and distribute this material for governmental purposes.

REFERENCES

Boyer, K. E., and J. B. Zedler. Forthcoming. "Cordgrass Damage by Scale Insects in a Constructed Salt Marsh: Effects of nitrogen additions." *Estuaries.*

Covin, J. D., and J. B. Zedler. 1988. "Nitrogen Effects on *Spartina foliosa* and *Salicornia virginica* in the Salt Marsh at Tijuana Estuary, California." *Wetlands* 8: 51–65.

Dahl, T. E. 1990. "Wetlands Losses in the United States 1780s to 1980s." Washington, D.C.: U.S. Department of the Interior, Fish and Wildlife Service.

Feierabend, J. S. 1992. "Endangered Species, Endangered Wetlands: Life on the Edge." Washington, D.C.: National Wildlife Federation.

Fink, B. H., and J. B. Zedler. 1989. "Endangered Plant Recovery: Experimental Approaches with *Cordylanthus maritimus* ssp. *maritimus*." In Proceedings of the First Annual Meeting of the Society of Ecological Restoration and Management. Edited by H. G. Hughes and T. M. Bonnicksen. Madison, Wisconsin. January 16–20, 1989.

Fong, P., J. B. Zedler, and R. M. Donohoe. 1993. "Nitrogen Versus Phosphorus Limitation of Algal Biomass in Shallow Coastal Lagoons." *Limnol. Oceanogr.* 38 (5): 906–923.

Gibson, K. D., J. Zedler, and R. Langis. 1994. "Limited Response of Cordgrass (*Spartina foliosa*) to Soil Amendments in Constructed Marshes." *Ecological Applications* 4: 757–767.

Jorgensen, P. D. 1975. "Habitat Preference of the Light-Footed Clapper Rail in Tijuana Marsh, California." Master's thesis, San Diego State University, San Diego.

Landin, M. C., E. J. Clairain, and C. J. Newling. 1989. "Wetland Habitat Development and Long-Term Monitoring at Windmill Point, Virginia." *Wetlands* 9: 13–26.

Langis, R., M. Zalejko, and J. B. Zedler. 1991. "Nitrogen Assessments in a Constructed and a Natural Salt Marsh of San Diego Bay, California." *Ecological Applications* 1: 40–51.

Macdonald, K. 1990. South San Diego Bay Enhancement Plan. Vol. 1: Resources Atlas. San Diego, Calif.: San Diego Unified Port District.

Minello, T. J., and R. J. Zimmerman. 1992. "Utilization of Natural and Transplanted Texas Salt Marshes by Fish and Decapod Crustaceans." *Marine Ecology Progress Series* 90: 273–285.

Moy, L. D., and L. A. Levin. 1991. "Are *Spartina* Marshes a Replaceable Resource? A Functional Approach to Evaluation of Marsh Creation Efforts." *Estuaries* 14: 1–16.

NRC (National Research Council). Committee on Restoration of Aquatic Ecosystems. 1992. *Restoration of Aquatic Ecosystems: Science, Technology, and Public Policy.* Washington, D.C.: National Academy of Science.

PERL (Pacific Estuarine Research Laboratory). 1990. "A Manual for Assessing Restored and Natural Coastal Wetlands with Examples from Southern California." California Sea Grant Report no. T-CSGCP-021. La Jolla, Calif.: California Sea Grant.

Parsons, L. S. 1994. "Re-establishment of Salt Marsh Bird's Beak at Sweetwater Marsh: Factors Affecting Reproductive Success." Master's thesis, San Diego State University.

Rutherford, S. E. 1989. "Detritus Production and Epibenthic Communities of Natural Versus Constructed Salt Marshes." Master's thesis, San Diego State University, San Diego, California.

Swift, K. L. 1988. "Salt Marsh Restoration: Assessing a Southern California Example." Master's thesis, San Diego State University, San Diego, California.

FWS (U.S. Fish and Wildlife Service). 1988. "Biological Opinion 1-1-78-F-14-R2: The Combined Sweetwater River Flood Control and Highway Project, San Diego County, California." Letter from Wally Steuke, acting regional director, USFWS, Portland, Oregon, to Colonel Tadahiko Ono, district engineer, Los Angeles District, Corps of Engineers, 30 March.

Winfield, T. P. 1980. "Dynamics of Carbon and Nitrogen in a Southern California Salt Marsh." Ph.D. diss. University of California, Riverside, and San Diego State University. San Diego.

Zalejko, M. K. 1989. "Nitrogen Fixation in a Natural and a Constructed Southern California Salt Marsh." Master's thesis, San Diego State University, San Diego, California.

Zedler, J. B. 1991. "The Challenge of Protecting Endangered Species Habitat Along the Southern California Coast." *Coastal Management* 19: 35–53.

Zedler, J. B. 1993. "Canopy Architecture of Natural and Planted Cordgrass Marshes: Selecting Habitat Evaluation Criteria." *Ecological Applications* 3: 123–138.

Zedler, J. B., and R. Langis. 1991. "Comparisons of Constructed and Natural Salt Marshes of San Diego Bay." *Restoration and Management Notes* 9 (1): 21–25.

Zedler, J. B., and A. Powell. 1993. "Problems in Managing Coastal Wetlands: Complexities, Compromises, and Concerns." *Oceanus* 36 (2): 19–28.

Zedler, J. B., P. Nelson, and P. Adam. 1995. "Plant Community Organization in New South Wales Saltmarshes: Species Mosaics and Potential Causes." *Wetlands* (Australia) 14: 1–18.

CHAPTER 15

Use of Corporate Lands

Brian J. Klatt and Ronald S. Niemann

Factors contributing to the worldwide loss of biodiversity have been enumerated by various authors, such as Diamond (1984); Pimm and Gilpin (1989); and Machlis and Forester (forthcoming). In general, the factors listed by most authors can be grouped into four main categories: (1) habitat alteration (such as habitat fragmentation or destruction due to agriculture); (2) overexploitation of species or ecosystems (such as clear-cutting of North American forests); (3) introduction of alien species (such as purple loosestrife (*Lythrum salicaria*) in upper midwest wetlands); and (4) disruption of ecosystem processes (such as elimination of fire in prairies and savannahs).

While the previous list may be expanded by subdividing the categories into more proximal levels of biological explanation, such as parceling habitat alteration into habitat fragmentation, habitat destruction, physicochemical alteration, and so on, we group the factors into the categories mentioned because they contain a commonality that gives them didactic purpose. The categories as presented are not biological mechanisms; rather they are classes of anthropogenic effects, all of which are directly or indirectly related to the acquisition of resources to meet the essential needs and nonessential wants of an ever-increasing human population. In short, the categories point out that the primary cause of the worldwide decrease in biodiversity is socioeconomic development associated with a burgeoning world population and resource consumption in developing nations. Consequently, the premise suggested by the title of this chapter—using corporate lands to help preserve, enhance, or restore biodiversity—may strike many readers as a conceptual contradiction, given the central role of corporations in socioeconomic development.

Those readers who have reservations concerning the potential role of corporations in preserving biodiversity have cause for such reservations. The reputation of corporations with respect to the environment is one of exploitation, not preservation; of pollution, not protection; of unwilling compliance with environmental regulation, not cooperation with agencies. Some of this perception is founded in fact, and there are certainly an unknown but substantial

number of corporate decision makers and employees who do not proactively seek to protect the environment. It is not a fact, however, that corporations — or, more precisely, the people who manage and operate them — are engaged in a Machiavellian plot to harm the environment.

It is undeniable that corporate activities such as clear-cutting, strip mining, organochlorine pesticide production, and shopping mall development have resulted in adverse environmental impacts such as increased siltation in spawning areas of salmon, production of acid mine drainage, decreased reproductive success of birds of prey, and filling of wetlands. However, the majority of this environmental damage was and is done in compliance with applicable environmental regulations; the methods being employed are simply the most cost-effective to the corporation at the time. It is extremely important to understand this point.

Most corporations are for-profit entities, and by definition have at least one stockholder who has contributed cash or something of value (their "investment") in exchange for shares of stock and who expects some return on the investment. Corporations attract stockholders by providing a return on investment that is competitive with other investment options available to the potential stockholder. That is, a corporation that offers a 10 percent return on investment will attract a greater amount of investment capital from stockholders than other corporations that offer only a 5 percent rate of return.

To maximize the rate of return, a corporation must maximize profits through a combination of minimizing costs, maximizing the amount of product sold, or maximizing the price of its product. All three of these factors are controlled to an extent through market forces and can also be affected by internal corporate decisions. It is these factors, and especially costs, that cause many corporations to engage in less-than-ideal environmental practices. By not replanting forests, by not restoring strip mined land to its original condition, by producing large amounts of organochloride residues from pesticide production and use, and by filling wetlands rather than buying additional property or building to avoid wetlands, corporations have been able to minimize costs in the past.

Hawken (1993) contends that if all the social and environmental costs a product caused could be accounted for and incorporated into the price of a product through a "green tax" (that is, if all of the costs could be internalized), consumers would automatically select the least environmentally damaging products due to the lower cost and lower price. This, argues Hawken, would result in a sustainable economy without adverse impacts to biodiversity. While the idea is intriguing, and the logic difficult to argue with, its implementation is complicated for a number of reasons.

Perhaps the greatest difficulty with implementing a complete cost internalization is our difficulty in capturing all of the costs associated with any

given product in order to equitably derive the "green tax." Using available ac-counting practices, it is very difficult to completely account for the actual pro-duction and overhead costs of a product, let alone its social and environ-mental costs (such as the increase in health insurance premiums due to the elevated number of cancer cases from production processes involving car-cinogens). Yet without the ability to accurately and objectively account for such costs, the appropriate amount for a "green tax" cannot be arrived at. If the "green tax" does not accurately differentiate the social and environmental costs of similar products from different producers, the green tax will not result in its intended goal of ensuring that the product with the lowest social and en-vironmental costs is also the least expensive.

Fortunately, we do not need to wait for the development of such cost-accounting techniques before enlisting the cooperation of corporations in the preservation of biodiversity. We can appeal directly to the corporation's self-interest. The establishment of a sound biodiversity management program for corporations that engage in land-intensive activities can result not only in sig-nificant environmental benefits but also in appreciable economic benefit to the corporation through enhanced long-term strategic land-use planning, better regulatory compliance, risk management, public relations, and regula-tory agency goodwill.

Thus, we feel that corporate lands can play a potentially significant role in preserving biodiversity and thereby make part of "the problem" part of the so-lution. To illustrate our point, we first review briefly the historical and current relationship between ecological resources and socioeconomic development in North America. We then describe a federal government initiative intended to promote biodiversity conservation among various nongovernmental orga-nizations and corporations and provide a few examples of corporate efforts growing out of this initiative. Finally and most important, we suggest some key elements that we feel will greatly increase the probability of successful biodiversity management program formation and implementation.

Reciprocity Between Biodiversity and Socioeconomic Development

Adverse effects on biodiversity due to socioeconomic development date back before the Industrial Revolution, to when populations were first large enough to overexploit local resources and to create habitat fragmentation due to urban and agricultural development. This statement is not meant to trivialize more recent anthropogenic effects, especially those of the industrial-era de-velopment in North America. Rather, the important point is that, in a world of finite land mass and resources, humans have been exerting a shaping force on biodiversity for a long time. While in some instances human activities may

have resulted in an increase in local diversity (such as burning of prairies and savannahs by indigenous peoples), such benefits were likely serendipitous or at least secondary to the main purpose of the activity (driving of game, clearing of travel routes, promotion of particular plant species, and so on). The effect of human activity on biodiversity has often been negative.

Whether pre- or post-historical, pre- or post-Columbian, or pre- or post—Industrial Revolution, the ultimate driving force behind human impacts is the same: acquisition of resources to meet essential needs (space, food, shelter) and to satisfy nonessential wants (increased social status, recreation, and so on). While the ultimate driving factors may not have changed, our capacity and propensity to exploit resources has. The effect of humans on the ecology of the North American continent during the last hundred years is undeniably greater than it has been throughout the previous hundred centuries. During the nineteenth century, wholesale impact on the environment via exploitation of natural resources (such as clear-cutting forests and plowing prairie) was not only accepted but encouraged. Indeed, most of us were probably taught in our grade-school history classes that during the 1800s the conquest and "settlement" of the entire continent was considered the manifest destiny of the United States.

While the adverse effects of industrial operations and socioeconomic development on biodiversity have been ongoing for centuries and are increasingly pervasive, it is only a recent development that the reverse is true—namely, that endangered species and natural resource issues have affected industrial operations and socioeconomic development. Due to an increase in environmental awareness stemming from the prophetic warnings of such authors as Rachel Carson (1962) and Paul and Anne Ehrlich (1970), various laws were enacted in an attempt to stem the tide of devastation. Thus, the 1960s and 1970s saw the passage of such federal laws as the Clean Water Act (CWA), the Endangered Species Act (ESA), the Surface Mined Land Reclamation Act (SMLRA), and the National Environmental Policy Act (NEPA). Some federal laws have been followed by a number of analogous state laws, such as the Wisconsin Environmental Policy Act and New York's State Environmental Quality Review Act, both of which are intended to serve the same function as NEPA, but at the state level (the function being consideration in the decision-making process of the potential environmental impacts of a project and development of mitigation measures to avoid or lessen those impacts).

While some of these laws apply directly to corporations (such as the SMRLA, the wetland protection provisions of the CWA, and the prohibitions on taking threatened or endangered animal species under the ESA), others exert their effect on corporate operations through internal regulatory agency review procedures. For example, NEPA and the state analogues of NEPA may

require that the agency issuing a permit for some corporate activity must prepare an environmental impact analysis regarding anticipated environmental and socioeconomic effects of the permitted activity. The agency in turn delegates the responsibility for preparing the impact analysis to the proponent of the action (the corporation). As part of the process of preparing the environmental documentation, mitigation measures may be suggested that avoid potential environmental impacts. Such measures may be negotiated between the regulatory agency and the corporation. For example, a corporation may be seeking a permit to construct and operate a facility (such as a landfill) that may negatively affect some resource or be considered potentially controversial from an environmental standpoint. As part of the environmental documentation, it is suggested that the corporation restore a degraded fen community located on the corporation's property as a mitigation measure offsetting the anticipated visual impacts of the landfill. Thus, while the agency had no direct statutory authority to require restoration of the fen, the agency nevertheless accomplishes this in an indirect manner by incorporating the suggested mitigation measure as a condition of issuing a permit to construct and operate the landfill. Thus, whether the mechanism is direct or indirect, rare resources already present significant regulatory hurdles to many corporate operations for which an environmentally related permit is required, especially if the operation is land intensive (such as mining, forest products, utility rights-of-way) or is considered controversial.

Biodiversity Conservation Initiatives on Private Lands

While some protection or even enhancement of endangered resources is accomplished through the laws discussed earlier, many resources are left unprotected. The limited effectiveness of the laws and their implementing regulations is attested to by the very need for a symposium of the type on which this book is based. Thus, while helpful, legislation and regulation do not always provide a sufficient incentive with respect to conservation of endangered resources or biodiversity in general. Additionally, some existing laws, such as the SMRLA and ESA, only come into play after an area has been affected or a species is at substantial risk of extinction. Private efforts by groups such as the Center for Plant Conservation, the Audubon Society, and the Nature Conservancy (some of which predate the legislative efforts) have helped to fill the gaps left by legislation and have helped to preserve genetic diversity, significant natural areas, and in some cases, entire species. Yet more can be done, and we feel that corporate lands could be much more

extensively utilized in efforts for biodiversity and endangered resource conservation.

In 1991, President Bush established the President's Commission on Environmental Quality (PCEQ). The PCEQ was made up of representatives of the business community, universities, and various nongovernmental organizations (NGOs). The purpose or goal of the PCEQ was to identify issues of critical environmental concern and, drawing on the diverse resources and perspectives of the group, to identify potential ways in which the members could act singly or in cooperation to address the issues. The PCEQ identified conservation of biodiversity as a key environmental issue to be addressed by the commission and established the Biodiversity on Private Lands Initiative. The goals of the initiative were (1) to demonstrate the feasibility of integrating biological diversity conservation into management of private lands; (2) to articulate the best practices that may be undertaken to achieve this integration; and (3) to foster increased understanding of the value of biological diversity. In an attempt to integrate environmental, economic, and quality-of-life issues and interests, the initiative was guided by a nineteen-member steering committee, on which corporations and nongovernmental organizations were approximately equally represented (ten corporations, eight NGOs, one government agency).

The committee recognized that 60 percent of the land area of the United States is privately held, with land uses and ownership varying from individual home owners with small lots to corporations holding millions of acres. Consequently, there was a need felt to develop a variety of strategies for conserving biodiversity that would complement the variety of land ownership and land use. PCEQ members were invited to submit to the committee projects demonstrating application of biodiversity conservation strategies to private land holdings. The following are a handful of brief descriptions of some of the demonstration projects resulting from the Biodiversity on Private Lands Initiative. These demonstrate the variety of strategies available for biodiversity conservation and endangered resource enhancement.

- On their 113-acre corporate headquarters in Fort Washington, Pennsylvania, the Johnson and Johnson/McNeil Consumer Products Company is reintroducing fire as a management tool to promote natural plant communities.

- A 200-acre research center in Granville, Ohio, is being managed by the Dow Chemical Company to restore and enhance biodiversity by reintroducing native vegetation on former farmland and managing existing wetland and woodland for wildlife.

- Pacific Gas and Electric is managing 8,700 acres of their 10,000-acre Diablo Canyon Power Plant in California as natural communities in-

cluding coastal scrub, chaparral, grassland, oak woodland, closed-cone pine forest, and riparian communities.

- Mead Paper Company is establishing a research and education program called Total Ecosystem Management Strategies on a 20,000-acre tract in Michigan's west-central Upper Peninsula. The aim of the program is to characterize the biological diversity of the area, then develop a strategy allowing for sustainable fiber production and conservation of the existing biodiversity.

- Champion International Corporation owns approximately 145,000 forested acres in the Adirondack Forest Preserve in upstate New York. Of this, 50,000 acres have been designated core to the company's operation and will remain in fiber production. The remaining 95,000 acres are considered nonstrategic to corporate goals and are being managed for multiple-use, including protection of unique ecological characteristics.

In addition to these parcel-specific projects, one demonstration project involved the authors of this chapter in the development of corporatewide guidelines and procedures for biodiversity management for WMX Technologies and Services (WMX) facilities (formerly known as Waste Management). WMX has a corporate policy that states that the corporation is dedicated to the conservation of nature and will manage its holdings so that there will be no net loss of biodiversity on the company's property.

To provide specific guidance to WMX facility managers in establishing a biodiversity management program, the Biodiversity Management Guideline (Klatt et al. 1993) was developed. The guideline explained the WMX corporate policy, then outlined a three-part approach at the facility level calling for a biodiversity inventory, development of a management plan, and establishment of a monitoring program. The guideline described specific procedures for conducting the baseline biodiversity inventory and developing management and monitoring plans based on a plant community–level approach (species-specific in the case of threatened or endangered species) that would ensure that corporate policy was met. The document was reviewed by a technical advisory committee consisting of six senior ecologists within RUST Environment and Infrastructure and was independently reviewed by Dr. Daniel Simberloff of Florida State University. Biodiversity inventories have been constructed at ten WMX facilities and follow-up monitoring began in the summer of 1995 to evaluate implementation of the management plans at the first facilities inventoried.

What is noteworthy concerning the WMX program and the other PCEQ demonstration projects just described is that none of them is required by regulation. Indeed, participation on the PCEQ was voluntary, as

was participation in the Biodiversity on Private Lands Initiative. While the corporations conducting demonstration projects certainly benefited from the publicity associated with these projects and can realize economic benefits from such efforts (as we will discuss later for the WMX program), it is encouraging that such efforts at conservation and cooperation with conservation organizations take place, regardless of the benefit to the corporation; and "win-win" situations such as these should be fostered.

Aside from the efforts connected with the PCEQ initiative, there have been other examples of rare species or community conservation and enhancement on corporate lands that were not specifically required by regulations, but were agreed to with regulatory agencies. For example, at a facility in Wisconsin a degraded oak savannah is being rehabilitated along with species-specific efforts for the Massassauga rattlesnake (*Sistrurus catenatus*) and Blanding's turtle (*Emydroidea blandingi*) at the site of a proposed landfill expansion. Though the state and corporation must ensure that permitting of the landfill expansion is protective of the turtle and rattlesnake due to considerations of the Wisconsin Endangered Species Act, the state has no explicit authority to require restoration of the oak savanna. Restoration of the savanna is a permit condition voluntarily agreed to by WMX and is being implemented through an ecological restoration plan for the facility.

At a clay-mining site in New Jersey, a population of the swamp pink (*Hellonias bullata*) is being intensively managed, and within the context of plant reintroductions is particularly noteworthy. Chemical Waste Management (CWM) purchased an existing clay-mining operation in New Jersey. Shortly after purchase it was found that the site contained a population of the swamp pink, a federally threatened species. Based on interpretation of historic aerial photos it is believed that the population of the swamp pink at the site may have once numbered ten thousand individuals. Mining operations at the site prior to purchase by CWM severely affected the forested wetlands that served as habitat for the swamp pink (a shade species) and caused the population to decline to fewer than twenty-seven hundred individuals. Upon discovery of the population, mining operations ceased, and restoration efforts began. Individual plants at immediate risk of mortality due to destruction of the canopy from mining operations, and consequent exposure to full sun, were moved from the modified habitat to greenhouses at Rutgers University for rehabilitation and eventual reintroduction to the site. Meanwhile the wetlands on-site are being reconstructed, with the prognosis for population recovery being good.

The above are but a few examples of positive efforts being made by corporations in biodiversity conservation and give a taste of the possibilities in this area.

Integrating Biodiversity Conservation into Corporate Land Management

Prior to the Leopolds, Roosevelts, and Shelfords, conservation efforts in the United States either were largely the result of philanthropic acts by individuals who could afford such efforts, or they were incidental to other human activities. Examples of the latter include benefits to nongame waterfowl and shore birds stemming from management of waterfowl for hunting; establishment of private hunting preserves; preservation of prairies along railroad rights-of-way; and preservation of large, relatively undisturbed tracts on military reservations. But biodiversity conservation is no longer be merely a luxurious or serendipitous pastime. As noted earlier, one of the primary goals of the PCEQ Biodiversity Steering Committee was to demonstrate the feasibility of integrating biological diversity conservation as an active element of private land management. In this section we present our views on some of the factors likely to maximize success for such integration on corporate lands; we suggest how this integration can be done; and we list some of the current limitations to integration. The views we present largely come from our experience as environmental consultants, and our frequent dealings with natural resource issues in the context of permitting and environmental impact analyses; they come especially from our involvement with the development of the WMX Biodiversity Management Guideline (Klatt et al. 1993).

Programmatic Framework for Corporate Biodiversity Conservation and Enhancement

There are many factors that can contribute to the success or failure of a corporate biodiversity program. In our view, the factors that lead to success can be combined into three primary requisites: (1) appropriate corporate culture must exist to create sufficient corporate-level support for the program; (2) commitment to the program must be continual and long term; and (3) the technical approach must be both scientifically credible and cost effective.

CORPORATE CULTURE

It should come as no surprise to anyone that corporations differ greatly from one another with regard to internal culture. The culture of a particular corporation is defined by many factors, such as the type of products it produces or services it provides, its geographic location, the background and experience of its management, the background and experience of its nonmanagerial staff, its corporate history, the status of its stock as privately or publicly held, the age of the corporation, the culture of its competitors, the public image it wishes to portray, the level of internal entrepreneurism, the current business climate in which it operates, and so on.

Given the fact that biodiversity management in general, and threatened

and endangered plant species management on private lands in particular, is mostly not a regulatory requirement for private firms, corporate culture is of paramount importance to the success of biodiversity conservation on corporate lands. Regardless of the exact forces at work, there is a suite of characteristics that we feel increases the likelihood of a corporation's success at integrating biodiversity conservation into its planning. Not surprisingly, single most important cultural factor is that the concept be supported and fostered by the corporation's most senior management. That support must be demonstrated not only to the other corporate employees but also to stockholders. As we discuss later, there are costs (and benefits) to active biodiversity management, and the exact cost/benefit ratio may be difficult or impossible to calculate. Consequently, senior management must support the concept in order to assure stockholder acceptance.

Senior management support should be embodied, at least implicitly if not explicitly, in corporate policies and goals. For example, WMX has established fourteen explicit corporate environmental principles (Figure 15-1), one of which states, "the Company is committed to the conservation of nature. We will implement a policy of 'no net loss' of wetlands or other biological diversity on the Company's property" (Principle 3). If the corporation is organized into subsidiaries each constituting its own operating unit, this policy should also be adopted by each unit.

In addition to policy setting, support for biodiversity conservation should also be incorporated into central business planning in order to fully realize the intangible or nonquantifiable benefits of such efforts. For example, a growing contingent of investors are expressing a preference for "green" companies, and various mutual funds have been established that invest only in environmentally conscientious companies. Failure to apply at least qualitative social cost–accounting considerations (at the very least, from a public relations standpoint and long-term risk management) are likely to result in sole consideration of short-term financial costs in corporate planning, which may decrease the likelihood of successful biodiversity management program implementation.

Personnel policies and procedures should also reflect the corporate commitment to biodiversity conservation. Responsibility and accountability for particular program activities should be assigned to specific individuals, with performance within these responsibilities evaluated like any other responsibility, with advancement and compensation being linked to performance. Indeed, for the authors of this chapter and for WMX employees, approximately one-fourth of their annual performance rating is based on support of environmental programs and regulatory compliance.

Formal incorporation of biodiversity into corporate policy, business planning, and personnel management allows for the possibility of external au-

diting of the corporation for compliance with its own policies. In the case of WMX, the Arthur D. Little Company performs independent audits for compliance with the fourteen environmental principles. By assessing performance in relation to stated goals and policy, the resultant audit reports represent an independent assessment of the corporation's commitment to the principles.

This last point—that there may exist external audit reports (available to the public under the rules of the Securities and Exchange Commission) on the actual corporate commitment to biodiversity conservation—touches on the final corporate cultural characteristic we consider, which is willingness to accept public scrutiny. As we describe, one of the benefits to a corporation of adopting and implementing a biodiversity policy and program is the goodwill it may foster with the public and with regulators. However, this can be a two-edged sword. Opening the program and specific projects to public or scientific inspection means that failures may become widely known. The case studies in this book clearly show that the science of biodiversity conservation and the state of its practices are young and imprecise. While some case studies reported clear successes, others were less than successful even when the effort was well planned. While the public and regulators are likely to be forgiving of reintroduction failures if they are performed by conservation groups or regulatory agencies, they are likely to be far less understanding if the endeavor was undertaken by a corporation, especially in a mitigation context. Consequently, while the corporation may gain substantial benefit from increased goodwill, reintroduction efforts may also carry a substantial downside risk compared to similar efforts performed by conservation groups or regulatory agencies.

LONG-TERM COMMITMENT

While business trends may come and go quickly, extant species and ecosystems did not evolve overnight. Similarly, the changes that may be required to accomplish significant biodiversity enhancement may require decades or even centuries. While we may be able to remove in a few weeks all the invading shrubs and nontypical trees from a Midwest oak savannah in which fire has been suppressed, it may require decades for the typical oak savannah understory to be re-established. Consequently, the corporate commitment to biodiversity management must also be long-term. In essence, to be truly effective, the commitment must be in perpetuity. Fortunately, many of the corporations with land-intensive operations also engage in long-term planning.

For example, in the case of forest products companies, stand rotation may be on the order of fifty to one hundred years. Obviously, management periods of such length offer a host of possibilities for biodiversity management. In the case of landfills, after the unit has reached capacity, the landfill unit itself and

FIGURE 15-1.   Fourteen corporate environmental principles of WMX.

Principle 1: The Company is committed to improving the environment through the services that we offer and to providing our services in a manner demonstrably protective of human health and the environment, even if not required by law. We will minimize and strive not to allow any releases to the atmosphere, land, or water in amounts that may harm human health and the environment. We will train employees to enhance understanding of environmental policies and to promote excellence in job performance on all environmental matters.

Principle 2: The Company will work to minimize the volume and toxicity of waste generated by us and others. We will operate internal recycling programs. We will vigorously pursue opportunities to recycle waste before other management practices are applied. The Company will use and provide environmentally safe treatment and disposal services for waste that is not eliminated at the source or recycled.

Principle 3: The Company is committed to the conservation of nature. We will implement a policy of "no net loss" of wetlands or other biological diversity on the Company's property.

Principle 4: The Company will use renewable natural resources, such as water, soils, and forests in a sustainable manner and will offer services to make degraded resources once again usable. We will conserve nonrenewable natural resources through efficient use and careful planning.

Principle 5: The Company will make every reasonable effort to use environmentally safe and sustainable energy sources to meet our needs. We will seek opportunities to improve energy efficiency and conservation in our operations.

Principle 6: The Company is committed to comply with all legal requirements and to implement programs and procedures to ensure compliance. These efforts will include training and testing of employees, rewarding employees who excel in compliance, and disciplining employees who violate legal requirements.

Principle 7: The Company will operate in a manner designed to minimize environmental, health, or safety hazards. We will minimize risk and protect our employees and others in the vicinity of our operations by

employing safe technologies and operating procedures and by being prepared for emergencies. The Company will make available to our employees and to the public information related to any of our operations that we believe cause environmental harm or pose health or safety hazards. The Company will encourage employees to report any condition that creates a danger to the environment or poses health or safety hazards and will provide confidential means for them to do so.

Principle 8: The Company will take responsibility for any harm we cause to the environment and will make every reasonable effort to remedy the damage caused to people or ecosystems.

Principle 9: The Company will research, develop, and implement technologies for integrated waste management.

Principle 10: The Company will provide information to and will assist the public in understanding the environmental impacts of our activities. We will conduct public tours of facilities, consistent with safety requirements, and will work with communities near our facilities to encourage dialogue and exchange of information on facility activities.

Principle 11: The Company will support and participate in development of public policy and in educational initiatives that will protect human health and improve the environment. We will seek cooperation on this work with government, environmental groups, schools, universities, and other public organizations.

Principle 12: The Company will encourage its employees to participate in and to support the work of environmental organizations, and we will provide support to environmental organizations for the advancement of environmental protection.

Principle 13: The Board of Directors of the Company will evaluate and will address the environmental implications of its decisions. The Executive Environmental Committee of the Company will report directly to the Chief Executive Officer of the Company and will monitor and report upon implementation of this policy and other environmental matters. The Company will commit the resources needed to implement these principles.

Source: WMI 1990.

the facility in which it is located are required to be maintained in accordance with a "post-closure" plan established as part of the operating permit. Post-closure plans cover a period of at least thirty years and, in all likelihood, will extend beyond that. In addition, the owner of the facility is required to provide financial mechanisms (such as trust funds, bonds, and so on) that ensure the availability of adequate funding to carry out the post-closure plan requirements if the company ceases to exist.

While the post-closure plan is a requirement and not a voluntary undertaking, biodiversity management considerations could be built into the post-closure plan. Such planning could be to the long-term economic benefit of the corporation if low–or no–maintenance natural communities are incorporated into the plan, even if initial closure costs are somewhat higher (such as establishment of forested wetlands in suitable areas of the facility). If long-term costs are minimized, the corporation can also reduce current costs by reducing the amount of the post-closure financial assurances that need to be established in the short-term.

Because long-term support of biodiversity efforts resulting from a socially conscientious corporate culture can be reversed by changes in corporate management, economic downturns, stockholder demands, corporate buy-outs, and so on, it is important that government agencies establish other incentives promoting endangered resource management or restoration. Some of these incentives might include facilitating of the transfer of restored or sensitive areas to public ownership, government subsidies for conservation management, mitigation banking or transferable credit for restoration efforts, and various tax incentives (see Bean, Chapter Sixteen).

A final consideration about corporate culture is how biodiversity management is accounted for internally. A number of the benefits a corporation receives from encouraging and practicing endangered resource management, or biodiversity management in general, are realized at the corporate level, yet the actual implementation of such efforts must take place at the facility level, and these efforts are unlikely to be evenly distributed across all corporate facilities. Consequently, there would be a greater probability of long-term commitment to such practices if corporate-level financial support were made available to individual facilities, thereby alleviating some of the burden on the individual profit centers (facilities) and amortizing the costs of such practices across facilities through corporate overhead costs.

TECHNICAL APPROACH

Biodiversity can be managed at different levels, including the genetic, species, and ecosystem levels. While a truly comprehensive biodiversity management program would directly manage all of these levels, practicality and limited resources prevent site managers from meeting this ideal.

Consequently, technical approaches to biodiversity management must be developed that, though not all-encompassing, are cost effective, ecologically meaningful, and scientifically sound.

A universally accepted definition of *biodiversity* has not been developed within the scientific community, and because of the multileveled, complex nature of the concept, no agreed-to set of parameters has been established to measure biodiversity. Machlis and Forester (forthcoming) present a list of parameters proposed by various authors as indicators of biodiversity, including species richness, population abundance and distribution, number of threatened or endangered species in an area, genetic variability, ecosystem functions, interactions, natural communities, successional stages, and ecological redundancy. Because biodiversity is both multileveled and multifaceted within each level, each of the preceding candidate parameters has utility as an indicator of biodiversity depending on temporal and spatial scales, level of organization under consideration, the particular question of interest, and/or the management goals. Despite the lack of a consensus as to specific measures of biodiversity, establishing a set of parameters as part of the development of a corporate biodiversity management program, appropriate to the scale and question at hand, is important for several reasons:

1. It allows for establishment of clear program goals (such as maximization of species richness)

2. While specific parameters may be debatable and subjective, program results can be evaluated objectively with respect to meeting or not meeting numeric criteria (but see caveat that follows)

3. It allows for hypothesis testing

4. It focuses data needs, thereby increasing efficiency of data collection and minimizing field costs

Despite our emphasis on the importance of selecting a specific suite of measurements for biodiversity, we do offer a caveat: As with any measurement of a complex biological system, the specific parameters must be biologically relevant and sensitive enough to answer the question at hand, rather than merely provide a false sense of rigor due to mathematical trappings.

In the case of the WMX program, the parameters established to determine if there is or is not a net loss of biodiversity on a facility (that is, compliance with the corporate policy) are based on floristics at the community and ecosystem levels. The vegetation of the facility is first quantitatively inventoried by plant community, according to the community classification system defined by the appropriate State Natural Heritage Program. From the inventory three specific parameters can be calculated: (1) the total number of plant communities represented on the facility, (2) the percentage of native plant

species found on the facility, and (3) Soerensen's Index of Similarity. The index is calculated from a list of the ten species considered indicative of each plant community (based on the literature or as defined by the Natural Heritage Program) and the ten species with the highest importance values (importance values equaling the combination of relative frequency, relative cover, and relative density) calculated from the inventory data. In general, if a facility exhibits a decrease in any one of the three parameters, it is considered to not be in compliance with the corporate policy (in other words, it has experienced a net loss of biodiversity). However, the management planning should recognize that under certain circumstances, the loss of an entire plant community in order to promote a different (and presumably rarer) community may increase biodiversity on a regional scale and should be encouraged. For example, in Wisconsin the elimination of wildfires has allowed many former oak savannas to undergo succession to mesic hardwood forests, yet many of the open-growth oaks remain. Removal of shrubs and reintroduction of fire may reverse the successional trend back to oak savanna, thus eliminating the mesic hardwood forest community but restoring the oak savanna community, a community considered by the Natural Heritage Program to be "globally endangered."

While the three parameters just described are the primary means of evaluating the level of biodiversity, other parameters are evaluated in the WMX approach via a geographic information system and trend analysis in order to identify the underlying causes of changes in any of the three primary parameters. These secondary parameters include the relative degree of habitat fragmentation (such as community patch size and contiguousness), the total area of each community, the relative abundance of bird species, and the relative abundance of mammal species (Klatt et al. 1993).

Regardless of the parameters selected, our experience with WMX leads us to suggest that a corporate biodiversity management program contain the following elements:

1. An appropriate array of parameters indicative of biodiversity

2. Specific data collection procedures (a biodiversity inventory phase)

3. A management plan based on the results of the inventory

4. A monitoring plan to evaluate the effectiveness of the management plan

5. A feed-back mechanism to allow adjustment of the management plan based on monitoring results

6. Written guidance with respect to the five previous points

Benefits, Costs, and Limitations

The experience of the PCEQ demonstrates clearly that some corporations are already integrating biodiversity preservation and enhancement into their land use management planning. If such planning is not a regulatory requirement, the question can be asked, Why are these corporations doing this? What is the benefit to the corporation? One reason why corporations are engaging in such planning is that some individuals managing corporations are, in fact, personally concerned about the environment and are in a position to act on those concerns. However, there can be additional, tangible benefits to the corporation (Klatt 1994).

Biological inventories of undeveloped land can reveal the presence of federal- or state-protected species or other rare resources that may be of great concern to regulatory agencies. From a corporate strategic-planning standpoint, knowledge of the distribution of rare resources among facilities may allow the corporation to adjust planned expansions among facilities, moving operations or planned expansions from facilities unusually constrained by protected resources. Identifying such species prior to project development can reduce costs substantially by allowing for timely permit application preparation and agency coordination. This in turn can help avoid project delays and allow for well-thought-out design modifications that maximize facility operations yet avoid impacts to the species or resource. Thus, a biodiversity baseline inventory not required by regulation may actually aid in regulatory compliance at a future date.

The establishment of a biodiversity program will also aid the corporation in demonstrating to the public and regulators its ability and willingness to manage its facilities in an environmentally sound manner. Such demonstration can foster an overall climate of goodwill and cooperation among the parties, especially if local conservation groups or universities are invited to participate in the program. Their participation may be in the form of input to the planning process, participation in the inventory effort, or utilization of study sites and research opportunities. Such cooperation can help put the corporate program into a regional planning context and thereby increase the ecological benefits of such efforts, such as maximizing the corporate holdings as part of a metapopulation.

While these are clear benefits to the corporation, it should also be kept in mind that there are also risks and costs to the corporation. These may be either direct costs such as commitment of staff time, expense of consultants, capital costs, supply costs, and maintenance costs—that affect the corporation's bottom line, or they may be costs that affect potential profitability due to lost opportunity, such as unusable lands and schedule impacts. In addition to direct and indirect costs, there may also be hidden risks such as increased

public scrutiny and the potential of negative publicity from a failed project. By reintroducing protected species onto their lands, corporations may subject themselves to increased regulatory burden in the future (see McDonald, Chapter Four). Well-intentioned regulators may incorporate voluntary long-term management efforts into permit conditions or closure/post-closure plans, thereby putting the corporation at risk of enforcement actions (court orders, fines, and so on) for deviation from a plan. Similarly, there is also a danger that when some corporations begin to make efforts of a voluntary nature, lawmakers may make such efforts a requirement for all corporations, a phenomenon known as "raising the baseline." However, some of these risks could be avoided by government enactment of the incentives (such as mitigation credits, public ownership of restoration sites, and so on) discussed previously in the section on long-term commitment.

Opportunity costs can be minimized by ongoing corporate conservation programs utilizing innovative approaches, such as using existing wetlands, regulatory setback areas, flood plains, and zoned conservancy lands for enhancement projects. Direct costs can be minimized by use of temporary staff (such as summer interns) and consultants in lieu of permanent staff. Long-term risks and costs due to maintenance and monitoring of sites can be minimized by deeding biodiversity preserve areas to conservation agencies, universities, or colleges for continued preservation or research areas, and by regulatory agencies agreeing not to increase the regulatory burden on facilities in consideration of voluntary efforts by the corporation. We are sure that other cost-containment techniques and economic benefits can be developed.

Conclusions

At the beginning of the chapter we stated that the concept embodied in the chapter title may strike some as a conceptual contradiction, given that changing land use due to socioeconomic development has been a primary cause for the loss of biodiversity. We hope we have convinced the reader that our chapter title is not a contradiction and that, despite the risk and costs described in the previous section, many corporations are playing an active role in conservation.

The opportunity for biodiversity conservation represented by corporately owned lands is significant. For example, in the seven-county area covered by the Southeast Wisconsin Regional Planning Commission (SEWRPC), 141,000 acres are used for commercial, industrial, transportation, communication, and utility activities (SEWRPC 1992). This represents 8.4 percent of the total land area available in the seven county area. While only some portion may be available for natural resource management, this is still a substan-

tial resource in a largely urban area. Similarly, over 97 million acres of U.S. forestland are owned by industrial concerns, and over 331 million acres are owned by nonindustrial private owners, for a total of 58.7 percent of all U.S. forestland (Society of American Foresters 1991). By anyone's measure, even a fraction of the hundreds of millions of acres of privately held forestland must be considered a potentially significant resource.

The widely held tenet that conservation and development are diametrically opposed concepts has sometimes resulted in rhetoric and isolationism between "ivory towers" and the "real world." We have tried to bridge that gap by demonstrating via example and argument that these concepts are not mutually exclusive. Along with Soulé (1986), we would argue that such disciplinary hierarchy and isolation are counterproductive, and that the only "real world" is the whole that comprises all of the parts. The different parts operate in different cultural milieus, and both conservationists and corporations may need to undergo a paradigm shift when working together (Cairns 1987), but when the seemingly different viewpoints are reconciled, the resulting synergism can constitute a major step toward biodiversity conservation and, in the larger sense, a sustainable economy.

REFERENCES

Cairns, J., Jr. 1987. "Disturbed Ecosystems As Opportunities for Research in Restoration Ecology." In *Restoration Ecology: A Synthetic Approach to Ecological Research.* Edited by W. R. Jordan III, M. E. Gilpin, and J. D. Aber. Cambridge, England: Cambridge University Press.

Carson, R. 1962. *Silent Spring.* Boston: Houghton Mifflin.

Diamond, J. M. 1984. "'Normal' Extinctions of Isolated Populations." In *Extinctions.* Edited by M. H. Nitecki. Chicago: University of Chicago Press.

Ehrlich, P. R., and A. H. Ehrlich. 1970. *Population, Resources, Environment: Issues in Human Ecology.* San Francisco: W. H. Freeman.

Hawken, P. 1993. *The Ecology of Commerce: A Declaration of Sustainability.* New York: HarperCollins.

Klatt, B. J., 1994. Biodiversity Management: A Comprehensive Approach to Natural Resource Management Issues at WMX Technologies, Inc. Facilities." In *WasteTech '94*, Technical Proceedings of the Annual Meeting of the National Solid Waste Management Association, January, Charleston, South Carolina.

Klatt, B. J., L. Neal, J. Merino, S. Rowe-Krumdick, and J. Beacham. 1993. *Waste Management Incorporated Biodiversity Management Guideline.* Available through RUST Environment and Infrastructure, 4738 North 40th Street, Sheboygan, Wis. 53083.

Machlis, G. E., and D. J. Forester. Forthcoming. "The Relationship Between Socioeconomic Factors and Biodiversity Loss: First Efforts at Theoretical and Quantitative Models." Presented at Biodiversity in Managed Landscapes: Theory and Practice. 13–17 July, 1992, Sacramento, Calif.

Pimm, S. L., and M. E. Gilpin. 1989. "Theoretical Issues in Conservation Biology." In *Per-
    spectives in Ecological Theory*. Edited by J. Roughgarden, R. M. May, and S. A. Levin.
    Princeton: Princeton University Press.

Society of American Foresters. 1991. *Task Force Report on Biological Diversity in Forest
    Ecosystems*. Bethesda, Md.: Society of American Foresters.

Soulé, M. E. 1986. "Conservation Biology and the 'Real World.'" In *Conservation Biology:
    The Science of Scarcity and Diversity*. Edited by M. E. Soulé. Sunderland, Mass.: Sin-
    auer Associates.

SEWRPC (Southeast Wisconsin Regional Planning Commission). 1992. "A Regional
    Land Use Plan for Southeastern Wisconsin—2010." Planning Report No. 40.
    Waukesha, Wis.: SEWRPC.

WMI (Waste Management, Incorporated). 1990. *1990 Annual Environmental Report*. Oak
    Brook, Ill.: Waste Management Incorporated.

CHAPTER 16

New Directions for Rare Plant Mitigation Policy

Michael J. Bean

—————◆◆◆◆◆◆—————

This chapter examines the role of endangered plant mitigation, particularly under the Endangered Species Act. As part of that examination, it first describes the origins of mitigation concepts and policy then examines the application of those concepts and policies to endangered plants. Concluding sections explore a variety of proposals now pending in Congress and offer several suggestions for new directions to guide endangered plant mitigation under the Endangered Species Act.

The Wetlands Origin of Mitigation Policy

The concept of environmental mitigation has been most fully explored in the context of regulating development in wetlands (Leslie 1990). As a result of the Fish and Wildlife Coordination Act and Section 404 of the Clean Water Act, most development activities undertaken or permitted by federal agencies in wetlands and other aquatic environments are subject to mitigation requirements aimed at minimizing or compensating for the adverse effects of such activities on wildlife or on the environment generally. The history of wetlands mitigation policy reflects changing perceptions of what works and what doesn't and, more fundamentally, what the very aim of mitigation should be.

In the wetlands context, mitigation quickly became discredited in the eyes of many because it provided cover for a continued erosion—or a "net loss"—of environmental values. One of the earliest mitigation strategies in the wetlands context was to require that the destruction of one area of wetlands be "mitigated" by the transfer of another area of wetlands to public ownership, usually to a state or federal conservation agency. The most apparent result of this strategy was a net loss of wetlands, because the mitigation didn't create or restore any wetlands to offset the loss; it simply transferred ownership of some wetlands in exchange for the destruction of others. The final result was that

the total remaining acreage of wetlands was reduced. This was particularly galling to those who believed that the various federal and state wetland protection laws were capable of ensuring substantial protection to wetlands in private ownership—in short, that no transfer to public ownership was necessary to ensure wetland protection.

While this mitigation strategy clearly did result in a short-term net loss of wetland acreage, its proponents could nevertheless argue with some force that in the long term it resulted in more acres of wetlands being protected—and often actively managed for their conservation value—than would otherwise have been the case. The premise of this argument was that, in fact, the regulatory controls over privately owned wetlands are inadequate to ensure their long-term protection. Legal loopholes, weak enforcement, political interference, and other factors invariably undermine the effectiveness of those regulatory programs. In short, the security they offer to wetlands in private ownership is largely illusory. This argument, regrettably, had considerable historical experience to support it.

Putting aside the question whether public acquisition of existing wetlands truly "mitigated" losses elsewhere, there remained the question of how much mitigation should be required in any given situation. Is it possible to articulate and maintain a principled set of standards by which to measure an appropriate level of mitigation? In practice, setting mitigation requirements has been only partly a matter of biology. It also has often been partly a matter of economics: How much mitigation expense is it reasonable to ask this particular landowner to bear? It has been partly a matter of precedent: How did we handle this the last time? And it has been partly a matter of political calculation: How much can we get away with? Rarely, if ever, are mitigation requirements considered without at least some of these decidedly nonbiological questions. Is it possible to construct an approach to mitigation that is more principled and more biologically based?

Today, proponents of wetlands mitigation policy nominally seek to do so by embracing a goal of "no net loss" of either wetland acreage or functional values. The goal of "no net loss" of wetlands can be traced to the 1988 report of the National Wetlands Policy Forum, a group of environmental, business, and governmental leaders convened by the Conservation Foundation (Conservation Foundation 1988). As that report recognized, such a goal can be met only in one of two ways. Either no destruction of any remaining wetlands can be allowed, or the destruction of such wetlands must be offset by the restoration or creation of other wetlands. Since a blanket prohibition against any further loss of existing wetlands does not exist and is not likely to be adopted soon, the role of mitigation is to minimize losses of existing wetlands while trying to compensate for unavoidable losses with restoration or re-creation. While considerable debate rages over whether long-term success in wetland

creation and restoration can be demonstrated (Erwin 1991; Roberts 1993), the goal of "no net loss" remains in vogue.

Wetlands Mitigation Banking

One of the more interesting wetland mitigation concepts currently receiving attention is that of "mitigation banking" (Environmental Law Institute 1993; Short 1988). In essence, this refers to the practice of performing mitigation, usually in the form of wetland restoration, in advance of carrying out the harmful activities that give rise to a mitigation obligation. The result is some number of mitigation credits (such as wetland acres restored) that can be put in a "bank" and used at a later time to mitigate subsequent development. A single company or government agency (such as a state highway department) can establish a mitigation bank for its own exclusive use. Alternatively, a private entrepreneur can create a bank in the hope of later selling credits in a free market to the highest bidder (which may include conservation or development interests). Banks offer a number of theoretical benefits. Mitigation is performed in advance of development, thereby providing an opportunity to assess the success of mitigation efforts before development is allowed. Banks also offer a means of pooling mitigation efforts so that a large-scale planned restoration effort can be undertaken as mitigation for many small, unconnected development impacts occurring at different times.

Mitigation banking, however, is not without its difficulties. It suffers from the same uncertainties about long-term restoration success as other forms of mitigation. The availability of banks may lead regulatory agencies to accept wetland losses that could have been avoided. Thus, the legally mandated "sequencing" of mitigation steps (first avoid, then minimize, and finally compensate) may be undermined. Finally, mitigation banks require a common "currency" by which the "credits" earned by mitigation efforts and the "debits" to be incurred as a result of development are measured. Of necessity, that currency can only roughly approximate the environmental values lost or gained. Unique or unusual wetland values cannot easily be reflected in such currency. Nevertheless, wetlands mitigation banking appears likely to receive substantially greater attention in the near future (Shabman 1993; Marsh and Acker 1992).

Mitigating Losses of Endangered Plants

The effort to apply wetland mitigation principles to endangered plants immediately encounters a formidable obstacle. Whereas a duty to mitigate

wetland losses arises from the authority of the federal government to prohibit wetland filling altogether, no comparable federal authority prohibiting the destruction of endangered plants exists. Under the Endangered Species Act, endangered and threatened plants on private land receive virtually no protection. A landowner is free to apply herbicides to them, disc them under, or otherwise destroy them. Only when a landowner's action is subject to a federal permit requirement or receives federal funding does the Endangered Species Act potentially restrain what the landowner can do. (The only exceptions to this statement are those few states, such as Hawaii, that prohibit landowners from taking or destroying endangered plants on their own property; in those states the Endangered Species Act restricts private actions in the same manner as the state law.) Even then, landowners are restrained only if the consequences of their actions are sufficiently serious to jeopardize the survival of the species. If the loss of some or all of the plants on the property does not cause jeopardy to the species, the landowner is under no duty to mitigate that loss.

This situation is quite different from that which applies to endangered animals. Any taking of an endangered animal is prohibited, subject to two exceptions (Bean 1983). If the taking occurs incidental to an action authorized, funded, or carried out by a federal agency, it may be authorized, subject to such reasonable and prudent measure as will minimize its impact. If there is no federal nexus with the activity, taking is prohibited unless a permit expressly authorizing it has been granted. To secure a permit for the taking of an endangered animal incidental to some development activity, the development proponent must prepare a conservation plan that "minimizes and mitigates" such incidental taking. None of these safeguards apply to plants, because there is no general prohibition against the taking of plants. The much more limited protection that endangered plants receive must be kept clearly in mind when considering possible mitigation strategies for endangered plants. Under the Endangered Species Act, a duty to mitigate the loss of endangered plants exists only if a federal action so detrimentally affects the species as to jeopardize its continued existence. In those circumstances, Section 7 of the Act requires a federal action to find some alternative that avoids the prohibited jeopardy. If the federal action is not so significant as to jeopardize the species, however, the federal agency has no further duty, even if some loss of endangered plants will occur incidentally to its action.

Not uncommonly, the project alternatives to avoid the prohibited jeopardy finding include relocation, enhanced protection of other populations on the project site, and artificial propagation—in short, the "alternative" becomes the project as planned, plus mitigation. Where populations of the affected species occur on private lands, their public acquisition or other appropriate protection may be considered as mitigation for the loss of populations (or individuals) elsewhere. To characterize this as mitigation may give rise to the

same sort of qualms that were voiced about wetlands acquisition as a mitigation strategy, but it shouldn't. Endangered plant populations on private lands enjoy no security. Public acquisition (if coupled with a commitment to conservation management) or other forms of protection confer otherwise absent security. As a mitigation strategy, therefore, acquisition is the most direct, most immediate, and most certain way to offset rare plant losses elsewhere.

Where other mitigation techniques are employed, particularly relocation or enhancement measures, it may be appropriate to consider setting performance standards. Such standards would require not merely that the mitigation action be undertaken (that is, that the plants be relocated) but that some predetermined measure of success be met. Project approval could be based on meeting those standards, with failure to do so either reopening the mitigation requirement or triggering the forfeiture of security bonds. There are no legal obstacles to designing mitigation requirements of this sort, but there are practical obstacles. Developers typically want to know what they must do to secure project approval; they rely upon the government to identify measures expected to mitigate adverse impacts. If they do what the government asks and those measures do not in fact work, they don't want to reopen the question of their mitigation responsibilities. Further, they don't want to postpone moving forward with their projects for the long period of time that many scientists would find necessary to draw meaningful conclusions about the success of mitigation. In short, mitigation agreements of the sort described here are likely to be exceedingly difficult to negotiate.

Pending Proposals Affecting Endangered Plant Conservation and Mitigation

Uncertainty about the future of mitigation policy for endangered plants is compounded by the uncertainty about the direction of future endangered species conservation policy generally. By 1993, controversy over the Endangered Species Act had reached unprecedented proportions. Conflicts between spotted owls and the timber industry in the Pacific Northwest, California gnatcatchers and real estate development interests in Southern California, irrigation farmers and aquifer-dependent species in west Texas, and other species and interest groups elsewhere had sharply polarized political debate about the act's future. Numerous proposals for major amendments to the act had been introduced in Congress. While the outcome of those various proposals cannot be foretold at the time of this writing, a variety of alternative approaches to mitigation are evident from the many proposals.

Although they agree on little else, both critics and supporters of the Endangered Species Act agree that they would like to create positive incentives for the conservation of endangered species on private lands. The need for

incentives is clear. At present, the Endangered Species Act relies exclusively upon the "stick" of penalties and prohibitions to deter harmful conduct; it offers no "carrots" to induce or reward beneficial conduct. Although the stick approach has accomplished some useful results, its limitations are real. First, where active management is needed to control introduced organisms, maintain a desired seral stage, or carry out other needed actions, the stick approach cannot compel such management. Second, the use of penalties and prohibitions has fueled an antiendangered species backlash that undermines public support for the effort and threatens the political future of the Endangered Species Act.

To create incentives, one approach is simply to authorize the Secretary of the Interior to enter into agreements with private landowners in which the secretary pays the landowner to carry out management measures not already required by law but necessary for the recovery of the affected species. The implications of such incentives provisions for endangered plant mitigation are significant. They would create, for the first time, meaningful opportunities for plant mitigation on private land. At present, when federal or federally permitted projects jeopardize the survival of an endangered plant, the mitigation that may be required as a means of avoiding that jeopardy is likely to be carried out exclusively on public land. This is because nothing in the Endangered Species Act offers any assurance that endangered plants will be maintained on private land. Without that assurance, it makes little sense to mitigate for endangered plant losses by relocating, enhancing, or otherwise managing endangered plants on private land. With the sorts of incentive provisions proposed, private landowners might be persuaded to offer the sorts of assurances necessary to make plant mitigation on private land worthwhile.

There is also congressional interest in encouraging "market-based incentives" for endangered species conservation. This includes experimentation with "transferable credit," "mitigation banking," and other ideas being developed in other environmental contexts. The transferable credit idea would allow one landowner to earn "credits" for commitments he makes to endangered species conservation on his land and to sell those credits to another landowner who needs them to mitigate endangered species losses associated with a development project that he undertakes. "Mitigation banking" is a variation on the same idea, allowing someone to save up, or "bank" credits now in anticipation of needing or selling them later. Both ideas could expand the potential for endangered plant mitigation on private land. Private landowners, who now have no incentive other than good citizenship to preserve the rare plants that occur on their lands, might be able to realize economic benefits under such approaches.

The difficulty of translating these market-based measures (which were developed in wetlands or traditional pollution contexts) to the endangered

species context is considerable. A ton of sulphur dioxide emissions is a more or less fungible thing. An acre of red maple swamp is less so, but at least there are models that purport to evaluate the habitat values associated with individual parcels. Endangered species and their habitats, on the other hand, seem far more likely to entail such site-specific values as to make the development of a common currency for facilitating trades exceedingly difficult. Nevertheless, such site-specific evaluations are inherent in making judgments about the adequacy of mitigation, whether the mitigation is done at the time of project evaluation or as part of a mitigation bank for future projects.

Another idea receiving congressional consideration would require mitigation for endangered plant losses incidental to federal actions. At present, if a federal action does not jeopardize the survival of an endangered plant, it can go forward without any mitigation requirements, even though some losses of protected plants will occur. As a result, the status of endangered plants can be steadily eroded away through these incidental losses, at least until the plant is so imperiled that any further losses cross the threshold of jeopardy. Thus, without any duty to mitigate such losses, federal actions can undermine the act's goal of recovery.

An amendment is under consideration that would require federal agencies to mitigate for the incidental loss of endangered plants in connection with actions they authorize or carry out. If a standard of "no net loss" to the survival prospects for the species governs such mitigation, then at least the status quo, if not the goal of recovery, can be served. Such an amendment would potentially open up major new opportunities for rare plant mitigation.

Several other ideas in Congress have potential implications for endangered plant mitigation. One such idea, for example, would emphasize captive breeding (and, implicitly, artificial propagation for endangered plants) as a preferred strategy for endangered wildlife. Some of the proponents of that idea, however, would also restrict opportunities for release back into the wild of captive-bred organisms. The net effect is to promote *ex situ* conservation strategies over *in situ* ones. Moreover, the presence of secure *ex situ* populations may undermine legal protection for remaining *in situ* populations. Already, the U.S. Fish and Wildlife Service treats reintroduced populations of red wolves and black-footed ferrets as nonessential, because growing captive populations of both exist. It is not a huge leap from that foundation to the conclusion that wild natural populations of a species that also exists in captivity are not "essential" and therefore subject to less protection than they would receive in the absence of captive populations.

Still other proposals in Congress would encourage a study of tax impacts on species conservation. Ideally, this might lead to tax breaks for landowners willing to commit to beneficial actions, including mitigation actions.

Although not directly concerned with mitigation, proposals to require

compensation to landowners for reduced land values stemming from endangered species regulation could have profound practical impact on the future of mitigation policy. The U.S. Constitution specifies that private property may not be taken for a public purpose without the payment of just compensation. Although in the first two decades of the Endangered Species Act's history there has never been an instance in which any court has found that a landowner's property has been taken by virtue of the Endangered Species Act, there are nevertheless many proposals in Congress to require that landowners be compensated when the effect of endangered species regulation is to reduce land value below some specified percentage of its fair market value. As a practical matter, such a requirement would likely fix the cost of permissible mitigation below the level at which compensation were owed. Thus, mitigation requirements would be constrained by legislatively determined cost considerations rather than determined by biological judgments about the actual impact of a proposed action on an endangered species.

Summary

An assessment of endangered plant mitigation experience to date suggests the need for the following policies:

1. If the present policy of not regulating private land use to avoid adverse impacts to endangered plants continues, the most effective mitigation strategy is likely to be public land acquisition.

2. If private land use regulation is to be pursued, it should be accompanied by incentive programs that reward landowners for beneficial actions, and not merely punish them for harmful actions.

3. Other chapters presented in this book make abundantly clear how little is known yet about the success of endangered plant reintroduction efforts. It is essential to start filling those information gaps with carefully designed research projects as part of endangered plant recovery plans.

4. Where mitigation requirements are imposed, they should clearly specify not only the actions to be undertaken but the results anticipated. If the results achieved do not match those anticipated, the mitigation requirements should be reopened. In this way, all parties have a stake in the success of the mitigation effort.

5. Again, because of the major uncertainties about mitigation techniques, mitigation ought not to be done unless it includes an experimental component, designed to test hypotheses about the efficacy of mitigation techniques.

REFERENCES

Bean, M. J. 1983. *The Evolution of National Wildlife Law*. New York: Praeger.

Conservation Foundation. 1988. *Protecting America's Wetlands: An Action Agenda*. The Final Report of the National Wetlands Policy Forum. Washington, D.C.

Environmental Law Institute. 1993. *Wetland Mitigation Banking*, Washington, D.C.: Environmental Law Institute.

Erwin, K. L. 1991. "An Evaluation of Wetland Mitigation in the South Florida Water District." Vol. 1. A report to the South Florida Water Management District. West Palm Beach, FL.

Leslie, M. 1990. "Mitigation Policy." In *Issues in Wetlands Protection*. Background Papers Prepared for the National Wetlands Policy Forum. Washington, D.C.

Marsh, L. and D. R. Acker. 1992. "Mitigation Banking on a Wider Plane." *National Wetlands Newsletter* Jan.–Feb., 8–9.

Roberts, L. 1993. "Wetlands Trading Is a Loser's Game, Say Ecologists." *Science*. 260: 1890–1892.

Shabman, L., P. Scodari, and D. King. 1993. "Expanding Opportunities for Successful Wetland Mitigation: The Private Credit Market Alternative." Staff Paper SP-93-5. Virginia Polytechnic Institute and State University, Blacksburg, VA. Department of Agricultural and Applied Economics.

Short, C. 1988. "Mitigation Banking." *Biological Report* 88 (41). 97 pp., US Fish and Wildlife Service.

FOCUS
Rare Plant Mitigation in Florida

George D. Gann and Noel L. Gerson

⟶✦◦✦⟵

The central question of this chapter is whether or not mitigation, as it is currently practiced in Florida, can function as a mechanism to provide substantial protection to natural communities or provide a venue for the successful reintroduction of endangered species.

In Florida, many endangered plant species are concentrated in relatively unprotected uplands. Most regulatory protection, however, is focused on poorly drained habitats such as wetlands. The ironic result is that development—including compensatory mitigation—is being pushed into the very areas that provide habitat for endangered species. We explore this dynamic and the general impact of land use controls with specific reference to Florida, although many of the issues we touch on will have broader applications. While many regulations have changed since the initial draft of this chapter, we believe that the underlying considerations for endangered species in a mitigation context remain the same.

Natural History of Florida's Rare Plants

Florida is among the more diverse states in terms of plant diversity, along with Hawaii, California, and Texas. Although there is no published flora of the state, several authors have estimated plant diversity. For example, Ward (1990) estimated the potential number of vascular plant species in the state as approximately 3,489, using J. K. Small's *Manual of the Southeastern Flora* (1933) and his *Ferns of the Southeastern States* (1938). Small, however, was a taxonomic "splitter," and many of the species recognized by him have since

been reduced to an infraspecific level or synonymy. On the other hand, new taxa have been described for the state since the time of Small, including newly discovered endemics, range extensions, and exotic species recently naturalized and added to the flora. A few species have also been extirpated. Utilizing a checklist of the state's species initiated in the early 1960s, Ward (1990) recognized 3,448 species of plants in the Florida vascular flora. Of these species, Ward classified 2,523 (73 percent) as native and 925 (27 percent) as introduced to the state.

Estimates of the number of endemic species in Florida have also varied. Based on Small's *Manual of the Southeastern Flora*, Harper (1949a, 1949b, 1950) listed 427 endemic species for the state, approximately one-sixth of the state flora. Muller et al. (1989) listed 235 endemic vascular plants and 40 near endemics for the state, or approximately 8 percent of the total flora and 11 percent of the native flora.

Like many other parts of the United States, the original flora of Florida has been negatively affected by anthropogenic disturbances. As a consequence, many species have become endangered. The State of Florida currently lists 255 species as endangered and over 300 species as threatened or commercially exploited (Coile 1993; Florida Natural Areas Inventory 1993; Wood 1994). Fifty-three of the state-listed endangered species are also listed as endangered or threatened under the U.S. Endangered Species Act (Coile 1993).

Myriad habitat types support endangered species in Florida. While the topography is relatively flat and somewhat monotonous to the untrained eye, Florida was historically composed of a variety of interesting and rare natural communities. Wetlands once constituted more than half of the state (Shaw and Fredine 1956), with swamps widely distributed throughout (Ewel, K. 1990) and marshes dominating in the southern half of the peninsula (Kushlan 1990). Pine flatwoods and dry prairies were the most extensive terrestrial ecosystem in Florida and were found virtually throughout the state (Abrahamson and Hartnett 1990). Along the coasts were coastal marshes and swamps and coastal uplands (beach-dunes, coastal strand, and maritime hammocks). Other major terrestrial ecosystems included high pine (clayhill and sandhill) and temperate hardwood forests in the northern three-quarters of the state (Myers 1990; Platt and Schwartz 1990). Scattered in small patches throughout the same area was scrub, a fire-maintained xeromorphic shrub community that is nearly endemic to Florida. South Florida rocklands (including pine rockland and rockland hammock) were found exclusively in the Florida Keys, along a thin outcropping of limestone in Dade County and within the Big Cypress Swamp (Snyder, Herndon, and Robertson 1990).

Development pressures have affected portions of northern Florida for over four-hundred years; they began affecting the interior of the state over two

hundred years ago and proceeded in the southernmost portions of the peninsula about one hundred years ago (Ewel, J. 1990; Smiley 1973). Major historic anthropogenic disturbances have resulted from logging and agriculture in the northern portions of the state, conversion to citrus in central Florida, and drainage in southern Florida (Ewel, J. 1990). Today, one can add urban and suburban development and a plethora of other activities, from phosphate mining to fish farming. The result of these pressures is that over 90 percent of the high pine community has been lost (Myers 1990), along with approximately 85 percent of the scrub (Stap 1994), over half of the wetlands (Ewel, K. 1990), and at least 20 percent of coastal uplands (Johnson and Barbour 1990). Perhaps as significant for endangered species is that natural communities within areas of concentrated endemism and rarity, such as the Lake Wales Ridge and the Miami Rock Ridge, have been all but obliterated.

While endangered species are found throughout the state (Table 17-1) and in every major habitat type (Table 17-2), certain habitats support far more endangered species than others. South Florida rockland, for instance, is a primary habitat for 30 percent of the state-listed endangered species, while scrub and high pine support 43 percent of the endangered endemics.

Legal Protection of Endangered Species

Two laws provide endangered plant species with direct protection in Florida (see Appendix). The Regulated Plant Index (Chapter 5B-40, Florida Administrative Code) lists endangered, threatened, and commercially exploited species. The Florida Department of Agriculture and Consumer Services requires both a permit and written permission from the landowner in order to remove endangered or commercially exploited plants or plant parts from any

TABLE 17-1. Number of state-listed endangered species by region. Grand totals are less than the sum of the components due to overlap in ranges.

Region	Non-endemic natives	Endemics	Total
South	84	36	120
Central	27	52	79
North	78	31	109
Grand total	162	93	255

Sources: Clewell 1985; Coile 1993; Godfrey 1988; Godfrey and Wooten 1979; Godfrey and Wooten 1981; Lakela and Long 1976; Long and Lakela 1971; Luer 1972; Wunderlin 1982

TABLE 17-2   State-listed endangered species by habitat. Subtotals and grand totals are less than the sum of the components due to overlap in habitats.

Habitat	Non-endemic natives	Endemics	Total
Uplands			
Coastal uplands	19	16	35
Hardwood forests	55	8	63
Pine flatwoods	5	4	9
Scrub and high pine	9	40	49
South Florida rocklands	56	20	76
Subtotal for uplands	119	69	188
Wetlands			
Edges or banks of lakes, ponds, rivers, and streams	29	7	36
Swamps and bogs	50	13	63
Wet hardwood forests	25	12	37
Wet pine flatwoods	4	21	25
Wet prairies and marshes	5	11	16
Subtotal for wetlands	75	36	111
GRAND TOTAL	162	93	255

Sources: Clewell 1985; Coile 1993; Long and Lakela 1971; Luer 1972; Wunderlin 1982

property. This law, however, does not prevent a landowner from destroying endangered plants on his or her own property. Moreover, the Regulated Plant Index must be used solely for the purposes specified in Florida Statutes 581.186(3)—to regulate the harvesting of plants—and may not be used for regulatory purposes by other state agencies. However, this does not preclude another agency authorized to protect endangered plants from including one or more of the species listed on the Regulated Plant Index under its own regulatory authorities. Finally, the index fails to list a significant number of the imperiled and critically imperiled species identified by the Florida Natural Areas Inventory (Coile 1993; Florida Natural Areas Inventory 1993).

The other major law specifically protecting endangered plant species in Florida is the U.S. Endangered Species Act, which provides protection for endangered plants on federal property or when federal funds are involved. All other laws that provide protection for endangered plant species in Florida do so as a secondary effect. Endangered plant species, for instance, may be protected through the establishment of a preserve for a protected animal (such as the Florida scrub jay [*Aphelocoma coerulescens*]), through the protection of a regulated plant community (such as pine rockland), or through the protection of natural vegetation for erosion-control purposes on barrier islands.

Thus much of the protection of endangered plant species is derived as an artifact of regulation of natural communities and the process of mitigation, which is required when all or a portion of a regulated natural community is developed on a site.

Compensatory Mitigation: Background

The concept of mitigation was originated (and developed) as a way to reconcile divergent societal views on land and water use. It is a complex and highly charged political issue that revolves around the trade-offs between short-term economic interests and the long-term (but equally important) need to protect our natural resource base.

In the mid 1960s, public interest in environmental protection grew as the escalating impact of residential and commercial development on land, water, and air became increasingly obvious. Public concern led to the enactment of several new environmental laws, including the National Environmental Protection Act (NEPA) in 1969. The NEPA "requires federal agencies to evaluate the potential impacts of major federal actions on the environment" (Sheldon 1993, p. 304). The term *mitigation* was defined in NEPA to include "a) avoiding the impact altogether by not taking a certain action or parts of an action; b) minimizing impacts by limiting the degree or magnitude of the action and its implementation; c) rectifying the impact by repairing, rehabilitating, or restoring the affected environment; d) reducing or eliminating the impact over time by preservation and maintenance operations during the life of the action; and e) compensating for the impact by replacing or providing substitute resources or environment" (40 Congressional Federal Register Part 1508.20[a–e]).

Based on NEPA and other authorities, mitigation policies have been developed by national, state, and local agencies to protect wetlands and, to a lesser extent, other ecosystems at risk from development pressures. To date, much mitigation work has been compensatory in nature (replacing or providing substitute resources), although this approach was originally intended as a last-resort option. Criteria calling for avoiding and minimizing impacts to meet government environmental regulations prior to taking compensatory measures are unfortunately often bypassed.

In Florida, a variety of natural communities is regulated by law (see Appendix to this chapter). These include most wetlands and some uplands. In general, wetlands are heavily regulated, while most uplands in the state have little or no legal protection. Theoretically, the regulation of natural communities and the institutionalization of programs to mitigate for development within these regulated communities should provide substantial protection for

both natural communities and their commensal endangered plant species. But does the regulation of natural communities and the mitigation process really provide substantial levels of protection for natural communities and endangered species?

Mitigation in Florida: Is It Successful?

Evaluating the success of mitigation is a difficult task. For our purposes, we divide success criteria into three major areas: administrative success (whether mitigation performs what the law says it must); "no-net-loss" success (whether the total acreage of natural areas remains constant or increases); and technical success (whether mitigation areas, through creation, enhancement, or restoration, are equivalent or superior to the natural areas that have been destroyed). These criteria are evaluated below.

Administrative Success
Statistically, both the Clean Water Act (Section 404) and the State of Florida mitigation programs have poor mitigation track records. This was documented in *Report on Effectiveness of Mitigation in Florida*, submitted to the Florida Legislature and the governor on March 5, 1991 (Florida Department of Environmental Regulation 1991). The mitigation requirement for one-third of all permits issued had never been initiated, and the ecological success rate for the projects that were completed was 27 percent. Success rates at the county and municipal levels are more difficult to determine, as no comprehensive review of projects on the substate level exists.

Two of the main elements contributing to the lack of administrative success are a lack of understanding of the law and a lack of compliance with the law even when it is understood. Mitigation and the process that regulates it are, at present, complicated and rapidly evolving. The "myriad laws and federal agencies with some responsibility for wetlands is confusing to the regulated public" (Environmental Law Institute 1993, p. 5). The National Wildlife Federation (1989), for instance, identified thirty-four major federal laws and regulations affecting wetlands. When multiple state, county, and municipal agencies, laws, and regulations are added to the equation, along with the regulation of uplands, the situation can seem endlessly complex.

Agencies rely on the private sector to know the rules, and generally the larger development and consulting firms do. Yet many individual landowners may not understand how the law affects their property and may violate the law unintentionally. In addition, some owners—especially new ones—may simply be unaware that the natural systems on their properties are regulated. The Environmental Protection Agency (EPA) has tried to address some of

these problems with a general public outreach program and a toll-free hotline number to answer questions concerning wetlands regulation. Other agencies have sent out letters to owners, but, in general, the notification process is still poorly developed.

Part of the reason a coordinated notification process has not been established is that there are still many unanswered questions about regulated ecosystems. For instance, wetlands boundaries change over time, and the definition of what exactly constitutes a wetland is the subject of national debate. The U.S. Congress has gone so far as to ask the National Academy of Science to develop a definitive definition for use as a scientific and legislative guideline. Many agencies have struggled over similar issues concerning regulated uplands. Regulated ecosystems must be defined in scientific, measurable terms, and complex questions such as what degree of disturbance makes a system "unnatural" (and therefore unregulated) must be answered.

Even if landowners are aware that their properties are regulated, violations of the law may occur. Regulations may cause "expensive delays in private parties' planned activities" (Environmental Law Institute 1993, p. 5) and preclude certain activities that landowners wish to perform. As a consequence, some landowners choose to ignore the law. According to the National Wildlife Federation (1989, p. 25), "Many wetlands are filled before the necessary permits are obtained," and some of these violations are committed by those who know the law. Gann has witnessed several upland violations in southeastern Florida where the landowner clearly knew the law and chose to ignore it.

Even assuming that a landowner understands the law and chooses to comply with it, the process of acquiring and implementing mitigation permits in Florida can be confusing and convoluted. A developer with a large project that potentially affects wetlands might have to obtain permits from a municipality, a county, the Florida Department of Environmental Protection, a Water Management District, and the U.S. Army Corp of Engineers, as well as go through development of a regional impact process. If any endangered or threatened animals are found on site, permits from the Florida Game and Fresh Water Fish Commission and the U.S. Fish and Wildlife Service might be needed also. Mitigation, if required, would have to satisfy all of them, and coordinating with and satisfying each agency can be quite difficult. Florida has recently passed a streamlining bill that, when implemented, may help to facilitate the permitting process for wetlands.

Also hindering administrative success is the fact that many agencies are understaffed, and the emphasis is on permitting rather than enforcement. For example, during the 1993 fiscal year, the Florida Department of Environmental Protection Division of Water Management employed 41.5 people to issue permits and 17.5 people to enforce the law (P. Brown, Florida

Department of Environmental Protection, personal communication, April 1994). Between October 1, 1991, and September 30, 1992, the Department issued 1,639 permits and denied 192 permits. Ninety-seven permits were withdrawn. Another twenty-eight hundred permit applications were determined exempt from the law following review (Florida Department of Environmental Regulation 1993). Assuming relatively constant levels of staffing and permitting, this means that each staff person handling permitting processes an average of 114 new permits anually, and each person responsible for enforcement handles an average of ninety-four new projects per year.

Insufficient funding plagues most agencies and makes it difficult for them to carry out their mandates. For instance, the Florida Department of Environmental Regulation (now Department of Environmental Protection) has "consistently requested the resources necessary to secure contracts" to use satellite imagery techniques to assess the loss or gain of wetlands in the state (Florida Department of Environmental Regulation 1993, p. 36). But as of 1993, the requests had not proceeded beyond the department's preliminary budget. In addition, the department stated that inadequate personnel for permit compliance restricted its ability to conduct an analysis of the actual acreage and quality of wetlands successfully created in the field (Florida Department of Environmental Regulation 1993).

Finally, agencies face a difficult challenge training personnel. Mitigation, like ecological restoration, is a new and rapidly developing science and craft. Little formal training exists on a national or state level to prepare students to become mitigation experts, and limited information is available on "technical guidance that integrates research with practical experience" (Davis 1993, p. 156). While some fields such as biology, engineering, landscape architecture, environmental studies, and even political science may provide some of the skills necessary to produce high-quality mitigation on-the-ground, these skills alone do not prepare agency personnel for their tasks. As a consequence, agencies must provide substantial in-house training. This process is complicated by the rapid growth, high employee turnover, and insufficient staffing levels experienced by many agencies.

"No-Net-Loss" Policies

In 1988, President Bush issued a "no net loss" of wetlands policy: as of 1993 the U.S. Environmental Protection Agency was still in the process of defining "no net loss" on a national level with state implications (Florida Department of Environment Regulation 1993). Here we offer a review of "no net loss" from the perspective of both wetlands and uplands.

From 1985 through 1992, 24,468 acres of wetlands were destroyed, and 53,984 acres of wetlands were created as a result of the State of Florida's wet-

land permitting activities (Florida Department of Environmental Regulation 1993). These numbers, however, do not mean that "no net loss" of wetlands actually occurred. First, several activities that destroy wetlands are exempt from environmental permitting on both federal and state levels (Florida Department of Environmental Regulation 1993; National Wildlife Federation 1989). Second, activities such as the hydrological connection of isolated wetlands to waters of the state are considered to be wetlands creation (Florida Department of Environmental Regulation 1993). Third, some projects are double-counted when permitting is handled by more than one agency (Florida Department of Environmental Regulation 1993). Finally, the numbers reported by the state only indicate the acreage of wetlands required to be created, not the acreage of wetlands that are actually built (Florida Department of Environmental Regulation 1993). Thus, it is impossible to determine at this time whether there has been "no net loss" of wetlands in Florida since 1985.

In terms of uplands, it must first be reiterated that uplands are poorly protected in Florida. Second, the primary laws that protect uplands are local, and even the most stringent of these allow for the destruction of a certain percentage of regulated uplands. Dade County, for instance, which has one of the most sophisticated uplands ordinances in the state (Article 3, Section 24-60, Dade County Code), still allows 10 percent of rockland hammocks and 20 percent of pine rocklands on a site to be cleared without compensatory mitigation. If the natural area is less than 5 acres in size, slightly more area may be cleared, although avoidance measures and/or compensatory mitigation may be required. Martin County, which also has relatively strong uplands protection, allows three-quarters of "common" uplands to be destroyed during the development process without compensatory mitigation and slightly lower levels for "unique, rare, and endangered habitats" (Section 9-47.f, Martin County Comprehensive Plan). Several other counties, including Alachua, Broward, Collier, Hillsborough, Lee, Monroe, and Palm Beach have similar legislation, but none provides "no net loss" levels of protection.

Unfortunately, most of Florida's rarest plants occur in poorly protected uplands (Table 17-3). One hundred and eighty eight (or 74 percent) of the state-listed endangered plant species utilize uplands as primary habitat. Of these one hundred and forty four (or 56 percent) are upland obligates. While there is no direct evidence to test this hypothesis, it is possible that strong regulatory protection of wetlands has caused a shift in development patterns toward upland habitats, thus causing increasing pressure on upland areas containing the majority of endangered species.

While regulation of uplands lags behind that of wetlands, attempts are being made to rectify this imbalance. Florida's 1985 Growth Management Act requires all local governments to adopt a comprehensive plan, including

TABLE 17-3   State-listed endangered species by upland and wetland habitats. Endangered endemics are in parentheses. Totals are less than the sum of the components due to overlap in ranges.

	Habitats			
Region	Wetlands	Wetlands and uplands	Uplands	Total
South	24(6)	24(6)	72(24)	120(36)
Central	16(11)	8(1)	55(40)	79(52)
North	39(15)	19(5)	51(11)	109(31)
Total	67(24)	44(12)	144(57)	255(93)

Sources: Clewell 1985; Coile 1983; Godfrey 1988; Godfrey and Wooten 1979; Godfrey and Wooten 1981; Lakela and Long 1976; Long and Lakela 1971; Luer 1972; Wunderlin 1982

an assessment of jurisdictional natural resources (including uplands) and a conservation plan. Plans have been completed for all 67 counties and the 458 local governments in the state. Comprehensive plans, however, place different importance on upland resources; while the trend is clearly to provide more protection of uplands, there is still a long way to go in Florida if uplands and their commensal endangered species are to be preserved.

Technical Success

There has been much debate within the ecological restoration community over the conceptual and practical underpinnings of mitigation (Hackeling, Leach, and Apfelbaum 1990; Higgs 1993; Zentner 1992). Part of this debate concerns the ecological quality of newly created "ecosystems." Biological assessments to date indicate that many artificially constructed systems are inferior to or different from the natural communities they are meant to replace (see Zedler, Chapter Fourteen; Florida Department of Environmental Regulation 1991; Reiger 1992; Weller 1989; Zedler 1988; Zedler et al. 1989; Zedler and Langis 1991). These findings, however, are not universally accepted (Lewis 1992; Reiger 1991; Sacco, Booker, and Seneca. 1988; Tilton and Denison 1992; Zentner 1992). Further complicating the analysis of the technical results of mitigation is the elusive definition of success (Dunwiddie 1992). Even the Florida Department of Environmental Protection considers the majority of mitigation projects in the state to be ecological failures (Florida Department of Environmental Regulation 1991).

In Florida, there are two basic types of mitigation projects: (1) *creation projects*, in which ecosystems are created de novo and (2) *enhancement* or *restoration projects*, in which damaged natural systems are improved. While cre-

ation projects were historically more popular, enhancement and restoration projects are gaining more popularity in Florida today.

One of the earliest models of wetland creation mitigation projects was known as "upland-to-wetland" conversions, in which wetlands were created in places where they did not exist before. In some cases, intact upland natural communities (such as pine flatwoods) were converted to artificial wetlands to satisfy mitigation requirements. More common now are mitigation projects that replace complex wetlands with systems that are much simpler, such as plantings along littoral shelves on the edges of artificial lakes or ponds. Mary Kentula, head of EPA's Wetland Research Program in Corvalis, Oregon, has collected and analyzed data to support this assessment. Her research revealed that the only wetland type that is increasing in acreage in the country is an open-water pond with a fringe of wetland vegetation (Roberts 1993).

Contributing to problems with the technical success of mitigation projects are poor standards for project design. Because the creation, enhancement, and restoration of ecosystems is such a new science, technical specifications for mitigation projects have, in the past, been oversimplified almost to the point of resembling developed landscapes. As one individual reviewing mitigation projects in California put it, "While the appropriate dominant plant species may be present, we often cannot shake the feeling of an artificial, landscaped habitat" (Baird 1989, p. 64). In two projects the authors have been involved with, mitigation for the destruction of natural communities has been partially compensated for through the planting of native trees in rows within an artificial landscape.

Plant diversity requirements for mitigation creation projects have often been low, and species selection inappropriate. Designs have rarely called for historic accuracy at the species level, much less at the population or genetic level. In some cases, projects specify native plants well outside of their historic range. For instance, one mitigation project near Miami International Airport specified southern red cedar (*Juniperus silicicola*), a species whose nearest natural population (Little 1978) is 150 miles to the north of the project site. If the goal of mitigation is truly to restore natural systems and to offset development, then, at a minimum, appropriate species should be used. There is little consideration of the possible impacts that expanding the ranges of various species or genotypes might have on the ecological structure, function, and integrity of mitigation sites or nearby natural areas. The genetic composition of introduced stock, for instance, can influence the behavior of individuals, which in turn may affect the dynamics of the community and disrupt or alter the coevolutionary pathway within the community (Millar and Libby 1989).

The emphasis in Florida's mitigation work has been on instant results, although agencies are now beginning to encourage more flexible time lines to the mitigation planning and implementation process (A. Redmond, Florida

Department of Environmental Protection, personal communication, February 1994). Historically, mitigation projects have been treated very much like construction or landscaping projects. Mitigation plans are developed, contracts are let out to bid, and work is completed within a relatively short period of time. This has affected several areas of mitigation design, one of the most problematic of which was the availability of appropriate plant materials. Contract growing, which allows for the cultivation of appropriate species using local germplasm, has been rarely used within the mitigation industry. Instead, mitigation contractors are forced by tight time constraints to go onto the open market and purchase whatever plant materials happen to be available during the contract period.

Once mitigation creation projects are completed, the follow-up to ensure their success is usually weak. The guarantee or required management period (usually one to three years) falls far short of the time needed to determine ecological success (see Pavlik, Chapter Six). In addition, mitigation projects are monitored as a "product" rather than as a dynamic ecological community. For example, project "success" is commonly based on percent survival of outplanted individuals or, minimum percent cover of native species (or, conversely, maximum percent cover of exotic or invasive species). There is an assumption at the design stage that there will be some natural recruitment, but generally recruitment is not monitored. A more accurate measure of a project's success might be obtained if monitoring reports required the inclusion of information on recruitment, wildlife use, and other indicators of habitat quality.

Almost all projects in Florida require control of invasive species during the guarantee period. Once this period is over, however, few institutional mechanisms exist to ensure that completed mitigation projects do not become highly disturbed and invaded fragments. No mitigation site will truly be restored to a state of ecosystem stability within the relatively short project guarantee periods. When projects are "released," the only institutional leverage is protection from future development of the site; no mechanism to continue removing exotics or fostering restoration of the site generally exists.

As a consequence, true long-term preservation and management of mitigation sites is rare. Once the guarantee period is over, no management of the mitigation site is legally required. As a result, mitigation sites in Florida will predictably be invaded by exotic plant species, and overall site quality will almost certainly deteriorate. Even large, relatively well-protected natural areas in Florida suffer from extensive exotic invasion problems. Everglades National Park, for instance, has been invaded by at least 148 species of exotic plants (Avery and Loope 1980). In a recent review of plant lists from nine fragmented sites in southeastern Florida, an average of seventy-two species of exotic

plants—representing an average of 25 percent of the flora —has been recorded for each site (Gann, unpublished data).

Endangered Species Within the Mitigation Context

The protection of endangered species in Florida within the mitigation context can only be considered as a byproduct of the overall process. First, recall that the Regulated Plant Index cannot be used by state agencies with the authority to regulate natural communities and require mitigation during the development process. Although these agencies may develop their own lists of endangered species, they are not required to list all of the species on the index. Second, that even the Regulated Plant Index fails to list a substantial percentage of the species identified as imperiled or critically imperiled by the Florida Natural Areas Inventory. As a result of these two realities, mitigation agencies are not forced to account for many endangered species within the mitigation process.

Populations of endangered plant species are handled during the mitigation process in two basic ways: through avoidance of impacts and through intentional reintroduction or establishment of new populations as part of the mitigation process. Partial avoidance can be accomplished by limiting development and establishing preserves as part of the mitigation process, as well as attempts to relocate endangered species populations from within the development area to existing or newly created preserves. Properly conducted, reintroduction of endangered species involves planting these species back into natural areas where they historically occurred prior to anthropogenic disturbance.

Both the establishment of preserves to protect endangered species and the relocation of endangered species suffer from many general problems associated with mitigation. While preserves may be established to protect endangered species, long-term management and monitoring of populations is often sporadic. In addition, the violation of preserve areas (and the concurrent destruction of endangered species) sometimes occurs. The authors, for instance, have been involved in one case of a preserve established as part of a mitigation process in pine rocklands that was partly bulldozed to remove debris following Hurricane Andrew. This site contained six species listed as endangered on the Regulated Plant Index and twelve species listed as imperiled or critically imperiled by the Florida Natural Areas Inventory. Similar problems are encountered with endangered species reintroduction. For example, in another case in which the authors were involved, several federally endangered deltoid spurges (*Chamaesyce deltoidea* ssp. *deltoidea*) were relocated to a "safe" location but with no long-term monitoring or management

requirements. Although short-term survival was over 50 percent, no one has revisited the relocation area in several years, and it is unknown if any individuals have survived or reproduced (Gann and Gerson unpublished data).

Endangered species reintroduction within a mitigation context in Florida is unusual and mostly occurs only on a voluntary basis. In preparation for this chapter, the authors conducted an informal interview to which twenty-seven companies involved with mitigation in Florida responded. Of these companies, not one had been required by an agency to reintroduce an endangered species as part of a compensatory mitigation process.

It is ironic that the technical knowledge does exist to reintroduce many endangered species within the mitigation process. Over one-fourth of the vascular plant species native to Florida are now in cultivation (Association of Florida Native Nurseries 1993; Ward 1990), including at least ninety one (36 percent) of the state-listed endangered species and thirty one of the species listed as endangered or threatened under the Endangered Species Act (Table 17-4). The techniques used to cultivate these plants, together with the experience gained through voluntary reintroduction efforts, could serve as the knowledge base for potential future efforts to successfully reintroduce endangered species in Florida.

Recommendations for Improvement

Applied biological conservation is not a simple task. Society is still learning how to manage the fragmented preserves that are already protected; we know even less about the science of restoring highly degraded natural communities and creating others de novo. Most wetland creation and restoration projects in the United States have been implemented within the last fifteen years (Spear, Lent, and Nunnery 1992). Furthermore, most evidence suggests that mitigation (and ecological restoration in general) is a trial-and-error process (Zedler 1988). Regardless of how many problems have been experienced with mitigation in the past, the opportunity exists to learn from past mistakes and improve the process in the future. The following paragraphs present some ideas about how mitigation, as both a process and an ecological reality, could be improved.

- Prohibit in perpetuity the destruction of any intact or moderately disturbed natural community as part of a mitigation project. This includes upland-to-wetland conversions.

- Charge a mitigation fee for the destruction of natural communities, pool the money, and use the available funds either to purchase remnant natural communities and place them under the public control and/or to

TABLE 17-4. State-listed endangered species in cultivation in Florida.

Species	Source	Species	Source
Acacia choriophylla	AFNN	Illicium parviflorum	AFNN
Acrostichum aureum	AFNN	Ipomoea microdactyla	FTG
Alvaradoa amophoides	AFNN	Jacquemontia curtissii	FTG
Amorpha crenulata *	AFNN, FTG	Jacquemontia reclinata*	AFNN, FTG
Aquilegia canadensis	AFNN	Justicia cooleyi*	AFNN, BTG
Asimina tetramera *	BTG, KBG	Lantana depressa	AFNN, FTG
Bonamia grandiflora *	BTG	Liatris ohingerae*	BTG
Bourreria cassinifolia	FTG	Liatis provincialis	BTG
Byrsonima lucida	AFNN	Licaria triandra	AFNN
Callirhoe papaver	AFNN, KBG	Magnolia acuminata	AFNN
Calycanthus floridus	AFNN	Magnolia ashei	AFNN, KBG
Calyptranthes zuzygium	AFNN	Matelea alabamensis	BTG
Canella winterana	AFNN	Myrcianthes fragrans	
Catesbaea parvifolia	FTG	var. simpsonii	AFNN
Cereus eriophorus		Nolina brittoniana*	AFNN, BTG
var. fragrans*	FTG	Opuntia spinosissima	FTG
Cereus gracilis	FTG	Opunita tricantha	FTG
Cereus robinii*	FTG	Okenia hypogaea	AFNN
Chionanthus pygmaeus*	AFNN, BTG	Paronychia chartacea*	BTG
Chyrsophyllum oliviforme	AFNN	Peperomia humilis	AFNN
Chrysopsis floridana*	AFNN, BTG	Peperomia magnoliifolia	AFNN
Clusia rosea	AFNN	Peperomia obtusifolia	AFNN
Conradina brevifolia#,*	BTG	Polygonella basiramia*	BTG
Conradina etonia*	BTG	Prunus geniculata*	AFNN, BTG
Conradina glabra*	AFNN, BTG	Psuedophoenix sargentii	AFNN, FTG
Conradina grandiflora	BTG	Remirea maritima	AFNN
Cordia sebestena	AFNN	Rhipsalis baccifera	FTG
Crossopetalum ilicifolium	AFNN	Rhododendron alabamense	AFNN
Crossopetalum rhacoma	AFNN	Rhododendron austrinum	AFNN
Crotalaria avonensis*	BTG	Rhododendron chapmanii*	AFNN
Deeringothamnus pulchelus*	BTG	Ribes echinellum*	AFNN
Deeringothamnus rugelii*	BTG	Roystonea elata	AFNN
Dennstaedtia bipinnata	KBG	Salix floridana	AFNN
Dicerandra christmanii*	BTG	Sarracenia leucophulla	KBG
Dicerandra immaculata*	BTG	Spigelia gentianoides*	BTG
Eryngium cuneifolium*	BTG	Staphylea trifolia	AFNN
Eugenia confusa	AFNN	Stewartia malacodendron	AFNN
Eugenia rhombea	AFNN, FTG	Strumpfia maritima	FTG
Euphorbia telephioides*	BTG	Suriana maritima	AFNN
Galactia smallii*	FTG	Swietenia mahagoni	AFNN
Glandularia maritima	AFNN	Taxus floridana	AFNN, KBG
Gossypium hirsutum	AFNN	Torreya taxifolia*	AFNN, BTG, KBG
Guaiacum sanctum	AFNN	Tournefortia gnaphalodes	AFNN
Harperocallis flava*	AFNN	Tripsacum floridanum	AFNN, FTG
Hypelate trifoliata	AFNN	Zanthoxylum coriaceum	AFNN, FTG
Hypericum edisonianum	AFNN	Zanthoxylum flavum	AFNN
Ilex krugiana	AFNN	Ziziphus celata*	BTG

*Listed as endangered or threatened under the Endangered Species Act
#Omitted from the Regulated Plant Index
AFNN Cultivated by one or more member nurseries in the Association of Florida Native Nurseries (AFNN 1993)
BTG Cultivated at Bok Tower Gardens, Lake Wales, Florida
FTG Cultivated at Fairchild Tropical Garden, Miami, Florida
KBG Cultivated at Kanapaha Botanical Gardens, Gainesville, Florida

restore degraded natural communities already publicly owned. This concept is already being used by Dade County, Florida, for both wetland and upland systems (see Bean, Chapter Sixteen).

- Restoration of degraded natural community remnants should be required as part of the mitigation process. Most remnant natural communities have been affected negatively by fire exclusion, drainage, invasion by exotic pest plants (and/or animals), and the extirpation of native species. Degraded natural communities on private or public lands should be restored by the permittee. If restoration on private lands is proposed as part of a mitigation trade, then the property should (at a minimum) be protected in perpetuity through a covenant or conservation easement or, better yet, transferred to a public agency or private conservation organization. Long-term management goals and procedures should be developed and funding secured. Once this is done, then the reintroduction of endangered plant species should be accomplished where appropriate. Projects of this type are already being implemented in Florida by the Nature Conservancy.

- Creation of new, "natural" communities should be eliminated as the primary objective of mitigation efforts, although it may be worthwhile to require habitat creation in addition to land preservation and/or natural community restoration. The best system would require the creation of new natural communities on at least a 1-to-1 ratio of created area to destroyed area in addition to restoration or enhancement of degraded natural communities. Newly created natural communities should be transferred to a government agency or a private nonprofit agency such as the Nature Conservancy for long-term restoration and management. Management funds should be budgeted for and made available through development fees and/or government grants.

- For projects that still require the creation of new natural communities, several fundamental changes are necessary:

    1. Adequate time must be provided for project development, and both restoration designers and contractors should be included in the earliest stages of project planning.

    2. Strict design standards must be developed that require that remnant natural communities be used as models for natural community creation projects. Only species native to the immediate locality of the mitigation site should be used, and diversity and composition specifications developed and enforced.

    3. Likewise, strict provenance requirements must be developed and adhered to. If in-kind, on-site, or near-site creation is proposed, then

propagules of native species should be collected from the natural area that is to be destroyed, and plants cultivated from these propagules should be used in the mitigation project. If not-in-kind or off-site creation is proposed, then propagules from nearby analogous natural community remnants should be collected and grown for eventual use in the creation project.

4. Finally, society must begin to think long-term about both monitoring and management of ecosystems. For the restoration of degraded natural communities and creation projects, three to ten years of management and monitoring are barely long enough to evaluate successful establishment and long-term viability of many species (see Sutter, Chapter Ten). Mitigation sites must be protected and managed in perpetuity. Exotic species and other human-induced external disturbances will continue to affect mitigation projects, and these influences must be compensated for with long-term management activities.

5. Concomitant with the effort to restore natural communities as closely as possible to model natural communities, plans for the eventual reintroduction of endangered species over their full natural range should be developed and executed. In some cases, endangered species should be reintroduced in the earliest stages of restoration or creation. In other cases endangered species should only be reintroduced once sufficient site stability has been achieved or conditions sufficiently altered to make the site habitable for endangered species. In all cases, endangered species should be reintroduced once long-term monitoring and management of the site have been planned for and secured.

Appendix

Regulations and laws affecting the Conservation of Endangered Plant Species in Florida:

Primary Laws

1. *Endangered Species Act (ESA)* (16 U.S.C. 1531 et seq.). Under the ESA, all federally threatened and endangered plant species are protected on federal lands or if federal funds are expended for a project. No protection of endangered plants is derived if these conditions are not met. Endangered plants, however, may derive secondary protection from the ESA if federally endangered or threatened animals are found on any site.

2. *Preservation of Native Flora of Florida* (581.185 Florida Statutes; 5B-40 Florida Administrative Code). This rule was established to restrict harvesting and willful destruction of Florida native plant populations. Under the rule, the relocation of endangered or commercially exploited plants or plant parts from one property to another requires both a permit from the Florida Department of Agriculture and Consumer Services and written permission from the landowner. Relocation of threatened plants requires written permission from the landowner. Additionally, landowners may not harvest endangered or commercially exploited plants from their own property for sale without a permit from the state.

Secondary Laws

1. Clean Water Act, Section 404 (33 U.S.C. 1344). Regulates the discharge of dredged and fill material into waters of the United States; established a permit program to ensure that such discharges comply with environmental requirements. The Section 404 program is administered at the federal level by the U.S. Army Corps of Engineers (Corps) and the Environmental Protection Agency (EPA). The Corps has the primary responsibility for the permit program and is authorized, after notice and opportunity for public hearing, to issue permits for the discharge of dredged and fill material. The EPA has primary roles in several aspects of the Section 404 program, including development of the environmental guidelines by which permit applications must be evaluated.

2. *Clean Water Act, Section 401* (33 U.S.C. 1341). Under the Clean Water Act, states have the power to restrict the discharge of "dredged or fill material" into wetlands through the Section 410 State Water Quality Certification provision of the Act.[1]

3. *Waters of the State Act* (Section 373.414(1)(a)2 Florida Statues). This act authorizes the Florida Department of Environmental Protection (DEP) to prevent harm to the waters of the state through a permitting program. In this program, aquatic- and wetland-dependent species receive some protection because they are considered to be part of the water resources of the state. Within this law, the DEP has the authority to deny a wetland alteration permit if the project adversely affects the conservation of fish and wildlife including threatened and endangered species.

4. *Coastal Zone Management Act of 1972* (16 U.S.C. 1451). This federal law establishes a regulatory program that sets national standards to guide

---

1. From National Wildlife Federation 1989

public and private use of land and water in the coastal zone. It also requires federal agencies to coordinate with state coastal management programs that follow national standards set by the act.

5. *Coastal Construction Control Line Permit Program* (161.053 Florida Statutes; 16B-33 Florida Administrative Code). Under this program, the Florida Department of Environmental Protection has the authority to protect beach-dune plant communities by requiring a permit from the department for any alteration of beach-dune topography or vegetation in the state.

6.a. *State Comprehensive Plan* (SCP) (187 Florida Statutes). The SCP was enacted to provide a strategic vision for Florida's future. Twenty-six goals are established to provide broad direction and long-range guidance to decision makers in the state; among these goals, the protection of coastal and marine resources and natural systems is addressed. The SCP provides general guidance for land use and conservation. It also serves as a basis and blueprint for the development of regional and local resource protection plans and ordinances.

6.b. *Local Government Comprehensive Plans* (163, Part II Florida Statutes; 9J-5 Florida Administrative Code). Directs each local government to prepare a comprehensive plan, consistent with the state comprehensive plan. Local ordinances are to be developed to enforce the policies established in the local plans. Endangered species protection and conservation are essential elements of the plans to determine land-use suitability within the local governments' jurisdiction.

6.c. *Development of Regional Impacts* (DRI) (380.06 Florida Statutes; 28-24 and 9J-2 Florida Administrative Code). Provides guidelines consistent with the state comprehensive plan to determine how any site over 50 acres is developed. Endangered species and habitat protection are key elements considered in the DRI process.

7. *County and Municipal Ordinances.* Many counties and municipalities have developed local ordinances protecting wetlands, uplands, and coastal resources. These ordinances must be consistent with the comprehensive plan and, in many cases, offer the only substantive protection to uplands and upland-dependent endangered species.

REFERENCES

Abrahamson, W. G., and D. C. Hartnett. 1990. "Pine Flatwoods and Dry Prairies." In *Ecosystems of Florida.* Edited by R. L. Myers and J. J. Ewel. Orlando: University of Central Florida Press.

Association of Florida Native Nurseries. 1993. *Plant and Service Directory*. San Antonio, Florida.

Avery, G. N., and L. L. Loope. 1980. *Plants of Everglades National Park: A Preliminary Checklist of Vascular Plants*. Report T-574. Homestead, Fla.: U.S. National Park Service, South Florida Research Center.

Baird, K. 1989. "High-Quality Restoration of Riparian Ecosystems." *Restoration and Management Notes* 7 (2): 60–64.

Clewell, A. F. 1985. *Guide to the Vascular Plants of the Florida Panhandle*. Tallahassee: Florida State University Press.

Coile, N. C. 1993. *Florida's Endangered and Threatened Plants*. Gainesville: Florida Department of Agriculture and Consumer Services, Division of Plant Industry.

Davis, M. M. 1993. "Wetlands Research Program Studying Restoration Techniques: Produces New Manuals." *Restoration and Management Notes* 11 (2): 156.

Dunwiddie, P. W. 1992. "On Setting Goals: From Snapshots to Movies and Beyond." *Restoration and Management Notes* 10 (2): 116–119.

Environmental Law Institute. 1993. *Wetlands Deskbook*. Washington, D.C.

Ewel, J. J. 1990. "Introduction." In *Ecosystems of Florida*. Edited by R. L. Myers and J. J. Ewel. Orlando: University of Central Florida Press.

Ewel. K. C. 1990. "Swamps." In *Ecosystems of Florida*. Edited by R. L. Myers and J. J. Ewel. Orlando: University of Orlando Press.

Florida Department of Environmental Regulation. 1991. *Report on Effectiveness of Mitigation in Florida*. Tallahassee, Florida.

Florida Department of Environmental Regulation. 1993. *1991–1992 Report to the Legislature on Permitted Wetlands Projects*. Tallahassee, Florida.

Florida Natural Areas Inventory. 1993 (May). *Special Plants and Lichens*. Tallahassee, Florida.

Godfrey, R. K. 1988. *Trees, Shrubs, and Woody Vines of Northern Florida and Adjacent Georgia and Alabama*. Athens: University of Georgia Press.

Godfrey, R. K., and J. W. Wooten. 1979. *Aquatic and Wetland Plants of the Southeastern United States. Monocotyledons*. Athens: University of Georgia Press.

Godfrey, R. K., and J. W. Wooten. 1981. *Aquatic and Wetland Plants of the Southeastern United States. Dicotyledons*. Athens: University of Georgia Press.

Hackeling, L. C., M. K. Leach, and S. Apfelbaum. 1990. "Ecological Restoration and Environmental Mitigation: A Relationship Worth Scrutinizing." In Proceedings of the *First Annual Meeting of the Society for Ecological Restoration, January 16–20, 1989*. Edited by H. G. Hughes and T. M. Bonnicksen. Madison, Wis.: Society for Ecological Restoration.

Harper, R. M. 1949a. "A Preliminary List of the Endemic Flowering Plants of Florida: Part I. Introduction and History of Exploration." *Quart. Jour. Florida Acad. Sci.* 11 (1): 25–35.

Harper, R. M. 1949b. "A Preliminary List of the Endemic Flowering Plants of Florida. Part II. List of Species." *The Quarterly Journal of the Florida Academy of Science.* 11 (2–3): 39–57.

Harper, R. M. 1950. "A Preliminary List of the Endemic Flowering Plants of Florida. Part III. Notes and Summary." *Quart. Jour. Florida Acad. Sci.* 12 (1): 1–19.

Higgs, E. S. 1993. "The Ethics of Mitigation." *Restoration and Management Notes* 11 (2): 138–143.

Johnson, A. F., and M. C. Barbour. 1990. "Dunes and Maritime Forests." In *Ecosystems of Florida*. Edited by R. L. Myers and J. J. Ewel. Orlando: University of Central Florida Press.

Kushlan, J. A. 1990. "Freshwater Marshes." In *Ecosystems of Florida*. Edited by R. L. Myers and J. J. Ewel. Orlando: University of Central Florida Press.

Lakela, O., and R. W. Long. 1976. *Ferns of Florida*. Miami: Banyan Books.

Lewis, R. R. 1992. Personal communication in "Zentner on Katz (and Zedler and Hiss)," J. Zentner, 1992. *Restoration and Management Notes* 10 (2): 113–116.

Little, E. L., Jr. 1978. *Atlas of United States Trees Vol. 5.: Florida*. United States Department of Agriculture, Forest Service, Miscellaneous Publication No. 1361. Washington, D.C.: U.S. Government Printing Office.

Long, R. W., and O. Lakela. 1971. *A Flora of Tropical Florida*. Coral Gables: University of Miami Press.

Luer, C. A. 1972. *The Native Orchids of Florida*. Ipswich: W. S. Cowell Ltd.

Millar, D. I., and W. J. Libby. 1989. "Disneyland or Native Ecosytem: Genetics and the Restorationist." *Restoration and Management Notes* 7 (1): 18–24.

Muller, J. W., E. D. Hardin, D. R. Jackson, S. E. Gatewood, and N. Caire. 1989. *Summary Report on the Vascular Plants, Animals, and Plant Communities Endemic to Florida*. Florida Game and Fresh Water Fish Commission, Non-Game Wildlife Program, Technical Report No. 7.

Myers, R. L. 1990. "Scrub and High Pine." In *Ecosytems of Florida*. Edited by R. L. Myers and J. J. Ewel. Orlando: University of Florida Press.

National Wildlife Federation. 1989. *A Citizen's Guide to Protecting Wetlands*. Washington, D.C.: NWF.

Platt, W. J., and M. W. Schwartz. 1990. "Temperate Hardwood Forests." In *Ecosytems of Florida*. Edited by R. L. Myers and J. J. Ewel. Orlando: University of Central Florida Press.

Reiger, J. 1991. "San Diego Bay Mitigation Study: A Response." *Restoration and Management Notes* 6 (2): 65–66.

Reiger, J. 1992. "Western Riparian and Wetland Ecosystems." *Restoration and Management Notes* 10 (1): 52–55.

Roberts, L. 1993. "Wetlands Trading Is a Loser's Game, Say Ecologists." *Science* 260: 1890–1892.

Sacco, J. N., F. L. Booker, and E. D. Seneca. 1988. "Comparison of the Macrofaunal Communities of a Human-Initiated Salt Marsh at Two and Fifteen Years of Age." In *Proceedings of a Conference: Increasing our Wetland Resources, October 4–7, 1987*. Edited by J. Zelanzy and J. S. Feierabend. Washington D.C.: National Wildlife Federation.

Shaw, S. P., and C. G. Fredine. 1956. *Wetlands of the United States, Their Extent and Their Value for Waterfowl and Other Wildlife*. U.S. Fish and Wildlife Service Circular No. 39.

Sheldon, K. P. 1993. "Wildlife." In *Sustainable Environmental Law*. C. Campbell-Mohn et al. St Paul, Minn.: West Publishing Co.

Small, J. K. 1933. *Manual of the Southeastern Flora.* Chapel Hill: University of North Carolina Press.

Small, J. K. 1938. *Ferns of the Southeastern States.* Lancaster, Pa.: Science Press.

Smiley, N. 1973. *Yesterday's Miami.* Miami: E. A. Seemann.

Snyder, J. R., A. Herndon, and W. B. Robertson Jr. 1990. "South Florida Rockland." In *Ecosytems of Florida.* Edited by R. L. Myers and J. J. Ewel. Orlando: University of Central Florida Press.

Spear, T., T. Lent, and K. Nunnery. 1992. "Study of 50-Year-Old Bottomland Forest Restorations Concentrates on Functional Attributes." *Restoration and Management Notes* 10 (2): 183–184.

Tilton, D. L., and D. L. Denison. 1992. "Colonization of Restored Wetlands by Invertebrates, Fish, Amphibians and Reptiles Studied." *Restoration and Management Notes* 10 (2): 187.

Ward, D. B. 1990. "How Many Plant Species Are Native to Florida?" *The Palmetto* 9 (4): 3–5.

Weller, 1989. In *Wetland Creation and Restoration: The Status of the Science.* Edited by J. Kusler and M. Kentula. Submitted to the U.S. Environmental Protection Agency.

Wood, D. A. 1994. (June). *Official Lists of Endangered and Potentially Endangered Fauna and Flora in Florida.* Florida Game and Freshwater Fish Commission. Tallahassee, Florida.

Wunderlin, R. P. 1982. *Guide to the Vascular Plants of Central Florida.* Tampa: University of Florida Press.

Zedler, J. B. 1988. "Salt Marsh Restoration: Lessons from California." In *Rehabilitating Damaged Ecosystems.* Edited by J. Cains, Jr. Boca Raton, Fla.: CRC Press.

Zedler, J. B., and R. Langis. 1991. "Authenticity: Comparisions of Constructed and Natural Salt Marshes of San Diego Bay." *Restoration and Management Notes* 6 (1): 21–25.

Zentner, J. 1992. "Zentner on Katz (and Zedler and Hiss)." *Restoration and Management Notes* 10 (2): 113–116.

P A R T   I V

Case Studies

Threatened species are being reintroduced in all regions, major biomes, and habitat types; among many different taxonomic groups; and by various agencies and organizations. This practice provides the basis for the development and empirical testing of the theory of restoration. The following case studies are a sampling of reintroductions that have been attempted across the country. It is not our intention to describe every possible circumstance for reintroduction of rare plants but rather to illuminate some relevant experience in reintroducing different life forms in various habitats and regions across North America. We have also selected these particular examples of reintroduction to illustrate some of the major themes of this book: the importance of monitoring, interagency collaboration, and careful genetic and community ecological design, as well as the influence of subsequent land management.

Each case study identifies the taxon reintroduced, along with its endangerment status, the threats to the population, and the conditions that led to the need for reintroduction. A statement of objectives of the reintroduction project and a description of the project follow, with funding sources, funding amounts, and cooperating partners identified. In addition, the policy and regulatory context of the reintroduction is included, along with comments about the project's bioregional context. This information provides a better understanding of the "who, what, where, when, and how" of a reintroduction program.

No two reintroduction projects are alike, and there are different lessons to be learned from each one (Table CSI-1). Funding levels varied widely, as did the institutional context and objectives. All species represented in the case studies are federally listed as threatened or endangered under the Endangered Species Act (ESA). The community restoration project in the Appalachians (Case Study Six) includes eight species, of which one is federally listed and four are state listed in North Carolina.

Threats to these species vary; however, habitat destruction is the dominant cause of decline in the species described herein. Additionally, collection is a threat for the cactus (*Pediocactus knowltonii*) and the orchid (*Isotria medioloides*).

For most of the taxa in these case studies, the major condition leading to the reintroduction was low population numbers and few sites of occurrence. One species, *Stephanomeria malheurensis*, had actually become extinct in the wild and could be reintroduced only because a genetically representative sample of seeds was available. One of the projects described, *Isotria medeoloides*, was part of a mitigation effort.

The reintroductions were conducted between 1985 and 1991. As such, they are among some of the first plant reintroductions conducted under the auspices of the U.S. Fish and Wildlife Service as outlined in ESA recovery plans.

Table CSI-1.   Comparative information from case studies.

Species	Life form	Initial project year	State	Number of partners	Funding*
Cirsium pitcheri	Perennial herb	1991	IL	5	$10,000
Conradina glabra	Shrub	1991	FL	3	$2,000
Isotria medeoloides	Perennial herb	1986	NH	3	$3,000
Pediocactus knowltonii	Succulent	1985	NM	7	$125,000
Stephanomeria malheurensis	Annual herb	1987	OR	5	$40,000
Styrax texana	Shrub/sm. tree	1987	TX	7	$17,900
Granite Outcrops	Perennial herbs	1990	NC	6	$35,200
AVERAGE				5	$33,300

*Funding does not represent full cost of reintroduction project, because in-kind services are not included.

All of the federally listed taxa described here had approved recovery plans at the time of the reintroduction, with the exception of *Conradina glabra*. With the exception of *Isotria medioloides*, the reintroduction projects examined here were undertaken in response to stated conservation objectives or research needs as part of the recovery process. For the most part, these reintroductions were carried out without the benefit of guidelines, standards, or planning models other than the species-specific tasks included in individual recovery plans.

These case studies illustrate the importance of partnerships to a successful reintroduction program. None of these projects was conducted by a single individual or organization; in fact, all involved at least three different organizations or agencies, with the average number of partners for this sample being five. The low level of available funding is one reason why partnerships are important: for these projects funding varied from a low of $2,000 to a high of $125,000, with an average of $33,300.

All of the projects described include a monitoring component; over half have more than six years of monitoring data. Monitoring is often the most costly long-term item in a reintroduction project; much of the expense for these projects is in the form of in-kind services of agency or organization personnel to perform ongoing monitoring.

Committed partnerships play an integral role in many aspects of a reintroduction project; however, partnerships are especially important in production of plant material to reintroduce into the habitat. Plant material for five of the seven projects was propagated by gardens within the Center for Plant Conservation network from plant material in the National Collection of Endangered Plants. Such cooperation is typical of rare plant work, since land-managing agencies often do not operate their own plant-propagation facilities.

With respect to the bioregional context of the reintroductions, none of these projects was designed explicitly as part of a larger landscape-level plan for ecosystem restoration. As the case studies illustrate, most rare plant reintroductions occur on a species-by-species basis for management and conservation. We hope that increased attention to the ecology of rare plant reintroduction will encourage more projects to be designed as part of master plans for restoring diversity on a bioregional basis.

Experimental Reintroduction of *Stephanomeria malheurensis*

Edward O. Guerrant, Jr.

<hr>

Taxon

Stephanomeria malheurensis Gottlieb (Gottlieb 1978); Malheur wirelettuce (Asteraceae).

Endangerment Status and Date of Listing

Federal: endangered November 10, 1982.
State of Oregon: endangered October 27, 1989.

Natural Distribution and Ecology of Taxon

Stephanomeria malheurensis is known only from a single location, in Eastern Oregon, USA (about 25 miles south of the town of Burns). It occurs in a high desert environment that supports a shrub-steppe vegetation (more habitat information can be found in later sections). This very rare species has enjoyed considerable scientific notoriety, in large part because it is thought to have originated recently near the northern range limit of its presumably ancestral taxon, *S. exigua* ssp. *coronaria*, with which it is sympatric (Gottlieb 1973).

The species is an obligately inbreeding, herbaceous annual. Most seeds germinate in late winter/early spring (April/May). Surviving seedlings form a rosette that bolts in June, flowers in July, and bears fruits that mature in August.

Threats to Population

The primary threat appears to be competition with the aggressive exotic annual grass *Bromus tectorum* (cheatgrass).

Prior to a fire that swept through the population in 1972, Malheur wirelettuce was most abundant in the open areas between shrubs (big sagebrush, *Artemisia tridentata*; and green rabbitbrush, *Chrysothamnus vicidiflorus*) and bunchgrasses (Great Basin wildrye, *Elymus cinereus*). Subsequent to the fire, cheatgrass invaded these formerly open areas, forming a nearly complete cover. Experiments have shown competition with cheatgrass significantly reduces Malheur wirelettuce fitness (Brauner 1988).

The apparent extinction in 1985 coincided with a series of very wet years, which also saw populations of the sympatric and previously much more abundant ancestral taxon *S. exigua* ssp. *coronaria* reduced from the tens of thousands to around ten individuals per year (Brauner 1988). Malheur wirelettuce was never estimated to number more than about 750 plants (Gottlieb 1979).

Conditions That Initiated Reintroduction

The reintroduction attempt was precipitated by two facts: (1) a federally listed endangered species known only from federal Bureau of Land Management (BLM) land, and that was being monitored by the BLM, had apparently become extinct; and (2) a genetically representative sample of seeds that had been stored off-site by Dr. L. D. Gottlieb were available for reintroduction.

The high quality of the *ex situ* sample of dormant seeds available for reintroduction of a presumably extinct taxon is probably unique. This was not a casual seed collection but one systematically assembled for scientific purposes by a geneticist interested in studying population genetic structure. Together with the fact that the species is highly self-pollinating and has relatively little detectable genetic diversity (Gottlieb 1973), it is highly likely that the collection represented the original gene pool remarkably well.

Objective of Reintroduction

The prime objective of the reintroduction was to provide information required to recover the species (Brauner 1988). A major focus of experimental reintroduction was to elucidate the role of competition with the aggressive exotic annual cheatgrass (*Bromus tectorum*) and other plants common to the area (Parenti and Guerrant 1990).

Description of Reintroduction Site

Reintroduction was attempted at a single locality, where the species occurred naturally before it had apparently become extinct in the mid 1980s. The site is on a hilltop composed of volcanic tuff surrounded by soils derived from basalt (see Gottlieb 1979). Within this single site are located four experimental reintroduction plots, which have been fenced to exclude rabbits. The type locality is within the South Narrows Area of Critical Environmental Concern, managed by the BLM. The surrounding habitat has been fenced to exclude domestic cattle.

Description of Reintroduction Project

This reintroduction is explicitly experimental, involving five treatments. Four separate plots were established, each delineated by rodent-proof wire-mesh enclosures extending both 32 inches above and 4 inches below ground level. Each of the plots is dominated by one of the major species mentioned previously that is common to the site: three are 5-by-5-meter plots of natives (*Artemisia tridentata*, *Chrysothamnus viscidiflorus*, and *Elymus cinerus*); the other is a 5-by-10-meter plot with two treatments of the exotic *Bromus tectorum*. All cheatgrass was manually removed from plots dominated by native species. Half of the cheatgrass plot was hand-weeded to 50 percent cheatgrass cover, and the other half, which was not weeded, had 50 to 100 percent cheatgrass cover.

The reintroduction itself took place in 1987 and consisted of transplanting one thousand seedlings, which had been produced by the Berry Botanic Garden, with seeds provided by Dr. L. D. Gottlieb. The plants were thoroughly watered at planting time and subsequently as needed for about four weeks. Extensive monitoring has occurred since that time. In the first year, one thousand plants yielded forty thousand seeds. Since then, the sagebrush plot was decimated by rabbits, and it had to be re-established. In 1990, which had a very cold, dry spring, no plants were seen. Plants appeared again in 1991 and have reappeared every year since. It is still too early to evaluate the ultimate success of the project.

Bioregional Context

This reintroduction is not to my knowledge part of a landscape-level plan. The four plots and surrounding area are protected from grazing by domestic livestock.

Funding

FWS and BLM have contributed funds, amounting to an estimated $40,000.

Partnerships

- Academic: Dr. L. D. Gottlieb, University of California, Davis, had collected and stored the seed originally.

- Public: BLM, Burns District, and the U.S. Fish and Wildlife Service.

- Private: Berry Botanic Garden and the Center for Plant Conservation.

Policy or Regulatory Context of Reintroduction

Final recovery plan, March 21, 1991: In an early report on the reintroduction effort, Dr. S. Brauner (1988) suggested that herbicides may be the only cost-effective means by which cheatgrass competition could be controlled. Grass-specific herbicides have been used effectively to reduce competitive pressure from exotic grasses in the reintroduction of *Amsinckia grandiflora* in California (see Pavlik, Chapter Six). However, this is not an option because of a legal injunction against the use of herbicides on BLM land.

That the species is known from only a single site may have contributed to the lack of effort to introduce the species to additional localities.

REFERENCES

Brauner, S. 1988. "Malheur Wirelettuce (*Stephanomeria malheurensis*) Biology and Interactions with Cheatgrass: 1987 Study Results and Recommendations for a Recovery Plan." Unpublished report to the Bureau of Land Management, Burns District, Oregon.

Gottlieb, L. D. 1973. "Genetic Differentiation, Sympatric Speciation, and the Origin of a Diploid Species of *Stephanomeria*." *American Journal of Botany* 60 (6): 545–553.

Gottlieb, L. D. 1978. "*Stephanomeria malheurensis* (Compositae), a New Species from Oregon." Madroño 24: 44–46.

Gottlieb, L. D, 1979. "The Origin of Phenotype in a Recently Evolved Species." In *Topics in Plant Population Biology*. Edited by O. T. Solbrig, S. Jain, G. B. Johnson, and P. H. Raven. New York: Columbia University Press.

Parenti, R. L., and Guerrant, E. O., Jr. 1990. "Down But Not Out: Reintroduction of the Extirpated Malheur Wirelettuce, *Stephanomeria malheurensis*." *Endangered Species UPDATE*, (8) 1: 62–63.

Knowlton's Cactus
(*Pediocactus knowltonii*) Reintroduction

Ann Cully

Taxon

Pediocactus knowltonii L. Benson; Knowlton's cactus.

Endangerment Status and Date of Listing

Federal: endangered November 28, 1979.
State: endangered 1985.

Natural Distribution and Ecology of Taxon

Knowlton's cactus is a narrow endemic limited to one known, viable population at the type locality in northwestern New Mexico near the Colorado border. This tiny, perennial cactus varies from single to multistemmed, with stem diameters in mature plants ranging from 5 to 25 millimeters. The plants are very low growing and occur primarily among alluvial gravel and cobbles on a low hill. The flowers are white tinged with pink, and the green-to-tan fruits split irregularly to release about twelve seeds (Benson 1982).

Threats to Population

Knowlton's cactus is threatened by its small population size and limited distribution. The only viable, natural population is restricted to a single hilltop

in northwestern New Mexico. The cactus is vulnerable both to natural disasters and human-made disturbance. Northwestern New Mexico is a major oil and gas production area, with attendant roads, well pads, and pipelines that directly threaten the existence of the cactus and its habitat. Another serious threat to the species is collection; because of its rarity and diminutive size, the cactus is much sought after by collectors. The site is currently in the ownership of The Nature Conservancy.

Conditions That Initiated Reintroduction

The establishment of new populations of Knowlton's cactus is desirable because of the limited distribution of the species. The single, viable population is vulnerable to extinction from natural causes (including disease, parasitism, genetic bottlenecks) and from human activities (oil and gas exploration and development; collecting; livestock grazing; destruction of habitat). At the time of the first reintroduction effort, population size at the type locality was estimated to be about five thousand to six thousand plants. At the time of the second effort, the population was estimated to be about ten thousand plants.

Description of Reintroduction Sites

There are two sites of Knowlton's cactus reintroduction. Multiple agencies (U.S. Department of the Interior, Fish and Wildlife Service, Bureau of Reclamation, and Bureau of Land Management; and New Mexico Department of Natural Resources, Parks and Recreation Division) agreed that for the first reintroduction the site should be (1) within the probable historical range of the species, (2) within a similar vegetation type, (3) at the approximate elevation and with similar topographic features and soil type, (4) in an inaccessible and federally protected area, and (5) protected from future oil and gas development. After screening several possible locations, the first site was selected in 1985. The second site was chosen using the same criteria in a cooperative effort by personnel from New Mexico Energy, Minerals, and Natural Resources Department; Forestry and Resources Conservation Division; Fish and Wildlife Service; and the Bureau of Land Management in 1991.

The two reintroduction sites are located within 8 kilometers (5 miles) of the type locality. All sites (including the type locality) are in San Juan County, New Mexico, at approximately 1800 meters (6000 feet) in elevation, at the southern end of the Colorado Plateau. They receive about 40 centimeters (15 inches) of precipitation each year. All sites are located in a plant community

dominated by piñon pine (*Pinus edulis*), Rocky Mountain juniper (*Juniperus scopulorum*), black sagebrush (*Artemisia nova*), blue grama grass (*Bouteloua gracilis*), a foliose lichen (*Parmelia physodes*), and other species. At the type locality, the plants grow on the top and slopes of a single hill composed of alluvial deposits of a mix of sandy loam and material ranging from pebble to cobble in size.

The first site, established in 1985, is about 800 m² in size. The area is under the management of the Bureau of Reclamation and the New Mexico Energy, Minerals and Natural Resources Department. The second site, established in 1991, under the management of the Bureau of Land Management, is about 600 m² in area.

Objectives of Reintroduction

In the recovery plan for Knowlton's cactus, reintroduction is listed as a priority 1 task for the recovery of the species (Heil 1985). Success will be measured by the survival of the reintroduced cactus and the presence of new, reproductive individuals at the reintroduction site. The desired future conditions are the survival and reproduction of the reintroduced plants and the presence of new, reproductive individuals within ten years of the first planting.

Description of Reintroduction Project

Site 1

Establishment of the new populations began in May 1985, when 250 cuttings were taken from wild plants at the type locality. The cuttings were taken to a greenhouse and placed in individual pots, where they established roots over the summer. Some of the cuttings (103) were planted in September 1985, in a pre-established grid system 30 by 20 meters, at 2-meter intervals within the grid. Each planting point was marked with a twenty-penny nail, which secured an identification tag. A hole was dug for each plant approximately 15 centimeters from the nail and tag, and the cactus and the contents of the pot were placed in the hole. Each plant received about 0.5 liters of water after planting. The diameter of each plant was measured after its placement in the ground so that its growth could be determined from year to year. In May 1986, forty-seven cuttings were added to the grid; they were also marked and measured so that their progress could be followed. The plot was visited each fall and spring from 1986 through 1991; from 1991 to the present, the plot has been visited each spring. During these monitoring visits, diameter measurements,

number of flowers and fruits, and notes on the general conditions of the plants have been taken.

In September 1987, agency biologists installed a seed reintroduction plot for Knowlton's cactus. The plot is located about 200 meters southwest of the transplanting plot. The seed plot is separated by a small drainage from the cuttings plot, in order to prevent mixing of seeds from the two different sources. The seed plot was arranged in a 10-by-10-meter grid, with planting points at every meter. At forty-eight grid point locations, six seeds were planted in the following way. At each point, a nail and an aluminum tag were placed to identify the grid point. A wire mesh template, with three 1-cm² apertures cut into it was placed over the point along the north axis to determine the precise spots for the seeds to be planted. Two seeds were planted on the surface, two at 0.5 centimeters below the surface, and another two at 1-centimeter depth. In all, 288 seeds were planted in September 1987. In May 1988, 120 additional seeds were planted in the same way. The seed grid has been visited each spring and fall until 1991 and then was visited each spring, during monitoring visits to the cuttings site nearby. The seeds for this project were obtained from the plants remaining in the greenhouse that were collected at the type locality in 1985.

In 1986, monitoring plots were established at the type locality. Information on stem and plant diameters, numbers of flowers and fruits, number of plants that died, and numbers of juveniles was recorded every year, while the monitoring of reintroduction sites took place in the spring. These data were collected to develop and understand the life history and population biology of the natural populations and to compare to data taken at the new locations.

At Site 1, from 1985 to 1992, the mean diameter of all plants increased from about 13 millimeters to 23 millimeters, and about 70 percent of the original plants have survived (Figures CS2-1 and CS2-2). At the type locality, with the exception of 1989 when not all plots were censused, cactus counts from study plots indicate that the population is increasing over time (Figure CS2-3). For several years, a higher percentage of plants produced flowers and fruits at the new population than at the type locality. The percentage of plants producing flowers and fruits has gradually become more equal at both locations over time (Figure CS2-4). Five years after planting, eight seedlings were observed in the seed grid in 1992.

Site 2

An additional 250 cuttings were taken from the type locality in the spring of 1991. Each of the parent plants used for the 1991 cuttings were marked, so that they could be monitored for any mortality that resulted from the cut. These cuttings were also taken to the greenhouse, where they established roots. In September 1991, a total of 149 of these rooted cuttings were planted at the

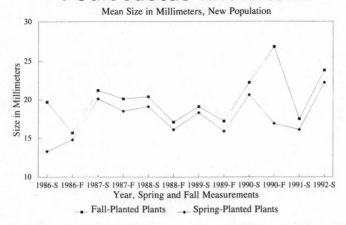

FIGURE CS2-1.   Mean Knowlton's cactus size in millimeters, site 1 population.

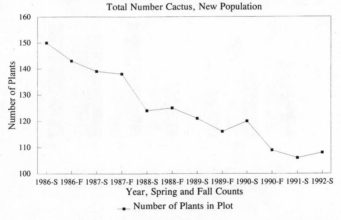

FIGURE CS2-2.   Total number of Knowlton's cactus, site 1 population.

second reintroduction site. This site differed from the first in method of planting and grid pattern. At this site, the transplants were placed in the ground with bare roots rather than with the pot contents. Ten clusters of five plants each (about 10 centimeters apart within the clusters) were spaced at 2-meter intervals along each of three grid lines. By planting closer together, agency biologists hoped to facilitate and increase pollination and seed set.

Agencies and organizations involved in planning and implementing the

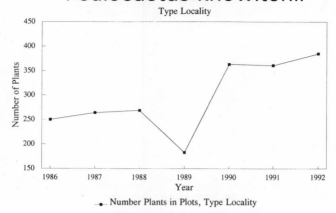

FIGURE CS2-3.   Number of plants in plot, type locality, Knowlton's cactus.

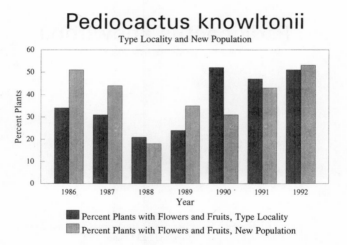

FIGURE CS2-4.   Percent plants with flowers and fruits, type locality and new population, Knowlton's cactus.

various components of this project at both sites include the U.S. Department of the Interior, Fish and Wildlife Service, Bureau of Land Management, and Bureau of Reclamation; the New Mexico Energy, Minerals, and Natural Resources Department, Forestry and Resources Conservation Division and State Parks and Recreation Division; and the Nature Conservancy. The primary monitoring responsibilities have been met by personnel of the Forestry and Resources Conservation Division, through cooperative agreements with the U.S. Fish and Wildlife Service.

Bioregional Context

This project is not part of a landscape-level plan. There are oil and gas and livestock grazing activities on adjacent lands that could affect the restoration project by impacting the reintroduction sites directly and indirectly and by limiting the amount of unoccupied habitat suitable for recovery.

Funding

Since 1985, the reintroduction project has been funded by the U.S. Fish and Wildlife Service. Some matching funds have been provided by the State of New Mexico. Logistical support has been provided by State Parks and Recreation and the Bureau of Land Management. The Fish and Wildlife Service anticipates funding will continue to be available for this species at least until 1995, and continued matching funds are expected from the State of New Mexico. The Bureau of Land Management is anticipated to continue to support the part of the project that is on lands under their management. Approximately $100,000 federal and $25,000 state have been spent.

Partnerships

- Fish and Wildlife Service: Peggy Olwell, Ann Cully, Charles McDonald

- Bureau of Land Management: James M. Ramakka, William Falvey, Jan Knight

- Bureau of Reclamation

- New Mexico Energy, Minerals, and Natural Resources Department:

  Forestry and Resources Conservation Division: Paul Knight, Robert Sivinski, Karen Lightfoot, Gregory Fitch

  Parks and Recreation Division: Mike Maddox, Larry W. Frederici

- The Nature Conservancy: John Egbert, Richard Johnson

- Mesa Gardens: Steve Brack

Policy or Regulatory Context of Reintroduction

The U.S. Fish and Wildlife Service Recovery Plan for this species was completed and approved on March 29, 1985. The establishment of new populations, or reintroduction, was a priority 1 task for the recovery of the species.

REFERENCES

Benson, L. 1982. *The Cacti of the United States and Canada*. Stanford, Calif.: Stanford University Press.

Heil, K. 1985. *Recovery Plan for* Pediocactus knowltonii *(Knowlton's Cactus)*. Albuquerque, N. Mex.: U.S. Department of the Interior, Fish and Wildlife Service, New Mexico Ecological Services Field Office.

Texas Snowbells
(*Stryrax texana*) Reintroduction

Charles B. McDonald

———————⟫•⟪———————

Taxon

Styrax texana Cory; Texas snowbells (Styracaceae).

Endangerment Status and Date of Listing

Federal: endangered October 12, 1984 (U.S. Fish and Wildlife Service 1984).
State of Texas: endangered January 9, 1987 (Poole and Riskind 1987).

Natural Distribution and Ecology of Taxon

Louis V. Cory discovered Texas snowbells on July 4, 1940, during one of his botanical explorations of the Texas Edwards Plateau. He formally named the species in 1943 (Cory 1943). Between 1940 and 1942, specimens were collected from six Texas sites: three in Edwards County, two in Real County, and one in Val Verde County. The next new site was discovered in Real County in 1980 (Mahler 1981). Two additional sites were discovered in Val Verde County in 1992 (T. James Fries, The Nature Conservancy of Texas, Texas Hill Country Bioreserve, Austin, Texas, personal communication, June 1992). Of the nine known sites, all are presently confirmed extant except the 1940s site in Val Verde County. The Edwards and Real county sites occur within about 32 kilometers (20 miles) of each other. The 1992 Val Verde County sites occur about 80 kilometers (50 miles) to the west.

Texas snowbells is a long-lived deciduous shrub or small tree that reaches 5 meters (16 feet) tall. Upper-leaf surfaces are bright green and lower-leaf

surfaces are conspicuously white with a fine, dense, silky tomentum. This contrast makes plants quite conspicuous when the slightest breeze rustles the leaves. Plants are even more conspicuous in the spring from a profusion of white blossoms that appear with the leaves. Because plants are large and easily seen from a distance, and because the Texas Edwards Plateau is botanically relatively well explored, it appears Texas snowbells is truly a rare species.

The Edwards Plateau is predominately formed from thick horizontal beds of limestone that in the eastern and southern parts of the plateau are deeply incised and form cliffs and steep-walled canyons. The region is locally known as the Hill Country. Areas that support Texas snowbells are lightly wooded vertical limestone cliffs, usually along rivers or streams. Large cracks in most cliffs provide plant-anchoring sites.

Threats to Population

Texas snowbells is quite palatable to wildlife and domestic livestock, and it is believed its present restriction to cliff habitats is a direct response to browsing pressure. Threats to Texas snowbells include:

1. *Domestic livestock, exotic game, and native wildlife browsing* . The Edwards Plateau has been grazed heavily for more than a century and is presently a major region for wool and mohair production. Sheep and particularly goats are the principal domestic livestock threatening Texas snowbells. Exotic game such as blackbuck, axis and fallow deer, Barbados sheep, and aoudad have been introduced on private land for sporting purposes. Many animals have escaped and are reproducing in the wild. White-tailed deer severely overpopulate much of the Edwards Plateau.

2. *Seed and seedling predation.* Texas snowbells produces abundant, highly viable seeds, but there is little recruitment into the populations. An experiment using exclusion cages indicates there is heavy small-mammal seed predation (probably from rock squirrels), rooting animal seedling depredation (probably from armadillos), and seedling browsing from the various animals mentioned above (Poole 1991).

3. *Flooding and erosion.* Floods, which are frequent in the region, wash away seeds that drop into streams or dry streambeds. Erosion poses a threat to some of the species' cliff habitat.

Conditions That Initiated Reintroduction

When the reintroduction was undertaken in 1987, only thirty-nine Texas snowbells plants were known in the wild at six sites. One site had twenty-five plants and one eight, but the other four had only one or two plants each. All plants were mature to old with no evidence of recent recruitment.

Objectives of Reintroduction

The Texas Snowbells Recovery Plan called for plant reintroductions into natural habitats to reduce the likelihood of extinction from natural random events and to shift the species' age-class structure toward younger plants (U.S. Fish and Wildlife Service 1987). The plan considered reintroduction to be an action that must be taken to prevent extinction or to prevent the species from declining irreversibly in the foreseeable future. The reintroduction was intended to reduce the short-term probability of extinction while solutions were being found to the underlying threats.

The reintroduction will be successful if the reintroduced plants can reproductively replace themselves in a self-sustaining population. However, it could take decades to determine the project's success or failure because of the longevity of individual plants. It will be difficult to sustain monitoring for this over a long period of time; however, once plants are established, there will also be considerable time for further management or experimentation.

Description of Reintroduction Sites

The reintroduction sites had to meet several criteria: (1) they must be within the historical range of Texas snowbells; (2) they must be generally similar to known occupied habitats; (3) they must be free from conflicting land uses; and (4) no sycamore-leaf snowbells (*Styrax platanifolia*) could be on or near the sites because of the possibility of hybridization, which is believed unlikely but still untested. Federal lands were the first choice, but none are in the area. State lands were surveyed and either had sycamore-leaf snowbells populations or provided no suitable habitat. Private land sites were selected based on the above criteria and contact with cooperative private landowners. Ultimately, two sites were selected: one each in Uvalde and Real counties. One reintroduction site was a limestone cliffside; the other was a steep hillside with limestone outcroppings. Neither site was known to have supported Texas snowbells in the past. Each site occupied less than 2 hectares (5 acres).

Description of Reintroduction Project

The reintroduction project began in the summer of 1985, when the San Antonio Botanical Gardens, a participating institution of the Center for Plant Conservation, collected seeds from what then were known to be the two largest Texas snowbells populations. The seeds were moist-stratified for six to eight weeks to simulate cold dormancy. The germination rate was high, and seedlings grew well in standard potting soil (Cox 1987). Pot-grown seedlings 1.5 years old and 12 to 20 centimeters (5 to 8 inches) tall were ultimately selected over seeds for reintroductions because of known seed predation in the wild and the vulnerability of germinating seedlings to natural environmental perturbations.

Twenty-five seedlings were planted at one reintroduction site, and twenty-four were planted at the other. The two original seed lots were kept separate, and material from a single source was used at each site. Because of the irregular terrain, no standardized planting pattern could be established. Plants were placed where soil and moisture conditions seemed to most favor plant establishment. Record rainfall in May and early June 1987, following the planting of the first site, washed out or damaged many plants. This site was replanted in July 1987. Although originally planned for the spring, the second site was not planted until July 1987.

A protective wire cage was placed around each seedling when it was planted. The cages were checked regularly to be sure they were still in place and to remove accumulated debris. Watering was also done as needed; no watering has been needed since 1992. As the plants grew, the original cages were replaced with larger ones.

A simple monitoring data sheet was developed to record general plant condition, phenological information, and watering dates. Monitoring was done bi-monthly for the first growing season, then monthly through 1992, when scheduled project funding ended. Monitoring since 1992 has continued irregularly on a volunteer basis.

Three plants died during the first summer, and one died during the first winter. Only seven of the original forty-nine plants have died since the project began. These plants were not replaced. The largest plants are now 1.4 to 1.5 meters (4.5 to 5.0 feet) tall and almost tall enough that their leaves are above the "browse line." Two plants flowered for the first time in 1992, five years after planting (T. Keeney, Southwest Texas Junior College, personal communication, June 1993).

Bioregional Context

The reintroduction was not part of a landscape-level plan. Expansion of the natural and reintroduced Texas snowbells populations will require regional

management of both domestic livestock and wildlife to reduce browsing impacts. Any regional management plan would be difficult to implement because most land is privately owned, and the browsing animals contribute significantly to the regional economy through either wool production or sport hunting.

Funding

This project was financed by the U.S. Fish and Wildlife Service. Funds expended were

$ 2,900   greenhouse propagation of seedlings
$ 7,500   population establishment
$ 7,500   monitoring for five years
$17,900   total

There is no future committed fiscal support for this project.

Partnerships

- U.S. Fish and Wildlife Service
- San Antonio Botanical Gardens (as a Center for Plant Conservation participating institution)
- Texas Parks and Wildlife Department
- Southwest Texas Junior College (through the participation of Toney Keeney)
- Two private landowners on whose lands the plants were reintroduced

Policy or Regulatory Context of Reintroduction

Texas snowbells is listed as endangered under the Endangered Species Act of 1973, as amended (16 U.S.C. 1531 *et seq.*). Conservation measures provided to listed species include recognition, recovery actions, requirements for federal protection, and prohibitions against certain practices. Recognition through listing encourages and results in conservation actions by federal, state, and private agencies, groups, and individuals. The act provides for possible land acquisition and cooperation with the states and requires that recovery actions be carried out for all listed species.

The U.S. Fish and Wildlife Service's Texas snowbells recovery plan was

approved July 31, 1987. Recovery plans delineate actions believed required to recover and/or protect listed species. Approved recovery plans represent the official position of the U.S. Fish and Wildlife Service but not necessarily the official positions or approval of other agencies, organizations, or individuals. Objectives in recovery plans are attained and funds expended based on budgetary and other constraints, and on the need to address other priorities. Recovery plans may be modified as dictated by new findings, changes in species' status, and the completion of recovery tasks.

Reintroduction is identified as a priority 1 task in the Texas snowbells recovery plan. (All tasks in recovery plans are assigned one of three priorities. A priority 1 task is an action the U.S. Fish and Wildlife Service considers must be taken to prevent extinction or prevent the species from declining irreversibly in the foreseeable future.)

REFERENCES

Cory, V. L. 1943. "The Genus *Styrax* in Central and Western Texas." *Madroño* 7 (4): 110–115.

Cox, P. 1987. "Chasing the Wild Texas Snowbells." *Plant Conservation* 2 (4): 1.

Mahler, W. F. 1981. "Status Report on *Styrax texana*." Albuquerque, N. Mex.: U.S. Fish and Wildlife Service.

Poole, J. M. 1991. "Reproductive Biology of Texas Snowbells (*Styrax texana*)." Section 6 project E-1, job 3-1. Austin, Tex.: Texas Parks and Wildlife Department.

Poole, J. M., and D. H. Riskind. 1987. *Endangered, Threatened or Protected Native Plants of Texas*. Austin, Tex.: Texas Parks and Wildlife Department.

U.S. Fish and Wildlife Service. 1984. Final rule to determine *Styrax texana* (Texas snowbells) to be an endangered species. *Federal Register* 49:40036-40038.

U.S. Fish and Wildlife Service. 1987. Texas snowbells (*Styrax texana*) recovery plan. Albuquerque, N. Mex.: U.S. Fish and Wildlife Service.

Apalachicola Rosemary
(*Conradina glabra*) Reintroduction

Doria R. Gordon

—————————————⇒•⊙•⇐—————————————

Taxon

Conradina glabra Shinners; Apalachicola rosemary (Lamiaceae).

Endangerment Status and Date of Listing

Federal: endangered July 12, 1993.
State: endangered September 20, 1993.

Natural Distribution and Ecology of Taxon

Apalachicola rosemary is endemic to Florida in xeric sandhills east of the Apalachicola River in Liberty County (Gray 1965). *C. glabra* was not described until 1962 (Shinners 1962), so its early distribution is unknown. Reconstructing that distribution is not possible, since most of the habitat had been converted to pine plantation by the late 1950s. *C. glabra* likely occupied the ecotones between sandhill and steephead ravines in the eastern Apalachicola River drainage. Prior to this reintroduction project, the Florida Natural Areas Inventory (in Tallahassee, Florida) had record of seven locations for *C. glabra*, six of which are on private timber company land, and one of which was in Torreya State Park and seems no longer to exist. Since the reintroduction, The Nature Conservancy (TNC) has discovered two additional locations on and adjacent to Apalachicola Bluffs and Ravines Preserve.

C. glabra is an obligately outcrossing (protandrous) perennial shrub that does not exceed 0.8 meters in height (Godfrey 1988).

Threats to Population

Threats to *C. glabra* come primarily from logging and from conversion of sites to sand pine silviculture, which produces an almost complete canopy. Apalachicola rosemary appears to require full sun to light shade (Kral 1983). Observations of this species' response to forestry site preparation indicate that no plants survive the common practice of roller chopping. The effects of other site preparation techniques (such as herbicide) are unknown. Conversion of habitat to pasture has also extirpated this species. Because fire is thought to have traditionally kept sandhill habitats open, fire exclusion may also threaten this species.

Records from the 1980s suggest that populations contain 25 to 300 plants; estimates for four sites are 25, 30, 100, 100, and 300, so over 555 plants existed at that time (data provided by the Florida Natural Areas Inventory). These numbers may include ramets because this species is thought to be clonal (Kral 1983).

Conditions That Initiated Reintroduction

The Florida Region of TNC decided to conduct the reintroduction for several reasons: (1) Apalachicola rosemary was endangered; (2) the species could no longer be located at the only known protected site; (3) the TNC preserve proposed as a reintroduction site was within the historical range and contained appropriate habitat for Apalachicola rosemary; (4) the species was present in two locations within 2.5 kilometers from the preserve; and (5) propagules were available from the Center for Plant Conservation's (CPC) collection at Bok Tower Gardens in Polk County, Florida. Additionally, the species could easily have been extirpated from the Apalachicola Bluffs and Ravines Preserve during silvicultural site preparation (Figure CS4-1). The reintroduction was for conservation purposes and not part of a mitigation effort.

Objectives of Reintroduction

No recovery plan for Apalachicola rosemary existed when the reintroduction was initiated. This reintroduction will be considered successful by TNC if

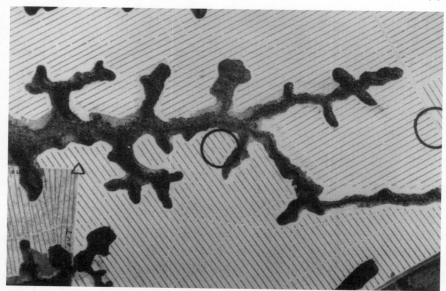

FIGURE CS4-1.   This 1959 aerial photograph of a portion of the Apalachicola Bluffs and Ravines Preserve illustrates the effect of windrow site preparation on the landscape (disregard symbols). The area that appears to be white is the exposed sandy soil that results from removal of the surface soil and vegetation to create competition-free planting sites for pine silviculture. The darker areas show locations where vegetation remains. Linear dark areas are where the soil and vegetation were piled when the clearing (windrows) occurred; irregularly shaped dark areas are the steephead ravines that support hardwood slope forest and drain into the Apalachicola River. Apalachicola rosemary is thought to have occupied the ecotones along the upper edges of the ravines. Reintroduced populations are in windrowed areas along those edges.

seedling establishment and reproduction is observed in all three sites when fire is reintroduced at natural frequencies. Establishment should balance mortality of adults such that the populations initially grow and then develop a stable age structure. Because we do not have information about the natural longevity, reproduction, and recruitment of individuals in this species, we are unable to estimate the timing for populations to stabilize. Early data on survival following fire and some seedling establishment suggest that the reintroduction will be successful.

Description of Reintroduction Sites

The Apalachicola Bluffs and Ravines Preserve (Liberty County, Florida), which is owned and managed by TNC, was selected as the reintroduction site because it is a protected property within the original range of *C. glabra*

containing the hypothesized original habitat. Management of the site in the 1950s for silviculture (prior to TNC ownership) involved the creation of windrows across the site as preparation for planting pines (see Figure CS4-1). This disturbance could easily have extirpated this species. Reintroduction sites were selected based on original habitat descriptions, uniformity of soils, orientation relative to steepheads, vegetation, and site management history and needs.

C. glabra was introduced to three sites at the preserve. All sites are within 1 kilometer of each other, on deep entisols. Elevation at all sites is roughly 61 meters (200 feet). Planting was within 45-by-80-meter grids at each site, with the longer axis along and parallel to the northern edge of a steephead ravine.

The sites are all xeric sandhill communities within the Northern Highlands physiographic region of the southeastern coastal plain. The tree canopy of this community is dominated by longleaf pine (*Pinus palustris*) and turkey oak (*Quercus laevis*), while the understory contains a diversity of herbaceous and woody species.

Climate in the region is characterized by wet, hot summers and cool, dry winters during which temperatures drop below $0°C$ twenty to thirty times per year. Summer rain may occur daily from convective thunderstorms from May through August, while winter rain, primarily from December through March, follows northern cold fronts. Annual rainfall ranges between 1,041 and 1,778 millimeters (Leitman, Sohm, and Franklin 1984).

Description of Reintroduction Project

When originally conducted, the translocation was considered an introduction because we had no records of populations of C. glabra at the Apalachicola Bluffs and Ravines Preserve. Discovery of a few individuals on and adjacent to the preserve in early 1993 indicates that this is more appropriately considered a reintroduction project.

Planting was conducted on February 17, 1991. Plants are currently ending their fifth growing season. Survival of these plants will be monitored annually. We planned to intensively follow the plants' growth rates the first season, somewhat less intensively the second season, and then only follow their survival into the future (no endpoint planned). New recruits into the population will be marked and survival tracked annually.

Planting material consisted of six-month-old cuttings, rooted from forty-eight propagated plants in the Bok Tower Gardens CPC collection. The collection plants were also propagated from cuttings from forty-eight plants in one of the natural populations within 2.5 kilometers of the preserve. Plants were transplanted with soil in February 1991 during a cold front. No watering

or other site amendments were added. *C. glabra* from randomly selected parents were planted in forty-eight clusters of nine plants each in each of the three sites. Cluster locations were randomly selected among the 5-meter intersections within the 45-by-80-meter grids. This design resulted in three relatively localized but clustered and randomly distributed populations. Forty-eight plots of nine plants each resulted in 432 plants in each of the three sites ($n=1,296$).

To examine whether competition would affect survival and growth, above-ground tissue of all vegetation within half of the plots at each site was clipped monthly through the first two growing seasons. One of the populations was subjected to prescribed fire in the second year to examine response.

All monitoring has been conducted by TNC staff and volunteers at Apalachicola Bluffs and Ravines Preserve. In the first year, survival of all plants and the height and number of woody branches were counted monthly in three plants randomly selected from each group of nine. The percent of branches on each plant bearing flowers was also recorded. Measurements were restricted to the beginning and end of the growing season in the next year. Only survival and recruitment of seedlings will be monitored in subsequent years.

Survival in the unburned plots was 95 percent after the first two years. Clipping had no effect on survival but significantly increased growth and flowering of plants. Several of the plants resprouted following fire. Thus far, seedling establishment is highest in the burned sites and almost entirely limited to clipped plots (Gordon 1995).

Sites will be managed consistently with the surrounding area. Prescribed fires (every two to five years) will be conducted during the lightning season as fuels allow. Over the long-term, we intend to restore other components of the over- and understory, starting with replacement of off-site slash pine (*Pinus elliottii* var. *elliottii*) with longleaf pine and reintroduction/augmentation of perennial bunchgrasses in the understory (such as *Aristida stricta*, *Sporobolus junceus*).

TNC will support continued monitoring of this species and site management through operational funds. In general, TNC interns conduct plant monitoring after training and assist with management. CPC (Susan Wallace at Bok Tower Gardens) and TNC were involved in planning the reintroduction.

Bioregional Context

This project is not within a landscape-level plan for this particular species, but the preserve management is moving toward that scale. Over the long term, we hope to increase the scale of prescribed fires and other processes and restore

vegetation across the uplands. No current threats to the translocated populations from activities on adjacent lands have been identified.

Funding

The original collection and propagation of Apalachicola rosemary was supported by Bok Tower Gardens and CPC. CPC transported the plants to the
site with TNC financial support. All further support for this reintroduction
has been from TNC's International and Florida offices. The Florida Program
continues this support (using intern and volunteer assistance). No accounting of funds or time spent on this project has been maintained. TNC is
committed to managing this preserve and monitoring these populations in
perpetuity as part of our other management and monitoring activities on site.

Partnerships

- CPC
- Bok Tower Gardens
- TNC and several TNC volunteers

Policy and Regulatory Context of Reintroduction

The U.S. Fish and Wildlife Service completed the recovery plan for C. *glabra*
in 1994.

REFERENCES

Godfrey, R. K. 1988. *Trees, Shrubs, and Woody Vines of Northern Florida and Adjacent
    Georgia and Alabama*. Athens: University of Georgia Press.
Gordon, D. R. In press. "Experimental Translocation of the Endangered Shrub *Con-
    radina glabra* (Apalachicola Rosemary), to the Apalachicola Bluffs and Ravines Pre-
    serve, Florida. *Biological Conservation*.
Gray, T. C. 1965. *A Monograph of the Genus* Conradina A. Gray *(Labiatae)*. Ph.D. diss.,
    Vanderbilt University, Nashville, Tennessee.
Kral, R. 1983. *"Conradina glabra"* In *A Report on Some Rare, Threatened, or Endangered
    Forest-Related Vascular Plants of the South*. USDA Forest Service Technical Publi-
    cation R8-TP 2.
Leitman, H. M., J. A. Sohm, and M. A. Franklin. 1984. *Wetland Hydrology, Tree Distribu-
    tion, Apalachicola River Flood Plain, Florida*. U.S. Geological Survey, Water-Supply
    Paper 2196-A.
Shinners, L. H. 1962. "Synopsis of *Conradina* (Labiatae)." *Sida* 1: 84–88.

Pitcher's Thistle (*Cirsium pitcheri*) Reintroduction

Marlin Bowles and Jeanette McBride

Taxon

Cirsium pitcheri; Pitcher's thistle

Endangerment Status and Date of Listing

Cirsium pitcheri was determined to be federally threatened in 1988; it is state threatened in Illinois, Indiana, and Michigan and endangered in Wisconsin (Harrison 1988).

Natural Distribution and Ecology of Taxon

Pitcher's thistle (*Cirsium pitcheri*) is endemic to dune systems of the western Great Lakes shoreline (Figure CS5-1). This thistle requires 70 percent open sand, which restricts its habitat to sand dunes where wind-generated disturbance processes maintain successional vegetation and open sand (McEachern 1992). Because of the dynamics of this habitat, metapopulations are required for persistence of *C. pitcheri*, allowing it to survive local population extinctions by colonizing new habitat from other populations (McEachern, Bowles, and Pavlovic 1994).

Cirsium pitcheri is a striking species. The stems and leaves of both juveniles and adults are wooly white and deeply pinnatifid, with larger spines between the lobes of the distal leaf margins' spines; flowering stems reach 1 meter and flowering heads are cream to light pink in color (Pavlovic et al 1993). Like many early-successional plants, *Cirsium pitcheri* is a monocarpic

FIGURE CS5-1.   Location of *Cirsium pitcheri* populations and habitats in the western Great Lakes and location of Illinois Beach State Park restoration sites (Bowles et al. 1993; McEachern, Bowles, and Pavlovic 1994).

perennial; plants flower after five to eight years and then die, with population maintenance dependent upon reproduction from seeds (Loveless 1984). This reproductive strategy presents a challenge for restoration, requiring establishment of staged cohorts that allow replacement of flowering plants in the population. Transplanting of greenhouse-propagated plants can accelerate cohort development and maturity, but may avoid environmental selection at the seedling stage and determination whether plants can actually complete their life cycle.

Threats to Taxon

Although Pitcher's thistle is adapted to a dynamic habitat, it is susceptible to disturbances that occur out of phase or with higher frequency, severity, or magnitude than natural dune processes (Bowles et al. 1993). For example, disturbances such as recreational impacts that consistently remove seedling cohorts or prevent seed production would lead to population decline (Dobberpuhl and Gibson 1987; McEachern 1992). Impacts such as sand mining, shoreline development, dune and shoreline stabilization, or disruption of shoreline currents that replenish eroded shorelines can eliminate thistle habitat and are now widespread throughout the range of the species (Pavlovic et al. 1993; Bowles et al. 1993). *Cirsium pitcheri* is also subject to potentially damaging seed predation, either in the seed head or after seed dispersal (Loveless 1984; Keddy and Keddy 1984), which can affect levels of seed production and seedling establishment that are critical for population maintenance.

Conditions That Initiated Reintroduction

Because of the narrow, linear Lake Michigan shoreline dune habitat in Illinois, former Pitcher's thistle populations there were probably extremely vulnerable. They apparently disappeared from Illinois shoreline dunes before 1920 due to the combined effects of increasing human activity, lake level fluctuations, collecting, and other chance events (Bowles et al. 1993).

Objectives of Reintroduction

With federal listing and recovery planning, restoration of a Pitcher's thistle metapopulation in Illinois became an important goal to test whether the species could be successfully restored, thereby improving its population status and protecting it from extirpation (Pavlovic et al. 1993). Protection of a large portion of the Illinois shoreline as a state park provided the only remaining Illinois dune system where *Cirsium pitcheri* restoration could be tested (Bowles et al. 1993). Because *C. pitcheri* persists as metapopulations, reintroduction into multiple habitats was required (McEachern et al. 1994).

Description of Reintroduction Sites

Illinois Beach State Park is located 70 kilometers north of Chicago on a low (up to 3 meters of relief), narrow (1.5-kilometer wide) sand deposit that

extends for over 20 kilometers along the Lake Michigan shoreline. The shoreline is dynamic, with sediment transport southward by the longshore current; former beach ridges form a compressed dune field north of the Dead River and a more widely spaced dune ridge and swale system south of the River (Figure CS5-2). In a detailed study, secondary dunes in a protected nature preserve south of the Dead River were found to replicate appropriate habitat for this species and appeared to be free from shoreline erosion and recreational impacts (Bowles 1991; Bowles et al. 1993). This area was recommended to the Illinois Endangered Species Protection Board as an initial site for reintroduction of Pitcher's thistle (Bowles 1991). To meet metapopulation requirements dunefield habitat north of the Dead River was also recommended for reintroduction of this species (McEachern *et al.* 1994).

Description of Reintroduction Project

With approval of the Illinois Department of Natural Resources and the Illinois Nature Preserves Commission, *Cirsium pitcheri* reintroduction to Illinois Beach began in 1991. Propagules for this restoration were derived from seeds collected by permit in 1990 from Indiana and southern Wisconsin. Seed collection in subsequent years also included southern Michigan populations. Seeds were moist stratified and greenhouse propagated the following spring. After hardening, seedlings were outplanted in early August. In 1991, seventy-seven Indiana and Wisconsin *C. pitcheri* seedlings were translocated into secondary dune habitat south of the Dead River. In 1992, a new cohort of eight Indiana seedlings was also introduced (Bowles et al. 1993). In 1993, a third cohort of seventy-nine Indiana, Wisconsin, and southwestern Michigan seedlings was introduced. In addition 853 seeds that had been moist stratified at the Morton Arboretum were planted in early April 1993.

In 1994, twenty-two seedlings were translocated, while 878 seeds were planted within the population south of the Dead River. To begin establishing a metapopulation, seedlings were also planted in dunefield habitat north of the Dead River in 1994. This area receives recreational disturbance, which may help maintain open habitat but might negatively impact thistle seedlings (Dobberpuhl and Gibson 1987), thus allowing an examination of recreational impact on thistles.

In 1992, twenty-five (32.5 percent) of the 1991 cohort had survived, with a higher over-winter mortality rate among Wisconsin plants (Table CS5-1). In 1993, four of the 1992 cohort survived, resulting in a population of over one hundred plants. By 1994, fifty-three of the *C. pitcheri* transplants for 1991, 1992, and 1993 were alive at Illinois Beach, representing 32.3 percent survival. Greatest mortality occurred over the first winter for each cohort and

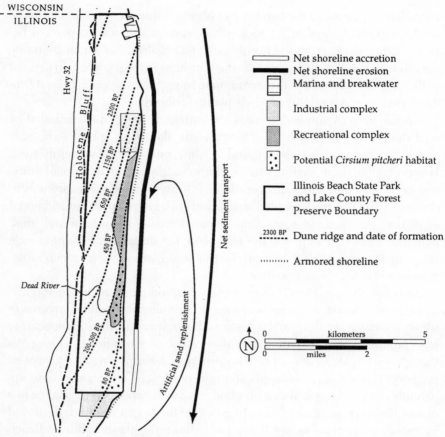

FIGURE CS5-2.   Locations of potential *Cirsium pitcheri* habitat in Illinois Beach State Park (Bowles et al. 1993; McEachern, Bowles, and Pavlovic 1994). Populations have been restored in habitat north and south of the Dead River.

TABLE CS5-1.   *Cirsium pitcheri* transplant survival at Illinois Beach. Percent cohort survival is in parentheses.

	1991	1992	1993	1994
1991 cohort	77	25 (32.5)	25 (100)	23* (92)
1992 cohort	—	8	4 (50)	4 (100)
1993 cohort	—	—	79	26 (32.9)
1994 cohort				22
TOTAL	77	33	108	75

*Two plants of the 1991 cohort flowered and died in 1994.

exceeded 60 percent in the two largest cohorts. Subsequent mortality was almost nonexistent. Among the 1991 cohort, survivorship was 92 percent between 1992 and 1994, leaving twenty-three plants alive. No winter mortality occurred among the 1992 cohort during transition to 1994, while 32.9 percent of the 1993 cohort survived the transition to 1994. Thus, we expect about two-thirds over-winter mortality of newly planted cohorts.

Germination of outplanted seeds was extremely low and confounded by seed dormancy (Table CS5-2). Of 853 seeds planted in spring 1993, only eleven (1.3 percent) germinated and became established by 30 July 1993. However, all of these seedlings survived into 1994, when an additional thirty-two previously dormant seeds germinated, thirty of which were from Wisconsin. Of the 878 seeds planted in 1994 only two germinated, but additional seeds may germinate in 1995. Although seed predation by birds and small mammals may have depleted the seed plots, the delayed germination indicates that some seeds survived. Germination among the greenhouse-germinated seeds was 22.1 percent.

In its fourth year, the *Cirsium pitcheri* reintroduction included seventy-three *ex situ*–propagated and translocated juvenile plants and forty-five *in situ*–germinated seedlings and juveniles. High rainfall during 1993 appears to have promoted plant growth, which allowed two plants to flower in 1994. Although insecticide treatment of flowering plants was planned to deter seed predators (Louda 1994), no applications were made. This treatment was apparently unneeded, as fourteen hundred seeds were produced among the two plants. However, the central flowering stem of the largest plant was destroyed by animal herbivory. The small number of flowering plants at Illinois Beach may have failed to attract seed predators, or seed predators may not be present. In comparison, eight flowering Wisconsin plants in a sand bed at the Morton Arboretum also produced fourteen hundred seeds. Seed production in this artificial population was over 20 percent higher in unbagged plants (26.1 percent of all ovules) than bagged plants (4.7 percent of all ovules), apparently due to outcrossing by pollinators. Although seed predation may have been unnoticed, it was apparently not a factor in the different levels of seed production at the arboretum. If animal herbivory continues, it could be a critical factor affecting growth of this small population.

Bioregional Context

Cirsium pitcheri possesses a low level of genetic variation, but it has a negative correlation between genetic similarity and geographic distance between populations (Loveless and Hamrick 1988). We found apparent genetic differences between Indiana and Wisconsin seed sources, with Indiana plants better adapted to habitat at Illinois Beach. A significantly smaller cotyledon

TABLE CS5-2.   Seed sources, seed numbers, and germination of
Cirsium pitcheri seeds introduced to Illinois Beach State Park.

	Wisconsin	Indiana	Michigan	Total
Seeds planted (1993)	648	205	—	853
Germinated (1993)	4	7	—	11
Germinated (1994)	30	2	—	32
Seeds planted (1994)	—	399	479	878
Germinated (1994)	—	0	2	2

size found in Wisconsin seedlings in 1991 (Bowles et al. 1993) was consistent during germination of seeds in 1993, and all Indiana and Michigan seedlings germinated in 1994 had similar cotyledons. Although Indiana plants had greater suvivorship (45 percent) than Wisconsin plants (16.2 percent) in 1992 (Bowles et al. 1993), there was no significant difference in survivorship between the 1992 Wisconsin cohort (36.1 percent) and Indiana cohort (37.9 percent) in 1994. The 1994 Michigan cohort had 16.7 percent survival in 1994.

Older cohorts of the Indiana plants have out-performed Wisconsin plants. At the time of planting in 1991, greenhouse plants from both sources had similar numbers of leaves and leaf lengths (Bowles et al. 1993). By 1992, Wisconsin plants were smaller in all aspects of morphology except number of leaves; by 1993, almost all plants from both sources had doubled in size, but Indiana plants were larger and had more leaves (unpublished data). Basal root-crown width appears to be a good indicator of plant growth (McEachern 1992). Only two plants, both from Indiana, had root-crown widths exceeding 1 centimeter in 1993, and these plants flowered in 1994. Population structure, based on root-crown width size-classes in 1994, was also dominated by Indiana plants (Figure CS5-3). Twelve of the Indiana plants had widths exceeding 1 centimeter in 1994, but only one Wisconsin plant was that large. This indicates that the Indiana plants have greater potential to flower or are more expedient in reaching flowering size.

Possible explanations for these potential genetic differences include geographic and ecological factors (Bowles et al. 1993). The Indiana seed source is 70 kilometers closer to Illinois Beach than the Wisconsin seed source, and its more southern habitat may have selected for adaptation to habitat conditions that are similar to Illinois Beach.

Funding

Federal funding for this project was coordinated by John Schwegman through the Illinois Department of Natural Resources, and by Amelia Orton-Palmer through the U.S. Fish and Wildlife Service.

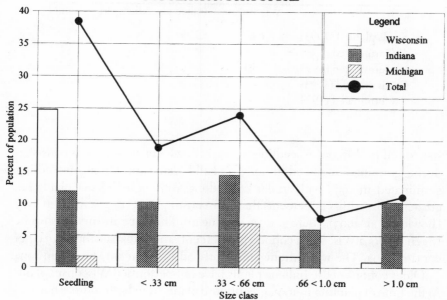

FIGURE CS5-3.   Population structure of 118 Pitcher's thistles at Illinois Beach Nature Preserve. Percentage of population is given for each of three seed sources among five size classes. Seedling size class indicates seedlings and one-year-old juveniles derived from on-site germination of seeds. Other size classes are 1994 basal root widths in centimeters of greenhouse-propagated plants translocated to Illinois Beach.

Partnerships

- Illinois Department of Natural Resources (through participation of Randy Heidorn and John Schwegman).

- Illinois Endangered Species Protection Board (through participation of Sue Lauzon).

- Illinois Nature Preserves Commission (through participation of Steve Byers).

- Illinois Beach State Park (through participation of Bob Grosso).

- Lake County Forest Preserve District (through participation of Ken Klick).

- U.S. Fish and Wildlife Service (through participation of Amelia Orton-Palmer).

Policy or Regulatory Context of Reintroduction

There are no direct policy or regulatory effects on the restored population. Because the plants occur within a state park and nature preserve, they are protected by Illinois state park and nature preserve statutes.

ACKNOWLEDGMENTS

We thank Cambridge University Press and the Natural Areas Journal for permission to use figures from their publications.

REFERENCES

Bowles, M. L. 1991. "Illinois Reintroduction Plan for the Federally Threatened Pitcher's Thistle (Cirsium pitcheri)." Lisle, Ill.: The Morton Arboretum.

Bowles, M., R. Flakne, K. McEachern, and N. Pavlovic. 1993. "Recovery Planning and Reintroduction of the Federally Threatened Pitcher's Thistle (*Cirsium pitcheri*) in Illinois. "*Natural Areas Journal* 13: 164–176.

Dobberpuhl, J. M., and T. J. Gibson. 1987. "Status Surveys and Habitat Assessment of Plant Species. I. *Cirsium pitcheri* (Torr.)" T. & G. Madison: Wisconsin Department of Natural Resources.

Harrison, W. F. 1988. "Endangered and Threatened Wildlife and Plants: Determination of Threatened Status for *Cirsium pitcheri*." *Federal Register* 53(137): 27137–27141.

Keddy, C. J., and P. A. Keddy. 1984. "Reproductive Biology and Habitat of *Cirsium pitcheri*." *The Michigan Botanist* 23: 57–67.

Louda, S. 1994. "Experimental Evidence for Insect Impact on Populations of Short-Lived, Perennial Plants, and Its Application in Restoration Ecology." In *Recovery of Endangered Species*. Edited by M. Bowles, and C. Whelan. Cambridge, England: Cambridge University Press.

Loveless, M. D. 1984. "Population Biology and Genetic Organization in *Cirsium pitcheri*, an Endemic Thistle." Ph.D. diss. University of Kansas, Lawrence.

Loveless, M. D., and J. L. Hamrick. 1988. "Genetic Organization and Evolutionary History in Two North American Species of *Cirsium*." *Evolution* 42: 254–265.

McEachern, K. 1992. "Disturbance Dynamics of Pitcher's Thistle (*Cirsium pitcheri*) in Great Lake Sand Dune Landscapes." Ph.D. diss.., University of Wisconsin, Madison.

McEachern, K., M. Bowles, and N. Pavlovic. 1994. "A Metapopulation Approach to Pitcher's Thistle (*Cirsium pitcheri*) Recovery in Illinois." In *Recovery of Endangered Species*. Edited by M. Bowles and C. Whelan. Cambridge, England: Cambridge University Press.

Pavlovic, N. B., M. L. Bowles, S. R. Crispin, T. C. Gibson, K. Herman, R. Kavetsky, K. McEachern, and M. R. Penskar. 1993. "Draft Federal Recovery Plan for Pitcher's Thistle (*Cirsium pitcheri*)." Minneapolis, Minn.: U.S. Fish and Wildlife Service.

Southern Appalachian Rare Plant Reintroductions on Granite Outcrops

Bart R. Johnson

———————➤·○·◄———————

Taxa

Geum radiatum Gray (Spreading avens)
Calamagrostis cainii Hitchc. (Cain's reedgrass)
Juncus trifidus ssp. carolinianus Hamet-Ahti (One-flowered rush)
Carex misera Buckl. (Wretched sedge)
Trichophorum cespitosum (L.) Hartman (Deerhair bulrush)
Sibbaldiopsis tridentata (Sol.) Ryd. (Mountain cinquefoil)
Krigia montana (Michaux) Nutt. (Mountain cynthia)
Polytrichum appalachianum And. (Appalachian haircap moss)

Endangerment Status and Date of Listing

See following table (Table CS6-1).

Natural Distribution and Ecology of Taxon

The vascular plant species described previously, along with other federal- and state-listed species (including *Houstonia montana*, *Liatris helleri*, and *Solidago spithamaea*, all federally listed), form the nucleus of an herbaceous plant community found only on exposed rock outcrops on isolated mountain summits and high ridges in the Southern Appalachians. Where the rare species occur, they typically dominate the vegetation of exposed areas of the outcrops. Plants are distributed in distinct patches across the outcrops, which

Table CS6-1.   Legal status of plant species used in restoration, including date of listing. Species are listed in order of federal and state rankings.

Scientific name	Family	Fed. Status (date)	N.C. Status (date)
Geum radiatum	Rosaceace	E (4-5-90)	E (1990)
Calamagrostis cainii	Poaceace	2 (2-21-90)	E (1989)
Juncus trifidus			
ssp. carolinianus	Juncaceae	3B(1993)	E (1987)
Carex misera	Cyperaceae	3C (1993)	SR (1993)
Trichophorum cespitosum	Cyperaceae	—	C (1989)
Sibbaldiopsis tridentata	Rosaceace	—	W1 (1991)
Krigia montana	Asteraceae	—	W1 (1991)
Polytrichum appalachianum	Polytrichaceae	—	—

E = endangered
2 = insufficient data for listing at this time
3B = name that does not meet definition of a species
C = candidate
SR = significantly rare
W1 = rare but relatively secure
3C = more widespread or abundant than previously believed

themselves are generally embedded in a matrix, or at least remnants, of heath bald or grassy bald. The bald component usually is embedded in either spruce–fir or northern hardwoods forest. Habitat patches typically occur as discrete islands of substrate in a "sea" of exposed rock and lichen. Alternatively, they occur in shallow soils at the margin of larger soil masses that are occupied by shrubs and that terminate at exposed rock or vertical drops (Figure CS6-1).

For restoration, species were limited to those known from the restoration site, and included those species that were considered important elements of the community because of their regional rarity, local abundance, or functional role as physical habitat structure. *Geum radiatum, Calamagrostis cainii,* and *Carex misera* are endemic to the Southern Appalachians and known from a total of eleven sites, three sites, and twenty-five sites, respectively. *Trichophorum cespitosum* and *Juncus trifidus* are disjuncts at the southern terminus of their ranges. The other three species used are Appalachian endemics that are not restricted to rock outcrops but also occur in grassy balds in the high mountains. The historical distribution of these species is not well known, although they appear to have been components of more extensive alpine communities in the high mountains during the Pleistocene (Ramseur 1960; Wiser 1994). More recently, *Geum radiatum* has been extirpated from five of sixteen historically known sites and has declined at another eight (Murdock 1992).

FIGURE CS6-1.   *Geum radiatum* blooms on the edge of a cliff (Craggy Gardens, Blue Ridge Parkway, N.C.). Photo B. Johnson.

Threats to Population

Threats to this community vary from site to site. They are the same as those that have been described for *Geum radiatum*, which include, but are not limited to, visitor trampling, rock climbing, recreational and residential development, and changes in surrounding vegetation (Murdock 1992). At most sites, rare plant populations have suffered serious decline, primarily due to visitor trampling. The outcrop habitat is particularly suited to visitor use, offering unobstructed views and easy maneuverability compared to adjacent woody vegetation (Johnson 1992). Large areas may be completely denuded of soil and vegetation (Figure CS6-2). Visitor trampling is concentrated, at least initially, on wide, level ledges rather than on steep rock faces. At some sites, however, rock climbing has caused major disturbance, even on vertical cliffs. A second major threat is woody plant succession. When shrubs occupy adjacent, deeper soils, they may encroach on the rare herbs, most of which appear unable to survive beneath the shrubs. While visitor trampling often begins in open, central areas of the largest outcrops, shrub succession typically moves in from the margins. In combination, the two processes can form a biological vise that is devastating to rare plant populations.

FIGURE CS6-2.   Heavy visitor use scours rare plant habitat to bare rock at many sites in the region (Grandfather Mountain, Avery County, N.C.). Photo B. Johnson.

Conditions That Initiated Reintroduction

Initial research began when the National Park Service became concerned by an apparent decline of rare plant populations at Craggy Gardens, Blue Ridge Parkway, North Carolina. At that time, the cause of the decline was unknown. Studies conducted through the Park Service Cooperative Studies Unit, located at the University of Georgia, concluded that visitor trampling was the most immediate cause of habitat loss. Subsequent experimental management studies led to a recommendation for an integrated strategy of design, interpretation, and restoration (Johnson 1992). The goal was to satisfy visitor desires for viewing, seating, and exploration while protecting fragile habitat through a combination of overlook design, interpretation, and trail rerouting. The implementation of these measures reduced visitor trampling on the summit by 97 percent, opening the way for restoration of disturbed habitat.

Objectives of Reintroduction

The project had three primary objectives. The first was to restore denuded habitat patches, as closely as possible, to conditions and species composition found prior to recent human disturbance. The second was to use experimental reintroduction plots to test hypotheses about species' microhabitats and community dynamics. The third was to explore the feasibility of creating

populations on exposed rock along parkway roadcuts and of reclaiming habitat by managing woody succession on outcrops overgrown by shrubs. The descriptions that follow focus on the first objective.

Description of Reintroduction Site

Craggy Pinnacle, elevation 1,754 meters (5,892 feet), is located on the Blue Ridge Parkway in Buncombe County, North Carolina. It was selected for restoration because it is considered one of the most botanically important sites on the parkway and because it was the focus of a concurrent National Park Service project to reduce visitor impacts on rare plant habitat. The area surrounding the outcrop habitat is somewhat over 1 hectare in size, of which the outcrops occupy only a small portion. Restoration was performed on thirty habitat patches, with a total size of 11 square meters (120 square feet). While small in area, this represents a substantial proportion of rare plant habitat found near the summit. In addition, it includes a disproportionately high number of the habitat patches found on relatively flat ledges, which often have different species composition than those found on steeper faces.

Description of Reintroduction Project

Experimental restoration and reintroduction plots were established in 1990 and monitored through 1992. Based on the results of these trials, both technical and ecological, restoration plots were installed in 1993. Plant stock was propagated from seed collected on site, with the exception of *Geum radiatum*. Poor germination of *Geum radiatum* seed from Craggy Gardens led to an investigation of whether inbreeding depression might be reducing seed viability. Plants propagated from a population that a genetic study (Hamrick, Godt, and Johnson 1991) indicated might be a good donor population were used for restoration. Evidence from the genetic study, from analysis of seed production at three sites, and from the transplant experiments currently is being weighed to make a final decision on whether genetic supplementation appears appropriate.

Because the habitat was scoured to bare rock, soil reintroduction and stabilization were necessary. Methods included the use of biodegradable fabric to make habitat modules filled with soil that were fitted to ledges and crevices on the outcrops (Figure CS6-3). Several mini-experiments were included to explore options for moss propagation, seed sowing, and planting time. Five hundred seventy-five plants were transplanted, of which 350 went to the restoration plots, and the rest were evenly distributed between the road cut and successional management plots.

FIGURE CS6-3.   Trial restoration plots were established by fitting soil-filled bags of biodegradable fabric to denuded habitat and then transplanting seedlings. Photo B. Johnson.

In the trial restoration plots, first-year survival ranged from 38 to 75 percent for the different species (Johnson 1995). Much of this mortality was expected, however, because species were planted both in microhabitats that resembled those under which they naturally occur, and in those where they did not. This planting design allowed comparison of the growth and survival of plants distributed according to the microhabitat assessment with systematic plantings of species across all microhabitats. The purpose was to address basic ecological questions about whether the natural species distribution primarily reflects physiological adaptation to specific environmental conditions or whether other factors, such as competitive interactions, dominate (Johnson, 1995). Second-year survival generally was good, except where erosion occurred: Under the intense movement of surface water across the rocks, the fabric on a number of plots shredded. Both the rapid degradation of the fabric and the slow establishment of mosses, which stabilize soil surfaces, contributed to patch breakdown. This led to a search for a biodegradable fabric that would last longer under extreme conditions and to consideration of alternative methods to establish mosses.

Ultimately, a sturdy, woven, coconut-hull fiber used both for stream bed stabilization and high-elevation restoration was located. In addition, final plots were grown under a mister in a greenhouse to allow propagation and establishment of native mosses, and vascular species prior to installation on-site

FIGURE CS6-4. To better establish mosses and vascular plants, final habitat modules were grown under a mister in a greenhouse prior to installation at the site. Photo B. Johnson.

(Figure CS6-4). Off-site fabrication required modular construction to allow transportation to the site, which included a half-mile hike up a trail, using army rescue cots to carry the modules. Thirty habitat patches were formed from over 120 modules. The size and shape of each prospective patch were measured and diagrammed, and this allowed development of a modular plan for each patch. The primary module was a rectangle, either 25 by 40, or 50 by 40 centimeters on a side. Irregularly shaped modules were custom-designed to create the appropriate shape of each patch.

Species selection for each patch was based on microhabitat conditions, results from the trial plots, and the species composition of nearby habitat. Structural features, such as the use of *Trichophorum cespitosum* to stabilize the leading edge of patches with *Geum radiatum*, mimicked natural patterns (Figure CS6-5). Each patch incorporated one to four species.

The final plots were installed in 1993, and evaluation of first-year survival is in progress. A substantial number of patches show high survival and vigorous growth, while others have experienced high mortality for certain species. Several of the original trial plots that were left intact appear to be adequately stabilized by both moss and herb growth. They are gradually taking on the appearance of natural habitat patches. It appears that soil stabilization requires three to five years for even initial evaluation. Other questions will require much longer waiting periods. Will rare species composition remain relatively

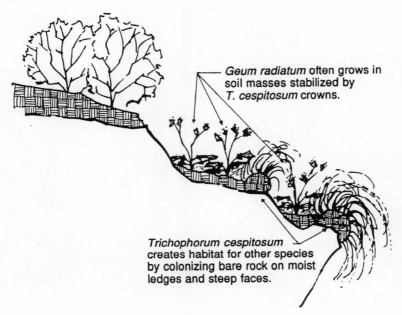

Geum radiatum often grows in
soil masses stabilized by
T. cespitosum crowns.

Trichophorum cespitosum
creates habitat for other species
by colonizing bare rock on moist
ledges and steep faces.

FIGURE CS6-5.   Habitat formation by *Trichophorum cespitosum*. Seedlings of *T. cespitosum* can colonize directly into moss on the edge of high ledges. As the plants grow, they form large crowns composed of organic matter, roots, and washed-in soil particles that rise above the rock surface and, in essence, become the habitat. As patch development continues, and, in particular, if multiple crowns grow together, other species, particularly *G. radiatum*, may colonize the *T. cespitosum* crowns, the soil mass left behind as they die, or soil held back by a *Trichophorum* "dam."

stable in the short term, and will it follow the long-term trajectory of natural patches? To what extent will patches be invaded by nontarget species? How will restored patches respond to a hundred-year drought? Our lack of knowledge about the basic biology and ecology of these species, such as expected rates of regeneration or replacement of individuals, may hamper some types of evaluation.

To date, restoration success has been evaluated as part of the ongoing research. The transition from research funding to in-house management will be challenging. The Blue Ridge Parkway currently monitors plots on natural habitat adjacent to the restoration plots and may be able to include simple monitoring of habitat stability, along with plant cover or abundance, in conjunction with this sampling. Restoration plots that include *Geum radiatum* also may be incorporated as part of long-term monitoring efforts under the Plant Conservation Program of the North Carolina Department of Agriculture.

Long-term monitoring is essential to the project. The final plan should be scientifically useful, with specific, measurable objectives. At the same time, it cannot be so complex or time consuming that it is discontinued. Concise doc-

umentation of methods, along with clear maps and directions to relocate plots will be essential. The current proposal is to create a quick checklist to indicate the physical conditions (such as erosion and evidence of trampling), total percent plant cover, and dominant species in each plot. This will serve as a "bottom-line" monitoring. A second level of detail will be to estimate the number of individuals and/or cover both of each target species and that of nontarget species. For the first five years, annual monitoring of all plots is recommended. After that, monitoring may be reduced in frequency. Periodic, more detailed monitoring may be integrated with less intensive but more frequent monitoring of selected plots. Previous results or unusual conditions (such as a severe drought) may trigger more frequent or intensive monitoring.

Ultimately, successful restoration at this site depends equally on the continued abatement of visitor impacts and on the actual restoration efforts. Hopefully, the two can be integrated so that interpretation, not only of the fragile habitat but of the restoration effort itself, will enhance habitat protection. Regardless of the technical success of the restoration, we will have gained important ecological knowledge about species and community dynamics by using restoration as a scientific tool of inquiry. Such an experimental and synthetic approach to restoration, as proposed by Jordan, Gilpin, and Aber (1987), is one way to integrate basic research with management efforts.

Bioregional Context

Most sites with this rare plant community are located in public natural areas and suffer visitor impacts similar to those at Craggy Gardens. The design and management approaches developed at Craggy Gardens are currently being implemented at several other sites in the region. The recovery plan for *Geum radiatum* (Murdock 1992) included habitat restoration, for which this project has become the model. Restoration of *Geum radiatum* habitat is proceeding at three additional sites on National Park Service and U.S. Forest Service land.

Funding

Funding included $7,000 from the National Park Service for restoration, along with another $10,000 for visitor management studies and $1,200 for overlook and trail design. A graduate research assistantship from the University of Georgia supported a broader set of studies that included restoration. Finally, the State of North Carolina, through a cooperative agreement with

the U.S. Fish and Wildlife Service, contributed $17,000 toward studies of *Geum radiatum*, of which restoration was one component.

Partnerships

The project has involved invaluable collaboration with a number of groups and individuals, in addition to those providing funding:

- Blue Ridge Parkway, National Park Service, Asheville, N.C. In addition to funding, the parkway has provided interpretive development, field assistance, and on-site management.

- North Carolina Arboretum, Asheville, N.C. The arboretum, including its staff and volunteers, has given extensive assistance in propagation, habitat module fabrication, and on-site restoration.

- Dr. James Coke, University of North Carolina, Chapel Hill Propagation.

- North Carolina Botanic Garden, Chapel Hill Propagation.

- Center for Plant Conservation. Consultation for initial project development.

Policy and Regulatory Context of Reintroduction

Both the recovery plan (Murdock 1992) and the management plan (Johnson, Murdock, and Frost 1993) for *Geum radiatum* specify reintroduction as a recovery tool. The recovery plan calls for population augmentation or reintroduction toward a goal of sixteen self-sustaining populations. The management plan for *Geum radiatum* outlines the need for different restoration strategies depending on the cause of population decline. Habitat restoration following visitor disturbance may require visitor management, followed by reintroduction and/or stabilization of soil to recreate habitat on denuded rock. Restoration of sites overgrown by woody succession may involve mimicking historical disturbance regimes to inhibit shrub establishment or growth, thus maintaining open habitat for herbaceous species. National Park Service policy extends protection efforts to state-listed species as well as federally listed species.

A further issue is whether small population size following recent declines may have reduced population viability through a loss of genetic variation. In some populations, initial evidence may warrant genetic and reproductive studies to investigate whether there is sufficient evidence of inbreeding de-

pression to warrant genetic augmentation with plants propagated from off-site stock or crosses with off-site stock.

Experimental plant introduction plots can be used for testing habitat manipulation and management techniques that initially may be risky to undertake on natural populations. Sites may include potentially suitable but unoccupied habitat; it may also include areas that require manipulation before they are suitable. Successional management—including pruning or prescribed burning—or genetic augmentation can be tested in this manner prior to implementation on natural habitat.

REFERENCES

Hamrick, J. L., M. J. Godt, and B. R. Johnson. 1991. "Levels and Distribution of Genetic Diversity in Four Rare Vascular Plant Species." Unpublished final report. Atlanta, Ga.: National Park Service, Southeast Region.

Johnson, B. R. 1992. "Mitigation of Visitor Impacts on High Montane Rare Plant Habitat: Habitat Protection Through an Integrated Strategy of Design, Interpretation and Restoration, Craggy Gardens, Blue Ridge Parkway, North Carolina." Master's thesis, University of Georgia, Athens. Printed as Technical Report for Southeast Region of the National Park Service.

Johnson, B. R. 1995. "The Ecology and Restoration of a High Montane Rare Plant Community." Ph.D. diss. University of Georgia, Athens.

Johnson, B. R., N. Murdock and C. Frost. 1993. "Spreading Avens Management Plan." Atlanta, Ga.: U.S. Fish and Wildlife Service.

Jordan, W. R., M. E. Gilpin, and J. D. Aber. 1987. "Restoration Ecology: Ecological Restoration As a Technique for Basic Research." In *Restoration Ecology, a Synthetic Approach to Ecological Research.* Edited by W. R. Jordan, M. E. Gilpin, and J. D. Aber. Cambridge, England: Cambridge University Press.

Murdock, Nora A. 1992. "Agency Draft Spreading Avens (*Geum radiatum*) Recovery Plan." Atlanta, Ga.: U.S. Fish and Wildlife Service.

Ramseur 1960. "The Vascular Flora of the High Mountain Communities of the Southern Appalachians." *Journal of the Elisa Mitchell Society.* 76: 82–112.

Wiser, S. K. 1994. "High-Elevation Cliffs and Outcrops of the Southern Appalachians: Vascular Plants and Biogeography." *Castanea* 59 (2): 85–116.

Small Whorled Pogonia
(*Isotria medeoloides*) Transplant Project

William E. Brumback and Carol W. Fyler

―――――――▸▪◦◂―――――――

Taxon

Isotria medeoloides; small whorled pogonia (Orchidaceae).

Endangerment Status and Date of Listing

Federal: endangered October 12, 1982.
State of New Hampshire: endangered 1986 (inception of the state law).

Natural Distribution and Ecology of Taxon

The small whorled pogonia is sparsely distributed in the Atlantic seaboard states from Maine to Georgia, with outlying occurrences in the Midwestern United States and Canada. Recent single colony occurrences have been located in Delaware, Tennessee, and Ohio. Historical records exist for localities within Vermont, Maryland, Missouri, Ohio, eastern Pennsylvania, and the District of Columbia. Recent efforts to relocate historical sites in New York, Vermont, and Missouri have been unsuccessful. The vast majority of extant sites are in New Hampshire and Maine. (U.S. Fish and Wildlife Service 1985, 1992).

This species is currently known from eighty-five sites in fifteen states with a total of approximately twenty-six hundred stems (1991 data; U.S. Fish and Wildlife Service 1992). This population level exceeds the number of occurrences known at the time of listing (seventeen extant sites); however thirteen

to fifteen sites are known to be extirpated, while as many as forty-one sites are considered historical (U.S. Fish and Wildlife Service 1992).

I. medeoloides occurs on upland sites in mixed deciduous or mixed deciduous/coniferous forests that are generally in second- or third-growth successional stages. Characteristics common to most sites include sparse to moderate ground cover, a relatively open understory canopy, and mostly highly acidic soils with moderately high moisture values. Light availability may be a limiting factor. (U.S. Fish and Wildlife Service 1992). No direct correlation with other plant species has been identified, and the plant remains rare in spite of thousands of acres of suitable habitat.

As with many orchid species, there is a relationship between this plant and mycorrhizal fungus (Ames 1922), which assists in the development of orchid embryos by providing nutrients. This relationship is poorly understood, and to our knowledge no cultivation of this species by seed has ever been documented. Furthermore, this species is capable of prolonged periods of dormancy. These dormancy periods necessitate long-term monitoring in order to establish an individual's survival.

Threats to Population

Threats to this species include actual and potential habitat destruction and collection. Although collecting can still be regarded as a factor, residential and commercial development, both directly and indirectly, is a primary factor in the destruction of the small whorled pogonia. Grazing by deer, formation of barriers to seed dispersal by fragmentation of habitat, and "people pressure" are also cited (U.S. Fish and Wildlife Service 1992).

Conditions That Initiated Reintroduction

At the time of the reintroduction (1986) approximately fifteen hundred to two thousand individuals were known to exist globally. A population of 147 plants was identified in New Hampshire on extremely valuable (expensive) land slated for subdivision development. No legal method to stop the project existed. Negotiations with the developers were initiated, and a land swap was proposed in which equal acreage of nearby land would be exchanged for the land where the plants occurred. Unfortunately the developers would only accept a two-for-one swap (two acres of land for every acre of orchid habitat they agreed to protect). This was not a feasible arrangement. Therefore, in a last-ditch effort to save the plants, the developers agreed to mitigate the loss of plants by financing the transplanting (and subsequent monitoring) of the or-

chids to a protected site where the species already occurred. As the first recorded transplant of this species, the project was controversial on both a national and regional scale. On the one hand, if the transplants survived, would a precedent not be set for moving a population each time its habitat came into conflict with future development? On the other hand, not many plants were known to exist globally in 1986, and the prospect of saving a large number of plants coupled with the chance to learn more about the species' biology eventually clinched the decision to transplant the orchids.

Objectives of the Reintroduction

Our objectives in this translocation were to remove as many plants as possible from the destruction site, to successfully transplant *Isotria medeoloides*, and to investigate transplant techniques in the event that future transplants were deemed necessary.

Description of Reintroduction Site

The reintroduction site was selected because it was protected (owned by state of New Hampshire), and at least one plant of *Isotria medeoloides* was found in the vicinity of the location where the transplants were eventually placed. Transplants of terrestrial orchids have, in general, been more successful when transplanted to an area already containing the orchid species, presumably because the proper cultural and fungal conditions already exist. Therefore, an extensive search of the reintroduction area was conducted before the transplanting was begun, and the transplant area was selected to include the existing plant at the site.

The transplants were eventually moved to eight different locations on the site, which is in Laconia, New Hampshire (precise location withheld). The land encompassed by the translocation totals approximately two acres of mixed coniferous and hardwood forest and appears identical in species composition and topography to the site where the plants were originally found.

Description of Reintroduction Project

One hundred forty-seven plants containing 157 total stems of *Isotria medeoloides* were moved from the area to be destroyed. Plants were dug with a root ball up to one foot in diameter and 8 inches deep if soil conditions allowed, and these were transplanted to the reintroduction site the same day as they

were dug. Half of the plants were transplanted in July 1986; the other half in September 1986 to determine if one time of year was more successful for transplantation. Transplants were watered immediately after transplanting. All plants were labeled, and their locations were precisely mapped within the eight separate groups.

Each individual plant was scheduled to be monitored for three years after transplanting, but because this species is capable of prolonged dormancy (Brumback and Fyler 1988; Vitt 1991, Ware 1990), monitoring continues yearly. Monitoring consists of four visits each year based on the method of Brumback and Fyler (1983). During the first visit (mid to late May), emergence of the plants from dormancy is recorded. Plants may be destroyed (eaten) by predators, and although plants may return the following year, this destruction can affect their reproductive status (Brumback and Fyler 1988). Therefore it is important to monitor emergence and any subsequent destruction that will likely affect a plant's future reproduction. A second visit in early June records reproductive status. The third visit in July records seed capsule formation (if any), survivorship of all individuals, and late emergence of some individuals. The late-emerging plants tend to be smaller and sterile. A fourth visit in October monitors the plants that formed seed capsules to determine if they survived to disperse seed.

No habitat management has occurred in the transplant areas since transplanting was accomplished. Volunteers assisted with the translocation, and approximately five persons were involved on each of the two transplanting days.

At the initial transplantation, it was decided that success would be measured by survival of plants, an increase in numbers of reproductive plants, and recruitment of new individuals through seed dispersal. Based on these criteria, this transplantation has been an almost total failure. Of the 147 plants that were moved, 96 percent of the plants are now dormant after eight years (Table CS7-1). Only 6 stems (out of an original 157 stems transplanted) have emerged in 1994 (June 1994 figures, unpublished data). Various studies have established that *Isotria medeoloides* individuals have successfully re-emerged after one to two years of dormancy (Ware 1990), two to three years of dormancy (Mehrhoff 1989), and up to four years of dormancy (Brumback and Fyler 1988). Other studies (Vitt 1991) found that most stems experienced dormancy for only a single year before re-emerging, while a very small percent were dormant for three consecutive years, re-emerging in the fourth. Given a maximum dormancy range of four years, and the fact that a number of plants have been dormant for more than four years, it seems likely that the vast majority of these transplants are dead. Similarly, the transplants have declined reproductively. Although 54 stems bloomed the year of transplanting, no stems have produced flowers in 1992, 1993, or 1994.

Based on this translocation effort (which is the only attempt on any scale to

TABLE CS7-1.   *Isotria medeoloides* Transplants—New Hampshire.

	1986	1987	1988	1989	1990	1991	1992	1993	1994
Plants moved: 147 (June 13, 1994)									
Stems emerged	157	121	111	95	77	53	10	10	6
Plants emerged	146	113	104	90	75	51	10	10	6
Plants dormant	1	34	43	57	72	6	137	137	141
Percent dormant		23	29	39	49	65	93	93	96
Plants moved: 61 (July 1, 1986)									
Stems emerged	65	53	50	43	34	29	4	6	3
Plants emerged	61	50	47	41	34	28	4	6	3
Plants dormant		11	14	20	27	33	57	55	58
Percent dormant		18	23	33	44	54	93	90	95
Plants moved: 86 (Sept 3, 1986)									
Stems emerged	92	68	61	52	43	24	6	4	3
Plants emerged	85	63	57	49	41	23	6	4	3
Plants dormant	1	23	29	37	45	63	79	82	83
Percent dormant		27	34	43	52	73	92	95	97

the best of our knowledge), transplantation of *Isotria medeoloides* should not be considered a viable alternative to protection of plants *in situ*.

Bioregional Context

At the time of the translocation there was no bioregional context in which to place the project, but transplantation was a part of the recovery plan for this species (see the section on Policy and Regulatory Context).

Funding

The translocation was conducted by the New England Wild Flower Society and was funded by the developers (including monitoring for three years) for $3,000. The New England Wild Flower Society continues to fund approximately ten person-hours per year for monitoring.

Partnerships

No formal partnerships between public and private agencies was formed for this project. Various agencies were involved with discussions with the developer, however, but no agency was formally involved in the transplantation.

Policy and Regulatory Context of Reintroduction

A recovery plan for the species was first approved in 1985 (U.S. Fish and Wildlife Service 1985). Section 6.3 of the step-down recovery outline discusses transplantation. In this section, transplantation is a "last option" alternative. "Only under those conditions in which the immediate destruction of an existing site will result in the loss of that population should this option be considered." Section 6.4 of the plan designates that an "experimental population" designation is available for translocated populations for which greater management flexibility is desired. Although the U.S. Fish and Wildlife Service and several New Hampshire state agencies were interested in the project, neither was involved in the transplantation in any official capacity.

Since then, a first revision of the recovery plan has been written (U.S. Fish and Wildlife Service 1992), and the early results of this transplantation are discussed in this plan. (It is interesting to note that transplantation does not appear as a task in this revised recovery plan.)

Since this transplantation, policies for introduction, reintroduction, and augmentation for the New England region have emerged from the creation

of a regional plant conservation program, the New England Plant Conservation Program (NEPCoP). The program is a voluntary collaboration of representatives from sixty-eight public and private organizations in New England. Although NEPCoP opposes the loss of any existing rare plant populations or their habitat, it will consider involvement with mitigation projects but only as a last resort and on a case-by-case basis under certain conditions, which are listed in its policies (New England Wild Flower Society 1992).

REFERENCES

Ames, O. 1922. "A Discussion of *Pogonia* and Its Allies in the Northeastern United States." *Orchidaceae* 7: 3–44.

Brumback, W. E., and C. W. Fyler. 1983. "Monitoring Study of *Isotria medeoloides* in East Alton, New Hampshire—1983." Unpublished report. Framingham, Mass.: New England Wild Flower Society.

Brumback, W. E., and C. W. Fyler. 1988. "Monitoring of *Isotria medeoloides* in New Hampshire." *Wild Flower Notes* 3 (1): 32–40.

Mehrhoff, L. A. 1989. "The Dynamics of Declining Populations of an Endangered Orchid, *Isotria medeoloides*." *Ecology* 70 (3): 783–786.

New England Wild Flower Society. 1992. "New England Plant Conservation Program." *Wild Flower Notes* 7 (1): 43–48.

U.S. Fish and Wildlife Service. 1985. *Small Whorled Pogonia Recovery Plan*. Newton Corner, Mass.: U.S. Fish and Wildlife Service, Northeast Region.

U.S. Fish and Wildlife Service. 1992. *Small Whorled Pogonia Recovery Plan—First Revision*. Newton Corner, Mass.: U.S. Fish and Wildlife Service, Region Five.

Vitt, P. 1991. "Conservation of *Isotria medeoloides*: A Federally Endangered Terrestrial Orchid." Master's thesis, University of Maine, Orono.

Ware, D. M. E. 1990. "Demographic Studies of the Small Whorled Pogonia, *Isotria medeoloides*, in Virginia." *Virginia Journal of Science* 41 (2): 70.

PART V

Guidelines for Developing a Rare Plant Reintroduction Plan

The reintroduction of any species is inherently complex. For endangered species, the complexity is exacerbated by a shortage of sound policies, effective models, and strong scientific underpinnings.

As is clear from several chapters in this book, the science of reintroduction is in its infancy; so too is the development of a policy framework to guide the use of reintroduction as a conservation tool. This policy vacuum is most evident with regard to the relationship between introducing new populations and conserving existing ones—a relationship that remains poorly articulated. For example, the policy of the International Union for Conservation of Nature and Natural Resources (IUCN 1987) describes translocations as powerful tools that can materially advance the diversity and viability of populations and habitat but notes that "like other powerful tools they have the potential to cause enormous damage if misused." (p. 1). The potential damage referred to—and much of the overall concern about reintroduction—is that its use will in some way displace the imperative to conserve existing populations and communities. The challenge, therefore, is to unlock the creative potential of reintroduction while guarding against its possible misuse. As the chapter on compensatory mitigation illustrates clearly, the ecological "meaning" of reintroduction depends largely on whether the natural populations still exist, or if they are somehow lost in the course of a project.

This chapter is intended to assist biologists and managers considering the use of reintroduction as a conservation tool. We have prepared these guidelines in the belief that well-planned and well-executed reintroduction can contribute materially to the goals of biodiversity conservation. While no single book can answer all relevant questions, the chapters in *Restoring Diversity* provide an excellent basis for advancing the understanding of the issues.

Re-establishment of populations is too variable to be reduced to simple formulations; there is no cookbook of reintroduction recipes. We have instead developed a series of questions that practitioners of reintroduction are likely to encounter. To form the framework for a reintroduction plan, the organizers of any well-planned project should be able, at a minimum, to provide coherent and well-researched answers to the following questions:

Is reintroduction appropriate?

1. What guidance can be found in existing policies on rare species reintroduction?

2. What criteria can be used to determine whether a species should be reintroduced?

3. Is reintroduction occurring in a mitigation context involving the loss or alteration of a natural population or community?

4. What legal or regulatory considerations are connected with the reintroduction?

How will reintroduction be conducted?

5. What are the defined goals of this reintroduction, and how will the project be monitored and evaluated?

6. Has available ecological knowledge of the species and its community been reviewed? What additional knowledge is needed to conduct the project well?

7. Who owns the land where the reintroduction is to occur, and how will the land be managed in the long term?

8. Where should the reintroduction occur?

9. What is the genetic composition of the material to be reintroduced?

10. How will the founding population be structured to favor demographic persistence and stability?

11. Are essential ecological processes intact at the site? If not, how will they be established?

These questions and the discussions that follow can help the restorationist cope with the likely event that the operative environments more closely resemble the imperfect world of compromise than they do the ideal. Our intention is, if not to provide definitive answers, at least to provoke good questions complementing other policy discussions (Falk and Olwell 1992; BGCI 1994). The guidelines are intended to be a template for further critical thinking about reintroduction. We hope that these ideas will serve as a foundation for scientists, agencies, non-government organizations, and others to develop specific policies and handbooks relevant to their own work.

We have discussed, in length, the semantics associated with reintroduction, introduction, and augmentation in the Introduction to this book. Restorationists should consider these concepts when setting goals or selecting the site or the source material. Guidelines 5 and 8 discuss the distinction between reintroduction, introduction, and augmentation; therefore, we have not redefined them here.

The biophysical aspects of a rare plant species—that is, the ecological

community, the ecological processes, and the environmental context into which the species will be placed—make up the core operational details of reintroduction planning. Each species is unique in its taxonomy, history, ecology, and biogeography; each reintroduction thus presents novel challenges. Several guidelines stress the importance of matching the ecological and physical characteristics (process and structure) of the rare plant in its native habitat with those in the reintroduction site. Some of these elements (such as site selection and selection of source material) are limited to certain phases of a reintroduction; others (such as genetics and ecological processes) are general aspects to consider during several phases of a project.

Is Reintroduction Appropriate?

1. What guidance can be found in existing policies on rare species reintroduction?

In an effort to assess the state of existing rare plant reintroduction policy, the editors and contributors surveyed a wide range of agencies, organizations, and corporations. We reviewed dozens of documents from international conservation organizations, U.S. federal agencies, state agencies, national conservation organizations, private corporations, native plant societies, and professional organizations. These policies, along with the chapters in this book and other published literature, served as the primary materials for development of the guidelines (Table 1).

Many policies were in draft form, reflecting the evolving state of the field. Some policies are broad formulations, while others (such as Gordon 1994) apply only to a single preserve system. United States federal agencies focused primarily on the legal aspects of reintroduction and the ways in which such activity relates to the Endangered Species Act.

In addition to commentary on various specific topics treated in the following sections, several "take-home" messages emerge in existing policies about reintroduction:

- *It is far better, where appropriate, to conserve existing populations and communities than to attempt the difficult and imperfect task of creating new ones.*

- *Reintroductions are fraught with uncertainty and difficulties and should be viewed as experiments.* As such, it is unwise to rely on "successful"

outcomes, given the risks of failure are significant (as is often the case in compensatory mitigation).

- *Determining the outcome of reintroduction takes time.* It certainly takes years, and probably takes decades, depending on species and community characteristics. For instance, Birkenshaw (1991) describes a detailed four- to five-year process for initial preparation, outplanting, and preliminary monitoring alone. As Sutter (Chapter Ten) points out, this means that for all practical purposes, monitoring should continue for the foreseeable future in most reintroductions.

- *Learning opportunities exist throughout the reintroduction process.* To reintroduce confidently, we need extensive and detailed knowledge about the species, its community, and the larger ecosystem. For most rare species this knowledge base is minimal and unevenly distributed among species or communities. Most projects will thus have to proceed on the basis of incomplete knowledge and preferably incorporate learning into the project design.

- *Documentation of outcomes of every reintroduction effort is extremely important.* Many journals accept data from reintroduction projects in progress; practitioners and scientists alike should publish preliminary results or progress reports, including negative outcomes (it is every bit as important to learn which techniques failed as it is to learn which ones worked.) If a project is well-conceived and executed, any outcome will yield useful ecological information.

- *Planning and long-term commitment are of utmost importance to the success of a reintroduction project.* Nearly all policy discussions agree that reintroduction is best when it is part of a comprehensive conservation and recovery strategy for the species and its community. If such a plan is developed, then reintroduction can be better incorporated into the larger objectives.

2. What criteria can be used to determine whether a species should be reintroduced?

Practically speaking, reintroductions are nearly always experiments. Accordingly, before beginning a reintroduction, organizations considering such projects should examine critically the reasons for conducting them. Reintroduction may not be the most effective or successful means to

TABLE 1. Policies and guidelines reviewed.

Agency/ organization	Type of organization	Form of document	Date of document
American Society of Plant Taxonomists	Professional organization	Resolution	1989
Botanic Gardens Conservation International	International conservation organization	Draft handbook	1994
Botanic Gardens Conservation International	International conservation organization	1988 workshop report	1990
Center for Plant Conservation	National conservation organization	Journal article (Falk and Olwell 1992)	1992
Florida Nature Conservancy	State conservation organization	Journal article (Gordon 1994)	1994
Illinois Endangered Species Protection Board	State government agency	Policy	1992
IUCN (International Union for Conservation of Nature and Natural Resources)	International conservation organization	Draft guidelines	1987 1992
National Park Service	Federal government agency	Policy	1988
Nature Conservancy Council (U.K.)	National conservation organization	Guidelines (Birkenshaw 1991)	1991
New England Wild Flower Society	Regional conservation organization	Policy	1992
Native Plant Society of Oregon	State native plant society	Policy	1992
The Nature Conservancy	International conservation organization	Draft policy	1992
U.S. Army Corps of Engineers	Federal government agency	Policy guidance letter	1991
U.S. Bureau of Land Management	Federal government agency	Policy	1992
USDA Forest Service	Federal government agency	Policy	n.d.
U.S. Fish and Wildlife Service	Federal government agency	Draft policy	1992
Waste Management	Private industry	Guidelines	1992
Wisconsin Department of Natural Resources	State government agency	Draft policy	1991

advance the conservation of an endangered species. Careful thought should be given to the reintroduction's potential effects on the future of the species and its community (Reinartz 1995). The expense and effort of reintroducing rare plants and establishing new populations should be undertaken for specific, defensible reasons, and not simply for opportunistic reasons, such as the availability of plant material.

By what criteria, then, can populations and species be selected as promising candidates for reintroduction? The following characteristics may render a species or population a good candidate:

- A species or population is extinct (or nearly so) in the wild. This depends on whether appropriate genetic material is available and whether threats can be managed.

- It has unnaturally few, small, or severely declining populations. Many new tools are emerging that can improve the traditional classification schemes used to identify the most endangered species. Among the most promising are those that use population viability analysis (PVA) to make quantitative, probabilistic predictions about the likelihood of a species becoming extinct (Mace and Lande 1991). While these methods are not without theoretical and pragmatic difficulties (Taylor 1995), they represent a potentially more powerful way to identify species that may be deserving candidates for reintroduction provided that other conditions listed below can be met.

- It has poor protection of existing natural populations.

- It shows evidence of problems with dispersal and/or fragmented habitat. Reintroduction may be a valuable conservation tool for overcoming the inability of some rare plants to disperse effectively to appropriate habitat, especially in fragmented natural habitats.

- It is anticipated to be affected adversely by climate change. Rapid climate change may place new demands and constraints on conservation of rare plant species (Kutner and Morse, Chapter Two; Morse, Chapter One); reintroduction may be part of the solution to these conservation challenges.

- It has available high-quality source material. This material should be genetically diverse, disease free, and of an appropriate provenance.

- It can be successfully propagated and established in experimental trials.

- Its reintroduction is supported by a recovery team. The team agrees that reintroduction will contribute positively to the conservation of the species.

Conversely, certain characteristics may render a species inappropriate for reintroduction:

- Reintroduction or establishment of new populations will undermine the imperative to protect existing sites.

- Feasibility of growing and establishing new populations has not been demonstrated, if the project involves loss of a natural site.

- High-quality appropriate source material is not available.

- Existing threats to natural (or other reintroduced) populations have not been controlled.

3. Is reintroduction occurring in a mitigation context involving the loss or alteration of a natural population or community?

Reintroduction of threatened species is most controversial when practiced in a context of compensatory mitigation. Mitigation directly challenges the relationship of restoration to conservation, in that it requires us to judge the value of existing nature against an artificial substitute. Recognizing that every mitigation case is different, we discuss some of the issues that should be examined.

A broad consensus exists among conservation biologists and planners that it is better to protect existing native populations and communities than to create new ones. Most policies that address compensatory mitigation emphasize the importance of protecting existing diversity. The highest possible priority must be given to avoiding or minimizing impacts to natural populations, especially where rare species or communities are concerned.

However, it is manifestly impossible to fulfill the mandate to "always protect existing sites from development" (Birkenshaw 1991, p. 4). If this were the case, mitigation would not have to occur at all. Compensatory mitigation represents a strategic gamble that the net goals of biological conservation will be furthered if resources of land development and commodity extraction can be diverted to protect some species and some habitats. In addition, legal protection for rare plants often does not apply on privately held lands (Bean, Chapter Sixteen; Klatt and Niemann, Chapter Fifteen). While many

private and corporate landowners voluntarily attempt to avoid damaging rare plants, they are often under no requirement to do so except in the case of wetlands and certain government-permitted activities. In such circumstances there may be no legal way to prevent a development-related translocation.

These realities suggest difficult questions: What, if any, are the characteristics of a good mitigation? Under what circumstances should species or communities be off-limits to any form of tradeoff? Are there circumstances in which mitigation-related reintroduction involving the destruction of a naturally occurring population advances the cause of conservation? Planners should consider the following:

SPECIES RARITY AND VULNERABILITY

Somewhere along the continuum of increasing abundance, an implicit judgment is made that a species is not of conservation concern, and that not every population of a species needs to be protected. For very rare organisms, by contrast, every individual may warrant protection. Somewhere between these two extremes lie the many species for which the fate of an individual population has an uncertain relationship to the future of the species. It is this middle zone of species for which mitigation policy is most important.

Any mitigation policy needs to state clearly that certain species and populations are categorically off limits to destruction. In particular, this applies to extremely rare species—those with very few populations, a small number of individuals, or an extremely restricted geographic range. Unfortunately, terms such as *few*, *small*, and *restricted* lack dimension and can thus be interpreted in several ways. For example, minimum viable population (MVP) standards could conceivably be used to ascertain the sustainable size of a population. But MVP analyses result in probabilistic statements about extinction or persistence, not absolute values. Similarly, rarity is a multidimensional quantitative attribute, not a simple categorical state (Fiedler and Ahouse 1992). The uncomfortable fact is that the threshold for tolerance of a possible destruction event is often difficult to define in intermediate cases of species viability.

There is no standard of rarity, numerical or otherwise, that can be applied across taxonomic and ecological lines. The conservative approach is thus to set limits high: only populations of abundant or stable species should be subjected to mitigation tradeoff. This places the burden of proof squarely on the mitigation proponent, where arguably it should be. The biological rationale for every case must be worked out individually, but mitigation should

proceed only when it can be demonstrated with acceptable certainty that there will be no irreparable harm to the species as a whole.

COMMUNITY OR HABITAT UNIQUENESS

Unique habitats are as important to save as are populations of rare species. Certain communities represent an irreplaceable combination of ecological history and function. Many also harbor populations of rare or habitat-restricted species. Mitigation tradeoffs of rare community types should be avoided altogether, if for no other reason than to prevent more species from becoming endangered (See Gann and Gerson, Chapter Seventeen; Zedler, Chapter Fourteen).

UNCERTAINTY AND THE DISTRIBUTION RISK

The natural processes of colonization and establishment are often very low-probability affairs. While some aspects can be made more predictable in a deliberate outplanting, a great deal of uncertainty surrounds any newly established population. Reviews of existing literature (Hall 1987; Fiedler 1991) indicate that failures—low germination and establishment rates, losses due to droughts or floods, massive herbivory events, and other obstacles to successful colonization—are more common than success (Howald, Chapter Thirteen; Case Studies). These difficulties may be only the *visible* evidence of failure. Less obvious problems may lurk in reduced gene pools; absent pollinators, dispersal agents, or mycorrhizae; and compromised functional parity with undisturbed natural systems (Zedler, Chapter Fourteen).

The threshold of acceptable certainty must be set substantially higher any time a natural population is proposed for destruction. Where reintroduction is practiced "proactively" (*sensu* New England Wild Flower Society 1992) as a conservation measure to heal past harms, this uncertainty may be tolerated because existing populations are not being placed at additional risk. When the equation involves the destruction of natural populations, however, the balance potentially shifts to the negative. Mitigation often involves trading off existing, naturally occurring habitat for created systems of unknown ecological value and an uncertain future.

One of the central problems with mitigation is the unequal distribution of risk in various parts of the process. For example, when a population is to be destroyed by construction activity and replaced by a newly established population elsewhere, the destruction is certain and immediate; it *will* happen. The replacement, however, faces an uncertain future; its prognosis fifty or even five years in the future cannot be predicted. The brunt of uncertainty, therefore, falls primarily on the replacement population. This asymmetry of risk constitutes a major problem for many proposed mitigation projects.

The condition of the reintroduction site poses an additional difficulty. If, as is often the case, the outplanting site is itself in a degraded or altered condition, then the prospects for successful establishment are reduced further. Altered or degraded sites will rarely offer suitable conditions for trading against any naturally occurring population.

Because mitigation efforts are so uncertain, they should be viewed as a last recourse in dealing with development impacts. Draft U.S. Fish and Wildlife Service policy states that propagation and reintroduction should supplement, not replace, conservation of existing populations (McDonald, Chapter Four). Some corporate policies also recognize avoidance or minimization of impacts to naturally occurring, sensitive populations as preferable (Klatt and Niemann, Chapter Fifteen). Only when impacts to rare species are genuinely unavoidable, after a good-faith effort, should compensatory mitigation be considered as an acceptable alternative.

THE MITIGATION TIME SCALE

Transplants take a long time to become part of a functioning ecological community, if they ever do (Pavlik, Chapter Six; Zedler, Chapter Fourteen). There is little research on establishment times for new populations under natural circumstances, let alone artificial outplantings. Whatever insight exists comes largely from the literature on postdisturbance recovery and succession, which suggests that community-level relationships can take decades to equilibrate.

As with the allocation of risk, the relative time scales of destruction and replacement are asymmetrical. Once a project begins, destruction of an existing population or habitat is more or less instantaneous. The "creation" of a new population or habitat, by contrast, is a matter of many years or decades. In combination with the high degree of uncertainty, the long time frame can make the promises of mitigation-related tradeoffs difficult to evaluate (Berg, Chapter Twelve; Zedler, Chapter Fourteen).

MITIGATING IMPACTS ACROSS BIOLOGICAL LEVELS

One commonly used compensatory mitigation technique involves salvage or rescue of individual rare plants that are about to be destroyed. In some cases entire populations consisting of hundreds of individuals are dug up and relocated. Under the best of circumstances, plants are taken in blocks of soil, in the hope of exporting site-level symbionts to the new location (Johnson, Case Study Six). Most of the time, however, what is removed from the site consists primarily of individual plants to be transplanted elsewhere.

As a form of mitigating impact, this practice obscures the different levels of biological organization affected in both sites. A natural, complex

ecological community is lost or destroyed, involving many species and their interactions with each other and with the abiotic environment. What are "saved" are a few individuals representing some fraction of a single population of a single species, with no supporting context. Even in the case of very rare species, this is not an acceptable exchange; if the species is of conservation concern, then so should be the habitat in which it exists.

ELIMINATING CAUSES OF DECLINE OR THREATS
A replacement population can be established only if the original causes of decline have been eliminated. These threats can include invasion by exotic weeds or feral herbivores, disease, suppressed or altered fire regimes, flood suppression, elimination of native pollinators or dispersers, or more pervasive effects such as weather or climate changes (Ledig, Chapter Eleven; Case Studies 2 and 6, this volume). If factors that led to the species' decline remain present, then there may be little reason for confidence in a replacement site. (See Pavlik, Chapter Six; and Pavlik, Nickrent, and Howald 1993) As with reintroduction into physically altered or degraded sites, a mitigation-related outplanting is unlikely to succeed in the long term if the threatening processes have not been eliminated.

4. What legal or regulatory considerations are connected with the reintroduction?
Legal protection for plants is far more limited than legal protection for animals. The most important legislation dealing with the reintroduction of rare plants is the U.S. Endangered Species Act of 1973. The act provides protection for federally listed plants on federal lands and in situations where federal funds, permits, or other actions are involved. The act does not protect endangered plants on private lands.

Reintroduced populations of federally listed plants on federal lands are automatically protected under the act (U.S. Fish and Wildlife Service 1988), and reintroduced populations are protected exactly as the other populations of the listed species, unless the reintroduced populations are listed under the act as experimental. Experimental populations are designated as either *essential* or *nonessential*. An essential population is protected as a threatened species, and a nonessential population is treated as a proposed species under the act. However, the experimental population designation has yet to be used for plants. If a federal agency is worried about reintroducing a listed species onto its lands because the reintroduction would limit their management actions, then an experimental population designation may be useful. In most cases the federal agency considers all sites, puts the reintroduced

population in an area with less conflict, and avoids the use of experimental population designation. To date, no experimental population designation has been used for a plant reintroduction.

If the reintroduction involves federal agencies, then a Section 7 consultation with FWS may be necessary. Permits may be obtained from the U.S. Fish and Wildlife Service (FWS) to collect propagules from lands under federal jurisdiction or to reintroduce a federally listed species on federal lands.

As McDonald (Chapter Four) indicates, draft policy guidance for the U.S. Fish and Wildlife Service states that "propagation programs will not be employed in lieu of habitat conservation (USFWS 1992, p. 1)." Protection of the species and its existing habitat is the foremost objective of a recovery program, with reintroduction being a tool to assist in the recovery of the species.

The lack of federal protection for plants on private lands simplifies the reintroduction process and may increase the likelihood of finding a private property owner who would allow a reintroduction on their property. In the case of *Amsonia kearneyana*, an endangered plant in southern Arizona, it was the goodwill of the owners of a canyon just east of its only known canyon locality who volunteered their property as the site for the reintroduction (Reichenbacher 1990).

State laws regarding endangered plants differ, and not all U.S. states have such legislation. Individual state laws should be checked to see which plants are covered, what activities are allowed, and how the permit process works. (For information on states with rare plant laws or contacts at federal or state agencies see the 1995 *Plant Conservation Directory* [Center for Plant Conservation 1995].)

How Will the Reintroduction Be Conducted?

5. What are the defined goals of this reintroduction, and how will the project be monitored and evaluated?

Once it is determined that reintroduction can help conserve a species or community, the planners must determine what the objectives are and how outcomes will be evaluated. As Pavlik (Chapter Six) notes, however, there is little consensus on standards of success. Moreover, reintroduction projects are so diverse that a single evaluation standard cannot be offered here. Consequently, the evaluation of each project will be based on some combination of standard measures and ad hoc criteria. Further project activities

can then be subject to adaptive management if the outcome does not meet those criteria.

Definitions of Success

Pavlik (Chapter Six) defines success at the population level as "meeting taxon-specific objectives that fulfill the goals of abundance, extent, resilience, and persistence." Pavlik is careful to distinguish between this definition of project success and biological success, which "only includes the performance of individuals, populations, and metapopulations of a targeted taxon."

Monitoring

Monitoring is essential for evaluating success in a reintroduction project. Sutter (Chapter Ten) observes that "Monitoring is the foundation of success . . . not a luxury." Sutter sets out four criteria for a reintroduction monitoring program: (1) monitoring data must have a known and acceptable level of precision; (2) data-collection techniques must be repeatable; (3) collection of data must be done over a long enough period of time to capture important natural processes and responses to management; and (4) the monitoring design must be efficient.

In addition, monitoring objectives must be specific and quantifiable and must define the framework for specific tasks. To evaluate outcomes, Sutter (Chapter Ten) suggests four elements of a reintroduced population that need to be monitored: (1) plants reintroduced to the site, (2) recruitment of new individuals, (3) condition and functioning of the community and ecosystem ,and (4) genetic variability of the population of reintroduced plants.

If we begin to think of all reintroductions as experiments, a vital step in any project will be to use the information from monitoring in managing the species or community. Such feedback is crucial because reintroduction should be an iterative process (Pavlik, Chapter Six). Agency plans must be flexible enough so that the original design can be modified to include information gleaned from the monitoring process (a process called adaptive management). However, those conducting reintroductions should not be too quick to change a monitoring or evaluation scheme simply because a project doesn't appear to be progressing as planned. The failure of a new population to establish provides important information about the biology of threatened and endangered plants and about the frequency of successful establishment in reintroduction programs (Pavlik 1994).

6. Has available ecological knowledge of the species and its community been reviewed? What additional knowledge is needed to conduct the project well?

Although most guidelines and policy formulations state that reintroduction should be based on a sound understanding of species and community ecology, there remains a general shortage of reliable information about many rare species. In such cases, should a project proceed or be delayed until an adequate (however defined) knowledge base exists? Moreover, it is commonly recommended that each reintroduction be treated as an experiment, in terms both of acknowledging uncertain outcomes and gleaning opportunities for learning. But designing an experiment to generate knowledge and designing an implementation project to maximize chances of short-term success may require different approaches. Can reintroduction be designed simultaneously as potential successes and as experiments?

BASIC KNOWLEDGE OF RARE SPECIES BIOLOGY

The field is wide open for research initiatives into rare species biology, especially for work involving the ecology of reintroduction and restoration (Wildt and Seal 1988; Falk and Holsinger 1991; Bowles and Whelan 1994; Schemske et al. 1994). The published literature will rarely be sufficient to answer all relevant questions about the ecology of a rare plant species proposed for reintroduction. Since these ecological relationships are especially germane to the process of reintroduction, it is unlikely that the practitioner will have the desired scientific basis in hand. This leaves reintroduction planners in the position of making more or less educated guesses about the response of the species, and makes the practice of restoration generally one of informed speculation. This predicament is most troubling in circumstances in which "failure" has significant consequences, such as critically threatened species, those for which very limited source material is available, or any situation involving a destructive tradeoff with an existing natural population.

TRANSLOCATIONS AS EXPERIMENTS

To some extent, simply documenting and publishing methods and outcomes will improve empirical understanding of reintroduction ecology. But if reintroductions are to serve as more refined investigations, they must conform to the criteria that would make them good experiments. This means including the basic elements of experimental design: controls, replication, a limited number of variables, and tests of statistical significance. These

conditions are not automatically satisfied in applied restoration work, where the proximate objective may be to succeed according to the terms of a contract, rather than to learn. There is no standard formula for achieving the correct balance of immediate results and expanding knowledge, although the two should be recognized as complementary in the long run (Zedler, Chapter Fourteen).

Any reintroduction project can contribute to the empirical knowledge base by recording baseline conditions. This includes carefully recording the number and type of individuals outplanted, their genetic diversity (if known), outplanting protocols (spacing, depth), soil treatments, management measures, and a detailed description of the receptor site and locality, preferably in a Geographic Information System (GIS). At the very least, this information should be recorded in the archives of a public or private conservation agency; a better practice is to offer the data for publication. Without such information, the knowledge surrounding reintroduction projects may be as ephemeral as the memories of those who conducted them; with proper documentation, projects can serve as empirical tests to which restoration ecologists can return decades later to interpret long-term outcomes. This orientation to the long term is vital to understanding natural ranges of variation and performing trend analyses for ecological responses such as population size and genetic variation. It is also probably the only means by which long-term empirical studies of time scales in the reintroduction process will be carried out. The key to all this is to capture baseline data early.

Some of the best research opportunities are for study of the ecological processes that reintroduction mimics: founder events, small population dynamics, establishment-phase competition, dispersal and disturbance ecology, and patch dynamics. Studies can be directed at colonization of ephemeral or disturbed habitats and at the effects of succession on population persistence, resilience, and stability over time. Cohort studies of reintroduced populations can provide data on the natural range of variation in survival, mortality, and recruitment. Reintroducing plants along gradients of key habitat parameters (moisture, light, elevation) will allow examination of the influence of these and other measures of habitat specificity.

Designing a study for research purposes may require a different approach than for maximizing short-term "success," at least in some instances. Research studies typically focus on a limited number of variables, and provide a wide enough range of conditions in the chosen variables to permit hypothesis testing. This means that some translocated plants may "fail" by growing, reproducing, or surviving at a lower rate than plants exposed to

other conditions in the test matrix. In other words, an experiment may "succeed" in explaining different outcomes, but "fail" to result in the establishment of a permanently viable population. By contrast, if the primary objective is to establish a viable population, then outplanting may need to be restricted to (or at least centered on) conditions known to offer the best prospects for survival. Hybrid approaches are possible, in which a somewhat larger number of variables are tested across a more limited magnitude of values, without the full range of conditions or controls. Under these circumstances, outplanting may provide some usable scientific information and create a viable population.

While the outcome of any individual reintroduction project is unknown at the beginning, the complex interactions of many factors offer an exciting and important opportunity for learning. As reintroduction progresses from trial-and-error to an adaptive ecological management tool, its design can increasingly accommodate needs for both science and conservation practice (Pavlik 1994). Over time, careful experiments will improve both the base of knowledge and the prospects for success.

7. Who owns the land where the reintroduction is to occur, and how will the land be managed in the long term?

Long-term funding and land management are important elements of any reintroduction plan. Many reintroductions are conducted without adequate planning for land use, management, or financial support. Since a reintroduction may take years or even decades to stabilize, inadequate planning can seriously compromise the long-term prospects for success. A reintroduction project that needs two decades of monitoring may outlast the tenure of most agency personnel.

Landowner Commitment to Permanent Site Protection

For rare plant reintroduction to be of permanent value to conservation, the habitat must be securely protected for the long term. Landowners and land managers of the sites must be brought into the dialogue early in the planning phase. The Knowlton cactus reintroduction project (Cully, *Pediocactus knowltonii* Case Study Two) began with considerable dialogue and commitment between seven federal and state agencies and one conservation organization. It is this continued dialogue and commitment that keep the project going ten years later. There are, unfortunately, many cases in which a rare plant was reintroduced onto a site and eventually forgotten because of personnel or priority changes. In some cases these "forgotten" sites were later

used for other purposes—parking lots, building sites or, mowed roadsides—not compatible with the population's survival (Hall 1987). Texas snowbells (McDonald, Case Study Three) illustrates of public and private landowner cooperation in a political climate that would otherwise be unsupportive.

A useful litmus test is to determine if the reintroduction is basic or incidental to the owner's primary interest or the managing agency's mission. Commitment by the landowner or manager to the project in general, and the use of the reintroduction site specifically, needs to be secured prior to initiating the project. Without a long-term commitment to protecting and managing the site, reintroduction projects are exposed to elevated long-term risk. A reintroduction plan should outline the main elements of the long-term program. This plan may take various forms, such as a recovery plan or land use document. Private lands can be secured by an easement or, as in the case of *Amsonia kearneyana*, by voluntary agreement with the landowner (Reichenbacher 1990). Reintroduction plans should cover a period as long as required by the institutional, legal, biological, and monitoring requirements of each species reintroduced; this will rarely be less than five years.

Funding for Long-Term Management

Two reviews of the outcomes of rare plant reintroduction work in California (Hall 1987; Fiedler 1991) found follow-up to be lacking in many projects. The main reason was the shortage of funds allocated to monitoring and ongoing habitat management. To design a legitimate reintroduction project, planners must define the time frame associated with measuring those outcomes and then incorporate these costs associated with the full life of the project. Such planning will make many reintroduction projects more expensive, but two benefits are realized: first, the process provides a more realistic assessment of the real long-term costs. Second, once such costs are made explicit as part of the agency's commitment, it may be more difficult for them to be rescinded than if they were "invisible" in the organization's budget.

Even if costs can be estimated for the entire life of the project, it may be difficult to secure financial commitment for a long-term project. Funds are usually appropriated to federal and state agencies on an annual basis, although programmatic commitments extend into decades. Funding for reintroduction projects undertaken in a mitigation context and paid for by a developer is often subject to even more stringent funding constraints, unless longer management is required by law. Institutional commitment and involvement in the project may improve prospects of funding for the full life of the project. In the extreme cases, conservationists may find it necessary to

refuse to undertake an outplanting project without long-term commitment and funding. Such a position would send a strong message about what is required to reintroduce rare plants successfully.

Bean (Chapter Sixteen) and Klatt and Niemann (Chapter Fifteen) suggest various ways for securing a long-term commitment. These include dedicated trust funds, surety bonds, and other irrevocable financial guarantees to be used for ecosystem management. Many statutory and contractual models exist for such guarantees, which can be adapted from construction and performance bonding or siting hazardous waste facilities. Although financial assurances for endangered species projects are generally not required at present, such guarantees could be included by regulatory agencies as a permit condition. Such up-front financial commitment is preferable if it can be obtained, in part because of the security afforded for the future.

HABITAT MANAGEMENT

A reintroduction site must be managed as an ongoing ecological unit long after the initial outplanting. Processes that need to be addressed include controlling exotic plants and animals, restoring disturbance regimes such as fire and floods, and reducing new sources of anthropogenic impact (Huenneke and Thompson 1995).

The reintroduction plan should also consider the bioregional context of the project. Since all reintroduction projects should aim to become part of landscape-scale conservation efforts, knowledge of current and future land use is imperative. By focusing on landscape-level management one can better ensure that the reintroduction is nested in a larger context and not subject to short-sighted decisions and policy changes.

OTHER LAND-MANAGEMENT CONSIDERATIONS

Populations of some rare species (such as *Betula uber* and *Asclepias meadii* have experienced what appear to be intentional acts of vandalism. Reintroduction in controversial locations (such as range allotments on public lands) may be similarly vulnerable. Public ownership can offer strong protection, although private nature preserves may be superior if available. Potential future landscape configurations should be considered in selecting the best site. For example, will land settlement or land management practices around the site result in a biological island for the rare plant?

Availability of superior sites meeting all criteria (see Guideline 8) will almost always be the limiting factor; the best site may be prohibitively expensive to obtain or administratively or practically unavailable.

8. Where should the reintroduction occur?

Selection of a reintroduction site is a central decision in any project; perhaps no other single aspect influences the eventual outcome as strongly. Site selection involves important long-term considerations of security, management, and monitoring physical/geomorphic, biological, and spatial-temporal considerations. And yet, as Birkenshaw (1991, p. 6) notes, "[E]xcept in the case of re-establishment . . . the selection of a translocation site is, to some extent, a shot in the dark."

Once a decision has been made to proceed with reintroduction, a key step is selecting the receptor site on which the new population will be established. Ideally, it is best to match physical and ecological conditions of the species in its native range with the reintroduction site. In practice, however, outplanting may have to occur in areas with introduced exotics, communities that differ from the species' native habitat, and the unpredictable challenges of climate change.

CRITERIA FOR SUITABLE TRANSLOCATION SITES

The ideal receptor site is not difficult to define hypothetically: it matches the habitat characteristics of the target species (such as biotic community, ecosystem function, and spatial context), and especially of those native populations closest and most similar to the potential receptor site. In practice, however, these conditions are rarely satisfied, for several reasons. First, for only a very few species can we define optimal (or even typical) habitat. The distribution of many rare species has been fragmented or altered, and existing populations often occur in habitat that is far from ideal. Superficially suitable reintroduction sites may prove to be unsuitable because of a cryptic ecological factor in soil chemistry, microhabitat or microclimate relations, absent pollinators, or mycorrhizae (Allen 1993). Second, even where appropriate conditions can be defined with some accuracy, available receptor sites that match these characteristics often do not exist. Available suitable sites may be too small or may lie on unprotected land, while land under reasonably secure management may not offer the appropriate biotic or abiotic environment. And third, sites that are ecologically suitable and protected may nonetheless fall outside of the species' known range. Or they may fall within its overall range but with no evidence that the species occurred at a particular location.

Site selection thus represents a series of tactical compromises. The process begins by determining areas encompassing tolerable variation in key biotic and abiotic parameters, which define suitable potential habitat (Figure 1). These include commonly used physical site indices (soil, available moisture, temperature regime, topographic position) as well as various

characteristics of the biotic community. This envelope of feasibility may be compared to an envelope of security, areas that meet the administrative and management criteria discussed previously. Among areas meeting these criteria, the restorationist can then select sites of known or suspected historical occurrence within a specified time frame. In this manner, potential receptor sites may be evaluated in terms of ecological, administrative, and historical suitability. Note that Figure 1 is a conceptual set diagram, not a "map" of a physical area; actual sites meeting the criteria may be patchy and dispersed across the landscape.

Three groups of sites emerge as possible reintroduction locations. Preferred sites are those that meet all three criteria. A second tier meets habitat and protection criteria but may fail (or not be demonstrated to meet) the criterion of historical occurrence. Ecologically and historically suitable sites on unprotected land would also fall into this category. Third-tier sites meet only the criterion of feasibility. Beyond this envelope, sites are neither biologically

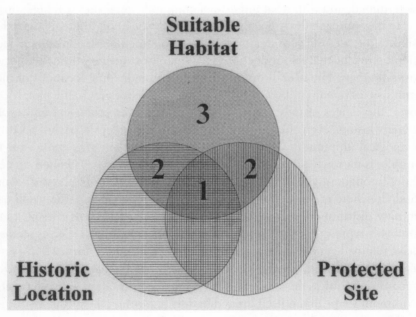

FIGURE 1.   Set diagram for evaluating potential reintroduction sites. Sites may be evaluated for the degree to which they meet primary criteria for habitat suitability, protection, and historic locality. Preferred sites (1) meet all three criteria to a high degree; secondary sites (2) meet two criteria, one of which must be ecological suitability. Tertiary sites (3) pass only the test of habitat suitability. All other sites are presumed to be nonviable for both ecological and conservation purposes.

nor administratively viable for the species in question. Any proposed reintroduction site may be classified according to these criteria.

THE HISTORICAL RANGE ISSUE

Existing policies vary in their approach to site selection and historical range. The U.S. Bureau of Land Management (1992, p. 11) permits release outside of historical range "for those threatened and endangered species for which remaining historical habitat has been destroyed or otherwise rendered unsuitable." The Botanical Society for the British Isles (Birkenshaw 1991) categorically restricts reintroductions to sites within 1 kilometer of a documented locality and considers all other outplantings to be introductions. The Illinois Endangered Species Protection Board (1992, p. 1) states unequivocally that outplantings "should not extend the historic ranges of distribution, the range of habitats in which a plant species is known to have occurred, nor exceed the pre-settlement [sic] abundance of a species in a community or in the state." While such policies are a good start, they beg the essential questions of defining spatial and temporal scales at which species distributions are to be determined, and with what fineness of grain individual sites will be defined.

In the approach described previously, more weight may be assigned to feasibility and security considerations than to documentation of historical locations, provided that reintroduction is reasonably occurring within the species' overall range. The overall historical range and individual locations of most rare species remain unknown for more than a few decades in the past. Moreover, the ranges of most species have changed over ecological and evolutionary history; even current distributions may not reflect a species' potential ecological amplitude. Giant Sequoia (*Sequoiadendron giganteum*), for example, is narrowly distributed at present, but the species is planted successfully throughout California's Sierra Nevada and in Europe, New Zealand, and elsewhere in a wide range of habitats. Giant Sequoia's tertiary and quaternary distribution were widespread in North America; its present range probably represents only a portion of its potential range. The species seems to have retained genetic potential for many habitats. Finally, even when overall range is known, individual past localities within that range may be difficult to reconstruct (unless the species has a very strong association with a habitat or community type that is itself limited in distribution).

Even more problematic are species with only one or two extant populations. In such cases, historical range becomes an almost irrelevant notion, and the emphasis must shift to search to sites that are ecologically and administratively feasible (Linder 1995).

In the face of such ambiguity, it may be difficult to apply criteria of historical range and documented localities in real-world practice. As discussed in the Introduction, we recommend instead an approach based on evaluating natural variation in range, distribution, and density of populations over time,

as more realistically reflecting the continually changing "history" of any species.

Defining potential habitat is itself difficult, because even naturally occurring populations may not reflect habitat and distributional optima. As the landscape becomes increasingly modified by human actions, populations of rare species are increasingly fragmented and pushed into ecological corners that may represent the fringes, rather than the center, of their historical distribution and niche space. Consequently, a search image based on current populations may reflect poorly the optimal conditions for their reintroduction. In southern Arizona, for example, Lemon lily (*Lilium parryi*) occurs in upper-elevation stream systems with moderate stream energy and periodically high flows. California populations are genetically similar but occur mostly in lower-elevation, low-energy cienegas and wetlands—habitat that has been largely destroyed in southern Arizona. A related rare species, *L. occidentale*, occurs in Northern California and southern Oregon on a handful of sites, virtually all of which appear to have been substantially altered by decades of fire suppression (E. Guerrant, personal communication, May 1995). In the midwestern United States, many formerly widespread prairie species (such as the royal catchfly [*Silene regia*]) now persist in only a few marginal sites, which may be wetter or drier (or of different soil composition or community type) than previous mean values.

ECOLOGICAL CRITERIA

At the level of community structure and process, the receptor site should provide resources and opportunities for key life history requirements: dispersal, pollination, germination and establishment, mycorrhizal associations, and other mutualisms. The site should also be similar to the rare plant's native habitat in floristic and faunal composition and structure, successional stage, functional parameters, and disturbance regime (Primack, Chapter Nine). If possible, avoid sites where disruptive exotics (including pathogens) persist, even if other conditions are appropriate. For instance, Rejmanek (1989) has demonstrated that some highly modified communities are no longer invasible by rare plants (Loope and Medeiros 1994; also Pavlik, Chapter Six). Although careful matching of native habitat and receptor site ecology is preferable, in many cases this approach will be difficult to implement (Schemske 1994). Given the paucity of information about many rare species, a directly experimental approach will often be necessary, using outplanted populations on an array of sites to evaluate outcomes over time.

THE SPATIAL CONTEXT

Landscape or spatial context is also important in the selection of the receptor site. White (Chapter Three) describes the importance of selecting sites that contribute to re-establishment of natural patterns of heterogeneity at the

landscape level. This may include the mosaic of successional habitats and the disturbance regime that will exist on the site.

The act of establishing a population creates a new source of propagules for the surrounding landscape. Depending on the dispersal ecology of the species, the amount of material translocated, and the characteristics of the surrounding vegetation matrix, each site-level introduction has the potential to influence the vegetation of surrounding areas.

Species characterized by metapopulation ecology require a careful definition of "site." For such species, the functional ecological unit may be a cluster of sites along a riparian corridor or among canopy gaps or edaphic islands. Each individual act of reintroduction should be intended to establish one component of this larger entity, in some cases simply filling in a gap within an existing metapopulation (Primack, Chapter Nine; Bowles and McBride, Case Study Five). Since gene flow is presumably higher among metapopulation sites than among differentiated populations, the genetic makeup of such subpopulations may be of special concern (see Guideline 9).

As with so many other aspects of biology and conservation, islands illustrate questions of spatial context intensively. The definition of a suitable landscape-scale area may be entirely bounded by a single island, even when potential habitat exists nearby. Rates of interisland migration are often unknown, and it may be difficult to ascertain if observed differences among island populations reflect significant ecotypic variation or simply chance events of dispersal, colonization, and survival. The most conservative policy is to restrict normal outplantings to the island that provided the source material (HRPRG 1992).

Whenever possible, reintroduction projects should account both for among-site or contextual factors, and for within-site criteria. Spatial and landscape characteristics of the receptor site should be compared with the species' native habitat, especially populations closest to potential reintroduction sites. Various characteristics of the landscape matrix can be evaluated, including corridors, patch configuration, buffer zones, fragmentation patterns, and watershed position (Naveh 1994).

CLIMATE CHANGE CONSIDERATIONS

Introducing some species outside of their known historical range may be a strategic hedge (or perhaps a response) to potential climate change (Kutner and Morse, Chapter Two). As vegetation zones shift, many rare species are predicted to be excluded from their current range and may face formidable dispersal barriers of both natural and anthropogenic origin. Intentional introduction outside of the envelope of known historical range could be part of the

conservation response to the effects of a change in climate (Peters 1985; Peters and Lovejoy 1992).

No species can be introduced, however, outside of its current envelope of ecological feasibility, no matter how urgent the need. Over the next fifty years, changing climates may just begin to affect many plant populations, primarily populations at the ecological margins of species distributions. Uncertainties about the local manifestation of climate change may hinder our ability to predict impacts. Because the actual physical and biological limits for rare plants are often unknown, predicting the movement of their suitable habitat is necessarily a highly speculative venture. For proposed receptor sites close to known localities, it is probably safe to assume some variation in the perimeter of distribution and perhaps even the occurrence of some disjunct or relictual populations over the species' natural history. In the published literature, we find few cases where the movement of a species just beyond its currently documented distribution can be proven to violate the area of previous colonization. Examples where this may be possible would be habitats (such as high alpine areas, riparian zones, or islands) that are by nature persistently patchy and isolated for ecologically significant periods of time. However, such habitats are often also closely defined by gradients in soil, temperature, or precipitation that correspond to the envelope of potential habitat.

Following the approach recommended throughout this book, we define the reasonable test discriminating a outplanting from a biologically new event to be whether the act of reintroduction exceeds the natural range of likely dispersal events over a specified period of time. Crossing a near-absolute dispersal barrier (2,000 miles of ocean, for instance, for most terrestrial species) would in nearly all cases constitute an introduction, while crossing a low range of hills with continuity in vegetation and climate might not. Because of the uncertainty of both future climate and species niche breadth, perhaps the best way to hedge bets on survival is to plan for buffering, resilience, and migration of communities over space and time, including many experimental outplantings that will offer a cushion against attrition.

9. What is the genetic composition of the material to be reintroduced?

Genetic composition and genetic processes at different phases of population growth influence short- and long-term population viability (Kress et al. 1994; Godt et al. 1995). The most important times to consider genetic aspects, however, are when (1) selecting the site, (2) developing and outplanting source material for initial planting and for supplementing the reintroduced population (such as replanting to replace initial mortality), (3) ensuring reproductive and dispersal adequacy of the new population, and (4) designing a

monitoring program (Guerrant, Chapter Eight). For excellent overviews, see
Fenster and Dudash (1994) and Schemske et al. (1994).

SITE SELECTION

When considering candidate sites, match known or inferred genetic elements with those in nearby healthy populations (Fiedler and Laven, Chapter
Seven). Ideally, sites should be avoided if they are surrounded by (that is,
within significant gene flow distance of) populations of nonlocal genotypes or
races capable of contaminating the reintroduced population. Sites should
also be avoided if populations are surrounded by widespread congeners capable of swamping the reintroduced gene pool via interspecific hybridization.
Conversely, if the species naturally occurs in a scattered metapopulation
structure, matching or creating this structure may be important both for genetic and demographic reasons (Primack, Chapter Nine). Choosing a site
that is large enough to accommodate a healthy native population may favor
maintenance of genetic diversity by maintaining large effective population
sizes.

SOURCE MATERIAL AND DESIGN FOR
INITIAL REINTRODUCTION AND SUPPLEMENTAL PLANTINGS

The objective in selecting and developing appropriate germplasm is to establish resilient, self-sustaining populations that retain the genetic resources
necessary to undergo adaptive evolutionary change (Guerrant, Chapter
Eight). One effective strategy to achieve this objective might be to do
whatever possible to maximize initial population growth and to minimize
short- and long-term extinction probability. Two aspects are most important:
(1) the genetic source of founders (and supplements) and (2) the diversity and
number of genetically effective individuals (that is, effective population
size). Regarding the source of founders, match germplasm to the reintroduction site by choosing native donor populations that are geographically close
and ecologically similar to the reintroduction site. However, the dimensions
and characteristics that define "local" are unique for each species—no direct
rule applies. "Local" is ultimately determined by the size of the genetic
neighborhoods of native populations, the environmental factors that condition selection gradients of genetic change over space, and the historical elements that influenced the evolution of the population's genetic structure
(DeMauro 1994). The conservative guide is to choose donors from closest
neighboring population(s) if those populations are relatively large, viable,
uncontaminated (by nonlocal genotypes of the same species or interspecific
hybridization), and healthy. In addition, conditions of the donor site should
match the ecological conditions of the reintroduction site. If neither of these
conditions applies, choose donors from more distant native populations, following the best ecological knowledge to match donor populations with sim-

ilar conditions to the reintroduction site and ensuring that collection does not harm donor populations (Lesica and Allendorf 1995). Only exceptional and urgent conditions would justify using germplasm of unknown origin. Even local germplasm grown for several generations in a nursery or botanical garden may be genetically different from native local gene pools (Kitzmiller 1990; Pavlik, Nickrent, and Howald 1993; Lippett et al. 1994).

Although few valid generalities exist regarding the optimal number of donor populations to use, the conservative guide is to use one if the population is healthy as described previously. Conditions other than these might favor using a mix of several native populations as donors. (See Guerrant, Chapter Eight, for discussion.) The guiding factor is to establish a population with appropriate genetic diversity to provide raw material for adaptation to the site.

In collecting propagules from natural or cultivated populations, the goal is to maintain high effective population sizes throughout the propagation process. This is achieved by maximizing the number of distinct founding genotypes (by collecting propagules systematically throughout the donor population), maintaining equal numbers of propagules from each founder through nursery stages to outplanting, and encouraging rapid early population growth (Lippett et al. 1994). Experience is insufficient to prescribe in general how large the founding population should be. Within obvious practical considerations, the default rule is "bigger is better" (Guerrant, Chapter Eight; Center for Plant Conservation 1991).

GENETIC CONSIDERATIONS FOR REPRODUCTION AND DISPERSAL

The general guideline for safeguarding genetic diversity during this phase of the reintroduction is to mimic the natural life-history characteristics of the rare species, including its pattern of dispersal (Primack, Chapter Nine). This minimizing inbreeding (except for inbreeders) and favoring natural dispersal patterns (numbers of propagules, dispersal distance, vectors). By so doing, genetic diversity is most likely to be maintained, and natural genetic structure will evolve in the new population. There are two ways to minimize inbreeding: (1) plant diverse genotypes scattered systematically over the planting site (that is, don't plant in groups of clones or close relatives such as selfed individuals, full sibs, or inbred propagules), and (2) plant with high stocking density to promote abundant cross-fertilization.

MONITORING GENETIC VARIATION

Genetic monitoring is a research field in itself and not straightforward in design, standards, or interpretation. The conceptual standard for genetic monitoring is that genetic diversity be adequate to maintain population viability (demographic stability and growth) and sustainability (long-term adapt-

ability and resilience to change). In practice, determining how much genetic diversity is enough to meet these goals is nearly impossible, even for species that are well understood genetically. Further, teasing apart genetic diversity from other factors that affect demographic stability and population viability is, with present knowledge and techniques, extremely difficult. The best practical guideline to ensure genetic integrity during the monitoring period is to begin with a good baseline genetic profile of the material to be introduced to the site and then over time to monitor levels and trends in overall genetic diversity. Other proxy data (such as demographic attributes and life-history parameters of population growth and viability) can be used to help interpret genetic status. If results from monitoring these traits indicate a population decline or a significant drop in viability, and genetic diversity similarly has dropped precipitously, then genetic factors may be contributing to the decline. Conversely, if overall levels of genetic diversity are maintained or increase gradually, and the population is viable and healthy, it can be assumed cautiously that genetic diversity is adequate. Abrupt changes in allele frequencies (that is, the appearance of unique alleles) may indicate gene contamination or interspecific hybridization and should be followed by careful inspection of neighboring populations.

The choice of genetic traits to monitor is debatable. *Marker traits* (allozymes, DNA), although expensive, are relatively quick to measure and indicate actual levels of genetic diversity. They are, however, almost always ambiguously and indirectly related to genetic loci controlling adaptive traits, which are usually the traits of interest in monitoring population viability. If used only to measure overall genetic diversity, and used in conjunction with other proxies, markers are probably best. *Quantitative traits* (such as plant height, fruit size, or seed weight) are of more direct interest. The role of genetic diversity in adaptive fitness requires either that the reintroduced population be originally designed as a common-garden test (rarely desirable for other reasons) or that common-garden experiments be undertaken periodically on propagules from the reintroduced population. An appropriate sampling design for monitoring genetic diversity should be followed regardless of genetic traits monitored (Center for Plant Conservation 1991).

10. How will the founding population be structured to favor demographic persistence and stability?

An immediate goal for reintroduction is to establish robust, self-sufficient, and reproductively effective populations. Attention must be paid to the early phases of a reintroduction, especially the demographic consequences of the outplanting materials chosen and the dynamics of early establishment and growth (Guerrant, Chapter Eight; Primack, Chapter Nine). Two important demographic goals are to maximize population growth and to avoid local ex-

tinction. Both stage-class (seed, seedling, juvenile, and so on) and age-class of founders affect subsequent population growth and extinction probabilities. How they specifically affect these values depends on the life history of the species. On average, simulations and empirical evidence show that populations expierence lower extinctions when mature plants are outplanted, rather than seeds or very small seedlings. Although larger plants appear to decrease extinction probabilities, the greatest gain in avoiding extinction is between seeds or seedlings and small plants. Therefore, when the goal is not mere persistence but rapid population growth, using the largest outplants appears to be best, since population growth rate generally increases continuously as plant size increases (Guerrant, Chapter Eight). These guidelines are based on simulation results and only tentatively offered practically, since there are some potential downsides of outplanting mature plants: selection under garden conditions; the time required to find out if the seed-to-seedling hurdle can be passed at a given site (McDonald, Case Study Three); and the lack of experimental or quantitative analysis of a significant demographic event (seed to seedling). Putting out seeds, mixed with whole plants, may circumvent these difficulties, especially if large numbers of seed are available (Guerrant, Chapter Eight).

There is no simple, standard answer to the question of how many plants are enough to constitute a viable founder population. Unfortunately, the actual critical values—minimum viable population size, founder population size to avoid an early extinction event, or even the age structure of a normal population—are unknown and probably nearly unknowable because of our uncertainty about future environments (see Pavlik, Chapter Six). The practitioner/researcher will have to explore the literature on the target species or congeners and make a series of best guesses based on the size and design of natural populations that appear ecologically comparable (Ruggiero et al. 1994; Schemske et al. 1994). We offer guidance here on some of the key questions that such exploration should attempt to address.

Founder Population Size

Reintroduction practitioners should become familiar with the literature applying analysis of minimum viable populations, population viability and vulnerability, founder events, and demographic stochasticity to problems in conservation biology (Pavlik, Chapter Six; Guerrant, Chapter Eight; Shaffer 1981; Gilpin and Soulé 1986; Soulé 1987; Menges 1991a, 1992). This work confirms both theoretically and empirically that, other things being equal, small populations are at greater risk of local extirpation due to demographic fluctuations than are large populations. However, the actual details about numbers are almost entirely a matter of speculation for most species, especially rare ones. Mathematically, predicted persistence time appears to be a function of the power of founding population size, but it is influenced as well by

predicted growth rates and many other factors (Menges 1990). Persistence over time is also a function of the effective population size, with regard to the maintenance of genetic variability and reproductive processes (Guerrant, Chapter Eight; Lande and Barrowclough 1987; Menges 1991b; Ryman and Lairke 1991).

Moreover, the true effects of demographic oscillations are only evident over many generations. Most demographic extinctions have been simulated in ecological models of populations over tens or hundreds of generations down the line. Chance extinction models based on birth-and-death processes (Goodman 1987) permit population size and growth rates to be correlated probabilistically with time (or number of generations) to extinction. Alternatively, the same parameters can be used to estimate the probability of population persistence.

Since minimum viable population values for individual species may be correlated with various life-history attributes, some insight may be gained by comparing the target species to others with similar characteristics (Pavlik, Chapter Six).

POPULATION GROWTH, RECRUITMENT, AND SURVIVORSHIP

Although rapid population growth is theoretically favored, management techniques to achieve it (such as fertilization or intensive culture) may not promote population sustainability. Most models suggest that population-level persistence can be enhanced by very large population sizes, high survivorship, or high growth rates; persistence probability can be expressed mathematically as a function of these and other factors. Ideally, the values for growth, recruitment, and survivorship will come from studies of natural populations or closely comparable congeners. In the absence of such baseline data in natural populations, the restorationist must estimate projected growth and survivorship and then monitor the outcomes closely.

Such trials may provide the best information over the long term if the project includes a series of successive outplantings over a period of years. In only a few cases will an initial outplanting be successful, in the sense of establishing a viable, self-reproducing population on the first try. The probability of extinction is very high in any given trial, especially those at the beginning of a reintroduction program. As such, the best strategy for achieving a stable population may be to treat the first attempts as ministudies that will provide qualitative information about survivorship, recruitment, growth rates, competitive interactions, pollination success, and other parameters.

If mature plants are placed on the site, every individual should be marked and mapped to allow survivorship, recruitment, and growth to be tracked accurately in subsequent years. If seed are used, a planting record and map

should indicate the amount of material released and its location. Over a period of several years, these methods will reveal a site-specific pattern of establishment and growth for the species. Survivorship studies must be continued for a long enough period to include the natural range of variation in weather and ideally some variation in related environmental factors (such as stream flow for riparian endemics). Laboratory-derived seed germination rates can provide a yardstick for evaluating response on the outplanting site, taking into account that germination rates can be an order of magnitude lower in the field due to suboptimal germination conditions, seed predation, and competition.

SIZE AND STAGE STRUCTURE
OF THE REINTRODUCED POPULATION

By itself, stage structure does not tell us all we need to know about a population's health. Species and populations may have a characteristic age structure reflecting multiple demographic, reproductive, and life-history factors, although many species are highly variable. For perennial species, the stage and size structure of comparable natural populations should be observed closely in designing the reintroduction program. Should the reintroduction program attempt to replicate this age structure, or should the population be permitted to establish its own demographic equilibrium over a period of years? In evaluating model simulations, Guerrant (Chapter Eight) argues that decisions about inital stage structure should include considerations of survivorship, cost, effort, and other variables beyond "pure" demogragraphy.

The simulations show that stage structure of the founding population can influence long-term extinction risk significantly, although the outcome depends on growth form (long-lived trees versus herbaceous perennials, and so on). The introduction of some plants of larger size classes (even relatively small or juvenile plants but one stage class only) dramatically reduced extinction risk compared to introductions using seeds. These simulation results suggest that using the largest founders practical may theoretically be the best, although using a diversity of stage classes as founders and multiple introductions may be safest practically. To the extent possible, reintroduction would treat these factors experimentally and track success by size class.

The multiyear outplanting approach r ecommended here will introduce a rudimentary degree of age structure diversity to the population. However, the resulting structure may not approximate the age-class distribution found in existing populations. In some cases, it may be possible to introduce literally a multiple age-class population by outplanting a combination of seeds, young rooted cuttings, and mature individuals of various sizes. Such an approach may be limited by the availability of material or time constraints for project completion, but if resources permit it may be worth considering (very little

empirical work with this approach has been conducted to date). Guerrant (Chapter Eight) discusses in detail the relative merits of outplanting various ages and types of plant material (see also Primack, Chapter Nine).

11. Are essential ecological processes intact at the site? If not, how will they be established?

Since an ultimate goal of reintroduction is not simply recovery of the rare plant but restoration of the ecological community, attention should be paid over time to ecosystem processes (White, Chapter Three; Sutter, Chapter Ten). In preceding sections, we have touched on several ecological processes important for successful reintroduction: interactions with ecological associates, symbionts, and mutualists such as mycorrhizae and pollinators; flower and fruit production; seed dispersal; gene-flow (within and among populations); disturbance processes (fire, flooding); and restoration of habitat relationships (Johnson, Case Study Six). Management actions for the reintroduced population should be adapted to promote key ecosystem processes such as nutrient cycling, disturbance and hydrologic regimes, watershed protection, wildlife corridors, and so on (Thomas 1994).

POLLINATION

Beyond the first outplanted generation of seeds or growing plants successful reproduction is vital to long-term success (Bond 1995; Weller 1994; Sipes and Tepedino 1995). Despite the obvious and fundamental importance of successful reproduction to persistence of the population, remarkably few reintroduction projects include any conscious effort to ensure that pollination can occur. For the many species that require animal vectors for pollen, the project will be short lived or will require hand-pollination, as does *Brighamia rockii* on Molokai, Hawaii (U.S. Fish and Wildlife Service 1995). If at all possible, the assistance of experts in pollination ecology for the taxon (or congeners) should be enlisted.

DISPERSAL

As with pollination, many plants rely on external agents to disperse propagules. In fact, there is evidence that dispersal failure accounts for at least some instances of recent decrease in range or population numbers of rare species (Primack, Chapter Nine; Primack and Miao 1992). The tendency toward ecological niche specificity in rare species amplifies the importance of dispersal. Moreover, many animal dispersal vectors (such as birds, mammals, ants, and bats) also prepare chemically for germination and may actually initiate the germination process.

In a real sense, reintroduction is an act of dispersal, at least for the first generation (Primack, Chapter Nine). Successful dispersal involves not only get-

ting seeds (or other material) to a suitable macrosite and community but also securing a variety of microsite factors: proper planting or sowing depth, litter cover, sun/shade position, soil moisture, and other parameters. Careful attention should be given to microsite characteristics and microclimatic variation, as they may affect the early establishment phase.

MYCORRHIZAL ASSOCIATIONS AND
OTHER SOIL MICROORGANISMS

The incidence and importance of root mycorrhizal symbiosis and other microorganismal interactions in rare plants are generally unknown. Allen (1993) has reviewed the role of mycorrhizae in ecological restoration efforts and concluded that their importance has probably been underestimated for long-term vigor and growth of established plants (Allen 1991; Weinbaum, Allen, and Allen in press) If reference natural populations or published literature do not provide data for the target species, the restorationist may have to look to congeners for clues. As an empirical alternative, some reintroduction projects bring soil from an existing population to inoculate the reintroduction site. While this practice may be effective, it may also introduce disease organisms or other undesirable elements into a new ecosystem; for this reason, one should be cautious about moving large amounts of soil.

DISTURBANCE

Few questions in ecology are as complex and controversial as the influence of disturbance on the distribution of species and communities. Many of these decisions will be delimited by the site itself, either because of its biological characteristics or land management regime (White, Chapter Three; Fielder and Laven, Chapter Seven; Guideline 8).

A guiding philosophy must be to understand and work with the dynamic nature of natural processes (Primack, Chapter Nine). Reintroduction should accommodate the reality of short- to medium-term (successional) changes, episodic processes (such as disturbance events), and long-term trajectories (climate change) and accept the stochastic nature of these dynamics (Falk 1990). Changes in landscape patterns due to human settlement and management (such as fragmentation) may have important implications for natural processes within the reintroduced population. In some cases, managers may have to intercede (by artificial pollination or dispersal, by clearing, or by prescribed burns) to promote important processes that are blocked due to highly altered landscape conditions.

Disturbance dynamics at a very fine spatial scale affect many of the more intimate aspects of population-level reintroduction. Many species have seeds that germinate only after fire scarification or contact with damp soil. Post-disturbance processes can directly affect germination and growth rates and

can alter profoundly the competitive interactions with other species on the site. Reintroduction is not simply a matter of bringing plant material to a site and then walking away; the disturbance regime is an important determinant of the pattern of distribution of species and communities on the landscape and on the long-term viability of the reintroduced population (Pavlovic 1994).

Recruitment and establishment of new individuals are critical measures of success for a reintroduced population (Pavlik, Chapter Six; Primack, Chapter Nine; Sutter, Chapter Ten). Mimicking known or inferred natural processes for the individual species is again the best guide: yearly recruitment is vital for annuals, while recruitment in long-lived perennials is often more sporadic. Since recruitment depends not only on adequate numbers of sound seeds but safe sites for germination and establishment, it is important to ensure that the reintroduction area can sustain disturbances that provide safe sites (Sutter, Chapter Ten).

Natural disturbances, such as fire, floods, windfalls, and insect and disease outbreaks often create gaps or areas of preferred habitat and should be permitted on the reintroduction site. Suppression of these processes, or introduction of artificial disturbance processes, may detrimentally affect creation of safe seed and seedling sites and thus inhibit recruitment and establishment. This illustrates the importance of ongoing habitat management (Guideline 7) even for single-species reintroductions (Gordon, Case Study Four).

ACKNOWLEDGMENTS

Many thanks to Ed Guerrant, Ann Howald, Bart Johnson, Brian Klatt, Charles McDonald, Linda McMahan, Loyal Mehrhoff, Ron Niemann, Bruce Pavlik, Joy Zedler, and Marlin Bowles, for comments and assistance. However, the editors assume all responsibility for the ideas expressed in this section.

REFERENCES

Allen, M.F. 1991. *The Ecology of Mycorrhizae*. Cambridge Study in Ecology Series, Cambridge University Press.

Allen, M.F. 1993. Microbial and Phosphate dynamics in a Restored Shrub Steppe in Southwestern Wyoming. *Restoration Ecology* 1(3): 196–205.

American Society of Plant Taxonomists. 1989. "A Resolution on Transplantation." Adopted at August 1989 annual meeting, University of Toronto, Canada.

Birkenshaw, C. R. 1991. *Guidance Notes for Translocating Plants as Part of Recovery Plans*. Cambridge, England: Nature Conservancy Council Report Number 1225.

Bond, W. J. 1995. "Assessing the Risk of Plant Extinction Due to Pollinator and Disperser Failure." In *Extinction Rates*. Edited by J. H. Lawton and R. M. May. Oxford, England: Oxford University Press.

BGCI (Botanic Gardens Conservation International). 1994. *A Handbook for Botanic Gardens on the Reintroduction of Plants to the Wild*. Richmond, Surrey, England: Botanic Garden Conservation International/IUCN Species Survival Commission.

BGCI (Botanic Gardens Conservation International). 1990. *Techniques for Germplasm Conservation of Wild Species by Botanic Gardens*. Report of a workshop, December 1988, in Maspalomas, Gran Canaria, Spain.

Bowles, M. L., and C. J. Whelan, eds. 1994. *Restoration of Endangered Species: Conceptual Issues, Planning and Implementation*. Cambridge, England: Cambridge University Press.

Center for Plant Conservation. 1991. "Genetic Sampling Guidelines for Conservation Collections of Rare Plants." In *Genetics and Conservation of Rare Plants*. Edited by D. A. Falk and K. E. Holsinger. New York: Oxford University Press.

Center for Plant Conservation. 1995. *1995 Plant Conservation Directory*. St. Louis, Mo.: Center for Plant Conservation.

DeMauro, M. M. 1994. "Development and Implementation of a Recovery Program for the Federal Threatened Lakeside Daisy (*Hymenoxys acaulis var. glabra*)." In *Restoration of Endangered Species: Conceptual Issues, Planning and Implementation*. Edited by M. L. Bowles, and C. J Whelan. Cambridge, England: Cambridge University Press.

Falk, D. A. 1990. "Discovering the Past, Restoring the Future." *Restoration and Management Notes* 8 (2): 71.

Falk, D. A., and K. E. Holsinger, eds. 1991. *Genetics and Conservation of Rare Plants*. New York: Oxford University Press.

Falk, D. A., and P. Olwell. 1992. "Scientific and Policy Considerations in Restoration and Reintroduction of Endangered Species." *Rhodora* 94 (879): 287–315.

Fenster, C. B. and M. R. Dudash. 1994. Genetic considerations for plant population restoration and conservation. Pp. 34–62 in Bowles *et al., op. cit.*

Fiedler, P. L. 1991. *Mitigation Related Transplantation, Relocation and Reintroduction Projects Involving Endangered and Threatened, and Rare Plant Species in California*. Final report to California Department of Fish and Game. Sacramento: Endangered Plant Program.

Fiedler, P. L., and J. J. Ahouse. 1992. "Hierarchies of Cause: Toward an Understanding of Rarity in Vascular Plant Species." In *Conservation Biology: The Theory and Practice of Nature Conservation, Preservation, and Management*. Edited by P. L. Fiedler and S. K. Jain. New York: Chapman and Hall.

Gilpin, M. E., and M. E. Soulé. 1986. "Minimum Viable Populations: The Processes of Species Extinctions." In *Conservation Biology: The Science of Scarcity and Diversity*. Edited by M. E. Soulé. Sunderland, Mass.: Sinauer Associates.

Godt, M. J. W., J. L. Hamrick, and S. Bratton. 1995. "Genetic Diversity in a Threatened Wetland Species, *Helonias bullata* (Liliaceae)." *Conservation Biology* 9 (5): 596–604.

Goodman, D. 1987. "The Demography of Chance Extinction." In *Viable Populations for Conservation*. Edited by M. E. Soulé. Cambridge, England: Cambridge University Press.

Gordon, D. R. 1994. "Translocation of Species into Conservation Areas: A Key for Natural Resource Managers." *Natural Areas Journal* 14 (1): 31–37.

Hall, L. A. 1987. "Transplantation of Sensitive Plants As Mitigation for Environmental

Impacts." In *Conservation and Management of Rare and Endangered Plants*. Edited by T. Elias and J. R. Nelson. Sacramento: California Native Plant Society.

HRPRG (Hawaii Rare Plant Restoration Group). 1992. *A System of Supplemental Outplanting and Propagation Facilities to Enhance the Conservation of Hawaiian Plants*. Honolulu, Hawaii: Center for Plant Conservation and Bernice P. Bishop Museum.

Heschel, M. S., and K. N. Page. 1995. "Inbreeding Depression, Environmental Stress, and Population Size Variation in Scarlet Gilia (*Ipomopsis aggregata*)." *Conservation Biology* 9 (1): 126–133.

Huenneke, L. F., and J. K. Thompson. 1995. "Potential Interference Between a Threatened Endemic Thistle and an Invasive Nonnative Plant." *Conservation Biology* 9 (2): 416–425.

Illinois Endangered Species Protection Board. 1992. "Establishing New Populations of Native Plants." Springfield, Ill.: Illinois Department of Conservation.

IUCN (International Union for the Conservation of Nature and Natural Resources). 1987. "The IUCN Position Statement on Translocation of Living Organisms: Introductions, Reintroduction and Restocking." Gland, Switzerland: IUCN.

International Union for the Conservation of Nature and Natural Resources/SSC Reintroduction Specialists Group—Plants. 1992. "Draft Guidelines for Reintroduction." *Reintroduction News* 4 (May). London: Species Survival Commission.

Kitzmiller, S. M. 1990. "Managing Genetic Diversity in a Tree Improvement Program." *Forest Ecology and Management* 35 (1,2): 131–150.

Kress, W. J., G. D. Maddox, and C. S. Roesel. 1994. "Genetic Variation and Protection Priorities in *Ptilimnium nodosum* (Apiaceae), an Endangered Plant of the Eastern United States." *Conservation Biology* 8 (1): 271–276.

Lande, R., and G. F Barrowclough. 1987. "Effective Population Size, Genetic Variation, and Their Use in Population Management." In *Viable Populations for Conservation*. Edited by M. E. Soulé. Cambridge, England: Cambridge University Press.

Lesica, P. and F.W. Allendorf. 1995. "When are Peripheral Populations valuable for Conservation?" *Conservation Biology* 9 (4) 753–760.

Linder, H. P. 1995. "Setting Conservation Priorities: the Importance of Endemism and Phylogeny in the Southern African Orchid Genus *Herschelia*." *Conservation Biology* 9 (3): 585–595.

Lippett, L., M. Fidelibus, and D. A. Bainbridge. 1994. "Native Seed Collection, Processing, and Storage for Revegetation Projects in the Western United States." *Restoration Ecology* 2 (2): 120–131.

Loope, L. L. and A. C. Medeiros. 1994. "Impacts of Biological Invasions on the Management and Recovery of Rare Plants in Haleakala National Park, Maui, Hawaii." Pp. 143–158 in Bowles *et al.*, *op. cit.*

Mace, G. M. and R. Lande. 1991. "Assessing Extinction Threats: Toward Re-Evaluation of IUCN Threatened Species Categories." *Conservation Biology* 5: 148–157.

Menges, E. S. 1990. "Population Viability Analysis for an Endangered Plant. *Conservation Biology* 4 (1): 62–62.

Menges, E. S. 1991a. "The Application of Minimum Viable Population Theory to Plants." In *Genetics and Conservation of Rare Plants*. Edited by D. A. Falk and K. E. Holsinger. New York: Oxford University Press.

Menges, E. S. 1991b. "Seed Germination Percentage Increases with Population Size in a Fragmented Prairie Species." *Conservation Biology* 5 (2): 158–164.

Menges, E. S. 1992. "Stochastic Modelling of Extinction in Plant Populations." In *Conservation Biology: The Theory and Practice of Nature Conservation, Preservation, and Management.* Edited by P. L. Fiedler and S. K. Jain. New York: Chapman and Hall.

National Park Service, Wildlife and Vegetation Division. 1991. "Endangered, Threatened, and Rare Species Management." *Natural Resources Management Guideline,* 77.

Native Plant Society of Oregon. 1992. "Policy Statement on Reintroduction of Extirpated or Rare Plant Species." *NPSO Bulletin,* January.

The Nature Conservancy Stewardship Program. 1992. "Draft Proposed Policy and Criteria for Species Translocations to Preserves." Arlington, Virginia.

Naveh, A. 1994. "From Biodiversity to Ecodiversity: A Landscape-Ecology Approach to Conservation and Restoration." *Restoration Ecology* 2 (3): 180–189.

New England Wild Flower Society. 1992. "New England Plant Conservation Program." *Wild Flower Notes* 7 (1): 40–48.

Pavlik, B. M. 1994. "Demographic Monitoring and the Recovery of Endangered Plants." In *Restoration of Endangered Species: Conceptual Issues, Planning, and Implementation.* Edited by M. L. Bowles and C. J. Whelan. Cambridge, England: Cambridge University Press.

Pavlik, B. M., Nickrent, D. L. and Howald, A. M. 1993. "The Recovery of an Endangered Plant. I. Creating a New Population of *Amsinkia grandiflora.*" *Conservation Biology* 7: 510–526.

Pavlovic, N. B. 1994. "Disturbance-Dependent Persistence of Rare Plants: Anthropogenic Impacts and Restoration Implications." Pp. 159–193 in Bowles, *et al., op. cit.*

Peters, R. L. 1985. "Global Climate Change: A Challenge for Restoration Ecology." *Restoration and Management Notes* 3 (2): 62–67.

Peters, R. L., and T. E. Lovejoy. 1992. *Global Warming and Biological Diversity.* New Haven, Connecticut: Yale University Press.

Primack, R., and S. L. Miao. 1992. "Dispersal Can Limit Local Plant Distribution." *Conservation Biology* 6: 513–519.

Reichenbacher, F. W. 1990. "Reintroduction Brings Kearney's Blue Star from Extinction's Edge." *Plant Conservation* 5: 3.

Reinartz, J. A. 1995. "Planting State-Listed Endangered and Threatened Plants." *Conservation Biology* 9 (4): 771–781.

Rejmanek, M. 1989. "Invasibility of Plant Communities." In *Ecology of Biological Invasions: A Global Perspective.* Edited by J. A. Drake, H. A. Mooney, F. diCastri, R. H. Groves, F. J. Kruger, M. Rejmanek, and M. Williamson. Chichester, England: Wiley and Sons.

Ryman, N., and L. Lairke. 1991. "Effects of Supportive Breeding on the Genetically Effective Population Size." *Conservation Biology* 5 (3): 325–329.

Ruggiero, L. F., G. D. Hayward, and J. R. Squires. 1994. "Viability Analysis in Biological Evaluations: Concepts of Population Viability Analysis, Biological Population, and Ecological Scale." *Conservation Biology* 8 (2): 364–372.

Schemske, D. W., B. C. Husband, C. Goodwillie, I. M. Parker, and J. G. Bishop. 1994. "Evaluating Approaches to the Conservation of Rare and Endangered Plants." *Ecology* 75 (3): 584–606.

Shaffer, M. L. 1981. "Minimum Population Sizes for Species Conservation." *BioScience* 31: 131–134.

Sipes, S. D. and V. J. Tepedino. 1995. "Reproductive Biology of the Rare Orchid, *Spiranthes diluvialis*: Breeding System, Pollination and Implications for Conservation." *Conservation Biology* 9 (4): 929–938.

Soulé, M. E., ed. 1987. *Viable Populations for Conservation*. Cambridge, England: Cambridge University Press.

Taylor, B. L. 1995. "The Reliability of Using Population Viability Analysis for Risk Classification of Species." *Conservation Biology* 9: 551–558.

Thomas, C. D. 1994. "Extinction, Colonization, and Metapopulations: Environmental Tracking by Rare Species." *Conservation Biology* 8 (2): 373–378.

U.S. Army Corps of Engineers. 1991. "Restoration of Fish and Wildlife Habitat Resources." Policy guidance letter number 24, March 7. Washington, DC: Department of the Army.

U.S. Fish and Wildlife Service. 1988. "Endangered Species Act of 1973 As Amended through the 100th Congress." Washington, D.C.: Department of the Interior.

U.S. Fish and Wildlife Service. 1992. "Policy and Guidelines for Planning and Coordinating Recovery of Endangered and Threatened Species." Washington, D.C.: Department of the Interior.

U. S. Fish and Wildlife Service. 1995. "Technical/Agency Draft Molokai Plant Cluster Recovery Plan." U.S. Fish and Wildlife Service, Portland, OR.

Waste Management, Inc. 1992. "Guidelines for Biological Diversity Conservation." Oak Brook, Ill.: Waste Management, Inc.

Weinbaum, B. S., M. F. Allen, and E. B. Allen. "Survival of Arbuscular Mycorrhizal Fungi Following Recipocal Transplanting Across the Great Basin, USA." *Ecological Applications*, in press.

Weller, S. G. 1994. "The Relationship of Rarity to Plant Reproductive Biology. Pp. 90–117 in Bowles *et al.*, *op. cit.*

Wildt, D. E., and U. S. Seal. 1988. "Research Priorities for Single-Species Conservation Biology." Proceedings of a workshop sponsored by the National Science Foundation and the National Zoological Park, November 1988, Washington, D.C.

Wisconsin Department of Natural Resources. 1991. "Draft Rare Plant Propagation and Distribution Policy." Madison, Wis.: Bureau of Natural Resources.

About the Contributors

Volume Editors

DONALD A. FALK is executive director of the Society for Ecological Restoration (SER), an international organization dedicated to advancing the science and practice of restoration. Prior to coming to SER he was the founder and director of the Center for Plant Conservation. Falk was awarded a Fulbright Short-Term Scholars Award in 1991, and is a Fellow of the American Association for the Advancement of Science (AAAS). He co-edited with Kent Holsinger *Genetics and Conservation of Rare Plants*, published by Oxford University Press. Falk is currently studying forest ecology as a Lockwood Fellow at the University of Washington.

CONSTANCE I. MILLAR received her bachelor's degree in forest biology from the University of Washington in 1977, her masters in forest genetics from the University of California, Berkeley in 1979, and her doctorate in genetics from the University of California in 1985. Subsequently she worked as project leader for the California Forest Genetic Conservation Project of the University of California. In 1987, she joined the Pacific Southwest Research Station of the U.S. Forest Service in Berkeley, California, as research geneticist and conservation biologist. Her research has focused on forest population genetics (genetic structure and systematics), evolutionary genetics (Tertiary and Quaternary impacts), and conservation genetics. Her recent research interests have dealt with the role of genes in ecosystems and in vegetation dynamics during the Holocene. Dr. Millar is active in policy development and implementation of ecosystem management in temperate forest systems. She was awarded a Pew Scholarship in Conservation and the Environment in 1991, and a Scientific Achievement Award from the International Union of Forest Research Organizations in 1995. Currently she serves on the congressionally mandated Sierra Nevada Ecosystem Project. This three-year project is charged to assess the health of all ecosystems in the Sierra Nevada of California and evaluate policy alternatives that maintain ecological sustainability. Miller has published over thirty papers on forest genetics, evolution, and conservation.

MARGARET OLWELL is endangered species coordinator for the National Park Service. She received her bachelor's degree in botany from the University of North Carolina at Chapel Hill and her masters in biology from Southern Methodist University. Prior to coming to the National Park Service, she was the conservation program manager for the Center for Plant Conservation where she developed policy and guidance for the National Collection of Endangered Plants. Olwell also worked as the regional botanist with U.S. Fish and Wildlife Service's Southwest Regional Office of Endangered Species where she reintroduced several endangered plant species as part of the recovery process. She is

currently chair of the Native Plant Conservation Initiative, a partnership of nine federal agencies and fifty state and private organizations implementing a national strategy to conserve native plants of the United States. Olwell is a member of The World Conservation Union's (IUCN) Plant Reintroduction Species Survival Commission and Cactus and Succulent Species Survival Commission.

Chapter Authors

MICHAEL J. BEAN is chair of the Wildlife Program of the Environmental Defense Fund in Washington, D.C. His work focuses on law and policy related to endangered species conservation.

KEN S. BERG is the national botanist and acting endangered species coordinator for the Bureau of Land Management in Washington, D.C. His professional interests include the conservation and management of rare and endangered plants.

MARLIN BOWLES is a research associate with The Morton Arboretum. His interests are in the conservation and restoration of endangered and threatened plant species, including participation in federal recovery planning for *Cirsium pitcheri*.

WILLIAM E. BRUMBACK is the conservation director of the New England Wild Flower Society. He has been involved in the propagation, cultivation, and reintroduction of endangered plants since 1980.

ANNE C. CULLY worked as an endangered species botanist with the New Mexico Energy, Minerals and Natural Resources Department, and the U.S. Fish and Wildlife Service in New Mexico. She is now working for the U.S. Fish and Wildlife Service in Manhattan, Kansas, where her current research and conservation interests are wetlands and plant species diversity in prairies.

PEGGY L. FIEDLER is associate professor of botany and conservation biology and director of the graduate program in conservation biology in the biology department of San Francisco State University. Her research focuses on the biology of rarity, the evolution and systematics of mariposa lilies (Calochortus: Liliaceae), and the ecology and restoration of wetlands in the American West.

CAROL W. FYLER has been a volunteer for the New England Wild Flower Society for over 15 years. She has personally located (and continues yearly monitoring of) over 1,000 plants of *Isotria medeoloides*.

GEORGE D. GANN is director of The Institute for Regional Conservation and president of Ecohorizons, Inc. in Miami, Florida. His research focus is on the conservation and restoration of regional biodiversity.

NOEL L. GERSON is research associate at The Institute for Regional Conservation and vice-president of Ecohorizons, Inc. in Miami, Florida. Her research focus is on biodiversity conservation and restoration policy as well as environmental education.

DORIA R. GORDON is the ecologist for the Florida Chapter of The Nature Conservancy and a courtesy assistant professor of botany at the University of Florida. Her research interests include rare plant demography, breeding systems, and conservation, species and habitat restoration, and nonnative plant effects and control.

EDWARD O. GUERRANT, JR. is the conservation director and seed bank curator at The Berry Botanic Garden, which is the participating institution in the Pacific Northwest of the Center for Plant Conservation. He oversees the operation of the Garden's Seed Bank for rare and endangered species and conducts demographic, genetic, and seed germination studies, usually in cooperation with federal and state land management agencies.

ANN M. HOWALD is a plant ecologist with the California Department of Fish and Game. Her responsibilities include all aspects of endangered plant management and protection.

BART R. JOHNSON teaches ecology in the landscape architecture department at the University of Oregon, Eugene. His primary focus is the integration of ecological analysis with landscape design, planning, and management initiatives; current research interests include biodiversity protection, ecosystem management and ecological restoration.

BRIAN J. KLATT is the associate director of the University of Michigan Matthaei Botanical Gardens. In addition to his participation in conservation planning efforts for private holdings, Klatt's conservation interests include public outreach and education, and advising on public policy.

LYNN S. KUTNER is managing editor of The Nature Conservancy's Biodiversity Book/CD-ROM Project. Drawing on data gathered by state natural heritage programs over the past twenty years, this project will examine the status of biodiversity in the United States, with an emphasis on rare species and natural communities.

RICHARD D. LAVEN is professor of forest and fire ecology and conservation biology in the department of forest sciences at Colorado State University. Laven specializes in the forest and fire ecology of the Central Rocky Mountains and the conservation of rare plants, particularly on federal lands.

F. THOMAS LEDIG is senior scientist at the Pacific Southwest Research Station, USDA Forest Service. His interests are in evolutionary biology and its applications to the conservation of genetic resources.

JEANETTE MCBRIDE is a research assistant at The Morton Arboretum, where she works with rare plant species recovery projects.

CHARLES B. MCDONALD is an endangered species botanist for the U.S. Fish and Wildlife Service, New Mexico Ecological Services State Office. His work involves the status assessment, listing, and recovery of endangered plants in New Mexico.

LOYAL A. MEHRHOFF is a plant ecologist with the U.S. Fish and Wildlife Service in Portland, Oregon. His research is focussed on patterns of extinction and rarity, Hawaiian bird–plant pollination systems, and species–area relationships.

LARRY E. MORSE is chief botanist for The Nature Conservancy, and is based at their international headquarters in Arlington, Virginia. His research emphasizes native plant conservation priorities and strategies, particularly in the United States and Canada.

RONALD S. NIEMANN is vice president of Rust Environment and Infrastructure. His interests include promotion of biodiversity conservation on corporate properties and ecological restoration on his farm in southern Wisconsin.

BRUCE M. PAVLIK is Gibbons–Young Professor of Biology at Mills College in Oakland, California. His research focus is on the ecology and physiology of plants native to western North America.

RICHARD B. PRIMACK is professor of plant ecology at Boston University and author of two textbooks, *Essentials of Conservation Biology* and *A Primer of Conservation Biology*. His research deals with the ecology and conservation of tropical forests in Asia and Central America and rare plant species in the eastern United States.

ROBERT D. SUTTER is the director of biological conservation for the Southeast Regional Office of The Nature Conservancy. In that position, he provides assistance to the Conservancy and federal agencies in the monitoring and management of populations and communities, teaches workshops on monitoring and conservation planning, and researches the demography of rare species, seed banks, and the dynamics of small nonalluvial wetlands.

PETER S. WHITE is professor of biology and director of the North Carolina Botanical Garden at the University of North Carolina at Chapel Hill. His research interests are vegetation dynamics, species richness, and conservation biology.

JOY B. ZEDLER is professor of biology and director of the Pacific Estuarine Research Laboratory, San Diego State University. Her research interests are in salt marsh ecology; structure and functioning of coastal wetlands; restoration and construction of wetland ecosystems; interactions of native and exotic species; endangered species, and use of experiments, mesocosms, remote sensing, and other scientific approaches to solve management problems.

Index